STRUCTURE AND DYNAMICS OF ATOMS AND MOLECULES: CONCEPTUAL TRENDS

This book is a companion volume to
Conceptual Trends in Quantum Chemistry,
edited by E. S. Kryachko and J. L. Calais, ISBN 0-7923-2621-0

Structure and Dynamics of Atoms and Molecules: Conceptual Trends

Edited by

J. L. Calais
Quantum Theory Project,
University of Florida, Gainesville, U.S.A.

and

E. S. Kryachko
The Bogoliubov Institute for Theoretical Physics,
Kiev, Ukraine

SPRINGER SCIENCE+BUSINESS MEDIA, B.V.

Library of Congress Cataloging-in-Publication Data

Structure and dynamics of atoms and molecules : concepts and trends /
edited by J.L. Calais and E.S. Kryachko.
p. cm.
"Companion volume to: Conceptual trends in quantum chemistry."
Includes bibliographical references (p. -) and index.

1. Quantum chemistry. I. Calais, Jean-Louis. II. Kryachko,
Eugene S.
QD462.S78 1995
541.2'8--dc20 95-4007

DOI 10.1007/978-94-011-0263-6

Printed on acid-free paper

CONTENTS

Foreword xi

Elements of Hydrogen Transfer Theory

R. Lefebvre 1

1. Introduction 1
2. One-Dimensional Tunneling 3
3. Spectroscopic Splittings 5
   3.1. 1-Dimensional Treatments 6
   3.2. N-Dimensional Treatments 8
4. From Quantum Beats to Relaxation 9
5. Calculation of Relaxation Rates 14

References 20

Building a Bridge Between Ab Initio and Semiempirical Theories of Molecular Electronic Structure

K. F. Freed 25

1. Introduction 25
2. Theoretical Bridge Between Ab Initio and Semiempirical Methods 29
   2.1. Derivation of the Exact Effective Valence Shell Only Equations 30
   2.2. Properties of and Computational Methods for H^V 32
3. The Ab Initio H^V Method 35

3.1. Description of the Ab Initio H^V Method 37
3.2. Illustrations of Recent Ab Initio H^V Calculations 43
4. Ab Initio Computations of Correlated Semiempirical-Like Integrals 49
4.1. Computations of One-Center H^V Parameters for Atomic Systems 53
4.2. Ab Initio Computation of Parameters in Pi-Electron Methods 56
4.3. Ab Initio Computations of the Effective Dipole Operator: Comparisons with Semiempirical Methods 62
5. Acknowledgment 64
6. *References* 64

Electron Delocalization in The Theory of Intermolecular and Intergroup Interactions: Cause, Effect, Prevention

W. H. Adams 69

1. Introduction 69
2. Unphysical States 71
3. Separability and Localizability of Electronic Wave Functions 73
3.1. Minimally Distorted Localized Wave Functions 74
3.2. Genuine Primitive Functions 78
4. Level Shift Operators 79
5. Hartree-Fock Approximation 82
5.1. Localized Orbital Equations 83
5.2. Huzinaga-Equation 87
6. Correspondences and Extensions 89
Appendix A 92
References 93

Quantum Dynamics of Diatoms in External Fields

J. Broeckhove, B. Feyen, and P. Van Leuven 97

1. Introduction 97
2. The Hamiltonian 98
   2.1. Born-Oppenheimer and Beyond 99
   2.2. The Dipole Interaction 101
   2.3. Diatoms in the Dipole Field 102
3. Quantum Time Evolution 104
4. Multiphoton Absorption in Hydrogen Fluoride 108
   4.1. Vibrational Motion 109
   4.2. Rovibrational Motion 113
5. Laser Assisted Dynamics in Boron Hydride 114
   5.1. The Excitation Pathway 119
   5.2. The Vibronic Picture 120
   5.3. The Rovibronic Picture 125
6. Conclusions 127

References 129

Dimensional Scaling in Quantum Theory

J. Avery 133

1. Introduction 133
2. The Jacobian Weighting Factor 135
3. The Large-D Limit 138
4. Three-Particle Systems 139
5. An Atom-to-Molecule Transformation 142
6. N-Particle Systems 148
7. Discussion 151

Acknowledgements 151
References 152

Probing The Collective and Independent-Particle Character of Atomic Electrons

R. S. Berry 155

1. Introduction 155
   1.1. Historical Background 155
   1.2. Alternative Quantizations and Constants of Motion 156
   1.3. The Tools of the Analysis 158
2. Symmetry-Based Interpretation 159
   2.1. Initial Explorations 159
   2.2. Multiplets and Supermultiplets 160
3. Interpretations Based on Wave Functions 162
   3.1. Graphic Representations of Helium 162
   3.2. Graphical Representations for Other Atoms: Alkaline Earths 163
   3.3. Overlap Criteria 167
   3.4. Oscillator Strength Criteria 168
   3.5. Quadrupole Moments 168
   3.6. Angular Correlations in Double Ionization 171
   3.7. Other Correlations and Dispersions 172
4. Interpretations Based on Classical and Semiclassical Models 173
   4.1. The Bohr-Sommerfeld Approach 173
   4.2. Orbits, Planetary Atoms, Langmuir States 174
5. Summary and Conclusions 176

References 176

Electronic Structure Models: Computations, Chemical Insights and Appropriateness

M. A. Ratner 183

1. Introduction 184
2. Model Hamiltonians 185
3. Response Properties, Wave Functions, and Levels of Theory 186
4. Example: The Frequency Dependent Hyperpolarizabilities of Push Pull Substituted Aromatics 191
   A. Model Hamiltonian 191
   B. Ground State 192
   C. Nature of the Excitations 192
   D. Some Selected Results 193

References 208

The Work Formalism: A New Theory of Electronic Structure

V. Sahni 213

1. Introduction 213
2. The Work Formalism 216
   2.1. General Framework 216
   2.2. The Work Formalism Hartree Approximation 220
   2.3. The Work Formalism Hartree-Fock Approximation 222
   2.4. The Work Formalism Configuration-Interaction Approximation 225
   2.5. Asymptotic Structure of Exchange-Correlation Potential 226
3. Comparisons with Other Local Potential Theories 227
   3.1. Comparison with Kohn-Sham Theory 227
   3.2. Rigorous Interpretation of the Local Density Approximation 232
   3.3. Comparison with Slater Theory 237
4. Results of Application to Atoms 238
   4.1. Structure of the Fermi Hole and Exchange Potential 238
   4.2. Structure of the Coulomb Hole and Correlation Potential 242
   4.3. Ground-State Properties in the Pauli-Correlated Approximation 247
   4.4. Ionization Potentials, Electron Affinities and Transition Energies 251
5. Conclusions and Future Work 255

References 257

Index 261

FOREWORD

The rivers run into the sea, yet the sea is not full

Ecclesiastes

What is quantum chemistry? The straightforward answer is that it is what quantum chemists do. But it must be admitted, that in contrast to physicists and chemists, "quantum chemists" seem to be a rather ill-defined category of scientists. Quantum chemists are more or less physicists (basically theoreticians), more or less chemists, and by and large, computationists. But first and foremost, we, quantum chemists, are conscious beings.

We may safely guess that quantum chemistry was one of the first areas in the natural sciences to lie on the boundaries of many disciplines. We may certainly claim that quantum chemists were the first to use computers for really large scale calculations. The scope of the problems which quantum chemistry wishes to answer and which, by its unique nature, only quantum chemistry can answer is growing daily. Retrospectively we may guess that many of those problems meet a daily need, or are say, technical in some sense. The rest are fundamental or conceptual. The daily life of most quantum chemists is usually filled with grasping the more or less technical problems. But it is at least as important to devote some time to the other kind of problems whose solution will open up new perspectives for both quantum chemistry itself and for the natural sciences in general.

When chemistry developed in the nineteenth century, its explanatory systems began to separate from those of physics, as peculiarly chemical theories of molecular structure and of reactivity were conceived. By the early twentieth century attempts were made to heal this rift and perhaps the most influential of these was the creation of the electronic theory of valency.

However it was not until the development of quantum mechanics in the late nineteen- twenties that it became clear how a proper reconciliation should be effected. It is in this context that quantum chemistry rises as a subject of its own and gets a very special place among the natural sciences. Quantum chemistry provides the conceptual apparatus by means of which chemical explanation is tied to the explanatory scheme of quantum mechanics, one of the most fundamental of all physical theories.

Until the nineteen-fifties - even though important numerical work was certainly carried out also before that time - the chief contribution of quantum chemistry to chemistry was in terms of concepts like atomic and molecular orbitals, resonance, hybridisation and many others. Since about nineteen- sixty the development of computers has led to a change in emphasis. The conceptual apparatus has remained largely that of previous generations but thanks to large scale sophisticated quantum chemical computations, quantitatively accurate results have made it possible to assist in an effective way the understanding of many phenomena in chemistry, physics, and biology. There can be no doubt about the tremendous quantitative successes that quantum chemistry has had in the last three decades. However, it may not be out of place to have a look retrospectively at different angles and to ask: Whether time might not be ripe for making a paradigm shift in the nature of the concepts that are used, for breaching the decorum to say what has not yet been said before because it has seemed too trivial to bother with or too embarrassing to own up to, and for showing us something that had been there all the time that we had not seen for ourselves.

That was our idea to launch "Conceptual Trends in Quantum Chemistry". This former collection of essays of a number of active quantum chemists has been appeared under the identical title in the early 1994. The present one has the same underlying idea, although it is slightly more specified by the purpose. We hope that both these

We are deeply indebted to the KLUWER Academic Publishers and personally, to Mrs. Wil Bruins and Dr. David J. Larner for all their interest and assistance in the preparation of this series.

JEAN-LOUIS CALAIS | EUGENE KRYACHKO

Uppsala | Kiev

ELEMENTS OF HYDROGEN TRANSFER THEORY

R. LEFEBVRE

Laboratoire de Photophysique Moléculaire, Bât. 213, Campus d'Orsay, 91405 Orsay, France

Also at Unité de Formation et de Recherche de Physique Fondamentale et Appliquée, Université Pierre et Marie Curie, Paris, France

1. Introduction

The purpose of this contribution is to describe the basic features of a theoretical treatment of hydrogen transfer. There is an evergrowing literature on the subject and this presentation is not meant to be a catalogue of all existing theoretical models. We will examine a few fundamental aspects which are described under the following titles:

Section 2 One-dimensional tunneling
Section 3 Spectroscopic splittings
Section 4 From quantum beats to relaxation
Section 5 Calculation of relaxation rates.

Section 2 is a reminder of the well known tunneling effect through a one-dimensional barrier. It insists on the sensitivity of the transmission probability with respect to barrier parameters (height and width). Section 3 is devoted to a discussion of the splittings of vibrational levels which is the result, for a molecule in the gas phase with two equivalent isomeric forms, of the permeability of a barrier to a light atom such as hydrogen. This section takes the example of malonaldehyde and stresses the need for a multidimensional treatment of the tunneling motion. Time is not involved in these two first sections, since an observation of the splitting in an energy resolved experiment reveals the existence of stationary molecular states with equal probabilities for the molecule to be in either one of the two wells. Sections 4 and 5 consider time: what happens to a molecule which is assumed to be prepared in one of the two wells? For an isolated molecule this situation would result in an oscillatory motion (quantum beats) which apparently have never been observed in relation to atom transfer. The medium can affect profoundly the dynamics. The molecule can be stabilized in one of the wells and departs irreversibly from an initial state on a time scale which is much longer than the oscillatory period. The challenge for the theory is to account for this modification of the motion and its slowing down. This can be done with rather simple models which identify the relevant mechanisms. Rather paradoxically most of the treatments of this problem introduce very little of the information about tunneling motion of the hydrogen atom. This is because in most situations one state only in each well is thermally accessible. Two parameters, an interaction constant between localized states and an energy bias, are the only fingerprints of hydrogen tunneling in the Hamiltonian. The relevant activation energy is not that of hydrogen motion but is associated with modes of lower frequencies. The theoretical treatments can be roughly classified in two types which correspond to Sections 4 and 5 respectively. Section 4: How are we to understand that the beat which in principle could be observed in the gas phase (but which is not observed because the period is much too brief) can be transformed into a relaxation process. A convenient way to

J. L. Calais and E. S. Kryachko (eds.), Structure and Dynamics of Atoms and Molecules: Conceptual Trends, 1–24.

study this is to introduce a random classical field expressing the effect of the environment. Section 5: Once we have accepted that the conditions are those of a relaxation, how are we to model the system in order to calculate a rate reflecting the behavior of the combined system hydrogen and its environment? The model considered in Section 5 introduces a single harmonic degree of freedom coupled to a 2 state description of hydrogen motion. There are various interpretations for this mode: it could be an intra- or intermolecular motion which assists hydrogen tunneling. It could also reflect the role of the solvent. The basic idea is anyhow the same: the environment of the hydrogen atom is affected by its position and the hopping of hydrogen requires a reorganization of the neighbouring atoms. This is also playing a role in the slowing down of the motion. The formula obtained with such a model is capable of accounting for instance for the rate of hydrogen transfer in the reaction $CH_3OH + CH_3 \rightarrow CH_2OH + CH_4$ (Hudson *et al.* (1977), Trakhtenberg *et al.* (1981)). As shown on Figure 1, a basic feature of hydrogen transfer rates is the existence of a plateau at low temperatures. The activation energy is much lower than that associated with excitation of local motion of hydrogen. The model developed in section 5 has the capacity for explaining this observation. The rate formula has a few parameters which can be fitted to experiment or tentatively (and optimistically) be deduced from ab initio quantum mechanical calculations. More sophisticated treatments are often needed, but a proliferation of unknown parameters can make the rate expression useless if there is only a moderate amount of experimental data to make a fit.

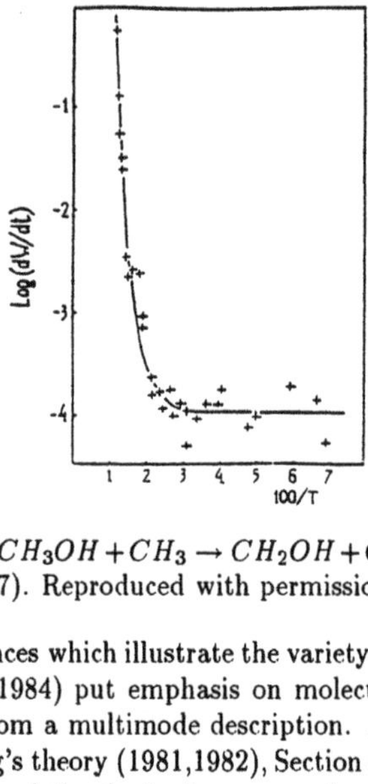

Figure 1
Rate constant of the reaction $CH_3OH + CH_3 \rightarrow CH_2OH + CH_4$ versus temperature. Experimental data from Hudson *et al.* (1977). Reproduced with permission from Trakhtenberg *et al.* (1981).

We now quote some references which illustrate the variety of procedures to approach the problem. Siebrand and collaborators (1984) put emphasis on molecular or supermolecular Franck-Condon factors which are obtained from a multimode description. A Franck-Condon approach is also the starting point of Trakhtenberg's theory (1981,1982), Section 5 being a simplified version of it. Silbey and Harris (1989) have reviewed the dephasing approach appropriate for the gas phase situation. They also develop a variational procedure for the condensed phase. Meyer and Ernst (1987) have generalized the random field approach described in section 4 (Bixon's treatment (1982)) and have also given (Meyer and Ernst((1990)) a purely quantum mechanical theory which, contrarily to a classical approach can account for detailed balance. The random field formulation is also the basis of the method adopted by Korst and Nikitin (1965) in their pioneering work on tunneling. Cukier and Morillo (1989) derive various limiting forms for the rate from a treatment of a 2-level system interacting with a bath. 2-level and 4-level systems coupled to a bath are studied by Skinner and Tromsdorff (1988) in order to extend the theory to double proton transfer. Borgis and Hynes (1989,1991) consider two modes coupled to the hydrogen transfer, one to represent an inter- or intramolecular rearrangement, the other to account for the solvent. They also use molecular simulation methods to examine the respective roles of these two types of interactions.

Finally we quote the important work of Leggett and collaborators (1987) who emphasize the great variety of solutions resulting from various assumptions made on the parameters of the Hamiltonian representing the coupling of a two level system to a bath of harmonic oscillators (the so-called spin-boson Hamiltonian). Other references will come as we develop the subject.

2. One-dimensional tunneling

This is a purely quantum mechanical effect which is described in every textbook (see for instance Merzbacher (1967), Chapter 6). Quantum mechanically a particle has a nonzero probability to go into a classically forbidden region with $E < v(r)$ (the energy is less than the potential energy). This corresponds to an imaginary wave number since for a particle of mass μ the wave number is

$$k(r) = \frac{\sqrt{2\mu}}{\hbar}[E - v(r)]^{\frac{1}{2}} \tag{1}$$

The wave function is damped in such a region. Within a barrier the two damped solutions and their derivatives propagated from the two classically allowed regions can be matched. In this way a particle impinging on the barrier can reappear on the other side. The simplest potential to describe this effect is the rectangular barrier (Merzbacher (1967)). Here we consider instead the Eckart barrier (Eckart (1930)) which is closer to those met in the molecular context. Because the transmission probability has an analytical form, this potential has been widely used in one dimensional descriptions of tunneling (Le Roy *et al.* (1980), McKinnon and Hurd (1983), Garrett and Truhlar (1979), McCurdy and Garrett (1986), Lefebvre and Taylor (1989), Lefebvre and Moiseyev (1990)). It will serve here to illustrate the sensitivity of the transmission probability versus two properties of this potential: its height and its width. A symmetrical Eckart potential with the top of the barrier at $r = 0$ has the form:

$$v(r) = \frac{Ay}{(1+y)^2} \quad \text{with} \quad y = \exp[\alpha r] \tag{2}$$

The barrier height is $V_m = A/4$. We start from the parameters used occasionally (Garrett and Truhlar (1979), McCurdy and Garrett (1986)) to give a one-dimensional view of the tunneling in the collinear gas-phase $H + H_2$ transfer reaction. The reduced mass of the collision is $\mu = 2/3m_H$, with $m_H = 1.00782$ a.m.u. α is 0.6 a_0^{-1} and A is 9233.48 cm^{-1} (barrier height: 2308.37 cm^{-1} or 6.6 kcal/mole). The transmission probability is:

$$T(E) = \frac{\cosh(2a) - 1}{\cosh(2a) + \cosh(d)} \tag{3}$$

with

$$a = 2\pi(\alpha\hbar)^{-1}(2\mu E)^{\frac{1}{2}} \ ; \quad d = 2\pi(\alpha\hbar)^{-1}\left(2\mu A - 0.25\alpha^2\hbar^2\right)^{\frac{1}{2}} \tag{4}$$

Two kinds of tests have been performed to study the changes of $T(E)$ when changing either V_m, the barrier height or α (which commands the barrier width).

(a) The energy is maintained at $E = 1154.18\,\mathrm{cm}^{-1}$ (half the barrier height of the $H + H_2$ model). V_m is changed from 1154.18 cm^{-1} to 3462.56 cm^{-1} (from 1/2 to 3/2 the barrier height of the initial model). Figure 2 shows that there is a change of $T(E)$ by 9 orders of magnitude. In the neighbourhood of the initial model, an order of magnitude change of $T(E)$ is obtained with 2182 cm^{-1} $\langle$ V_m $\langle$ 2332 cm^{-1}. The change in V_m amounts to 0.7 kcal/mole.

(b) The energy is still E=1154.18 cm^{-1}. The barrier height is 2308.37 cm^{-1}. The parameter α which is related to the width at half-height Δ by $\Delta \simeq 3.53\,\alpha^{-1}$ is changed from $0.5a_0^{-1}$ to $5a_0^{-1}$. The transmission probability changes by 15 orders of magnitude. Close to the initial model, an order of magnitude change of $T(E)$ is obtained when α goes from $1.43a_0^{-1}$ to $1.77a_0^{-1}$. This is illustrated in Figure 3.

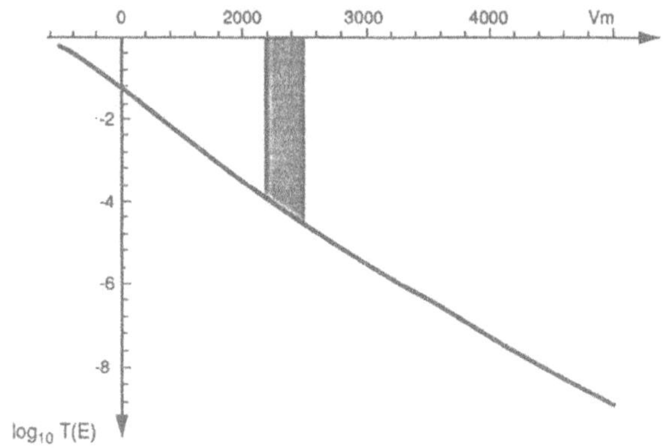

Figure 2
Logarithm of the transmission probability as a function of barrier height for an Eckart barrier with α and μ of the collinear $H + H_2$ transfer reaction (Garrett and Truhlar (1979)). The shaded area corresponds to one order of magnitude change in $T(E)$ when V_m varies in the vicinity of the model value.

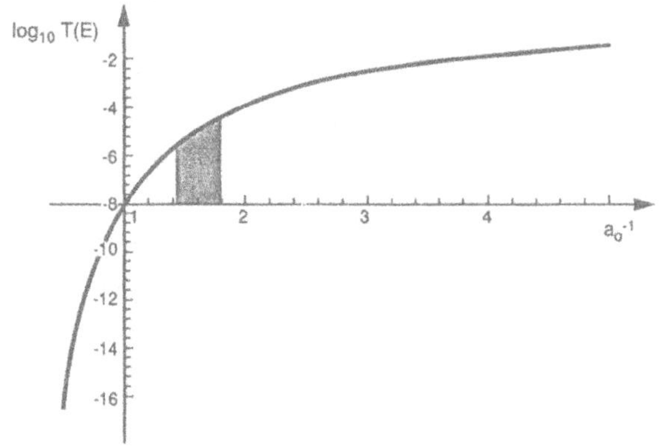

Figure 3
Logarithm of the transmission probability as a function of the parameter regulating the barrier width for an Eckart barrier with V_m and μ of the collinear $H + H_2$ transfer reaction (Garrett and Truhlar (1979))]. The shaded area corresponds to one order of magnitude change in $T(E)$ when α varies in the vicinities of the model value.

Before leaving the subject of tunneling through a one dimensional barrier, it is important to stress that when no analytical formula for $T(E)$ is available, the semi-classical expression:

$$T_{SC}(E) = \left(1 + e^{2\theta}\right)^{-1} \quad \text{with} \quad \theta = \hbar^{-1}\sqrt{2\mu}\int_{r_1}^{r_2} dr[v(r) - E]^{\frac{1}{2}} \tag{5}$$

can give excellent results (Garrett and Truhlar (1979)). In the integral giving θ, r_1 and r_2 are the two turning points at the edges of the non classical region. For the Eckart potential this integral can be calculated exactly, with the result:

$$\theta = \frac{1}{2}(f - 2a) \tag{6}$$

with a given by Eq.(4) and

$$f = 2\pi(\hbar\alpha)^{-1}(2\mu A)^{\frac{1}{2}} \tag{7}$$

Formula (6) can be used even above the barrier (Garrett and Truhlar (1979)), even though the turning points become imaginary.

Table 1 shows the quality of the semi-classical approximation both for energies below and equal to the top barrier energy, or above this energy.

Table 1: Exact and semi-classical transmission probabilities across an Eckart potential used to model the $H + H_2$ collinear transfer reaction (Garrett and Truhlar (1979), McCurdy and Garrett (1986)). The energy E is $pV_m/10$. Parentheses are powers of 10. The top barrier energy is obtained when p = 10.

p	T_{exact}	$T_{semi-classical}$	p	T_{exact}	$T_{semi-classical}$
1	0.1643(-11)	0.1451(-11)	8	0.1655(-1)	0.1465(-1)
2	0.3043(-9)	0.2689(-9)	9	0.1277	0.1145
3	0.1673(-7)	0.1478(-7)	10	0.5310	0.5000
4	0.4904(-6)	0.4331(-6)	11	0.8879	0.8750
5	0.9615(-5)	0.8494(-5)	12	0.9807	0.9782
6	0.1417(-3)	0.1252(-3)	13	0.9967	0.9974
7	0.1680(-2)	0.1484(-2)	14	0.9999	0.9999

3. Spectroscopic Splittings

One of the manifestations of the permeability of a potential barrier to a quantum particle is the occurrence of energy splittings for a symmetric double well potential. The most famous example is NH_3 which evolves between two pyramidal forms with the planar geometry as the transition configuration. The ground level is split by 0.66 cm^{-1} (Herzbzerg (1967)). Another well studied case is malonaldehyde where a hydrogen atom can be transferred from one oxygen to the other:

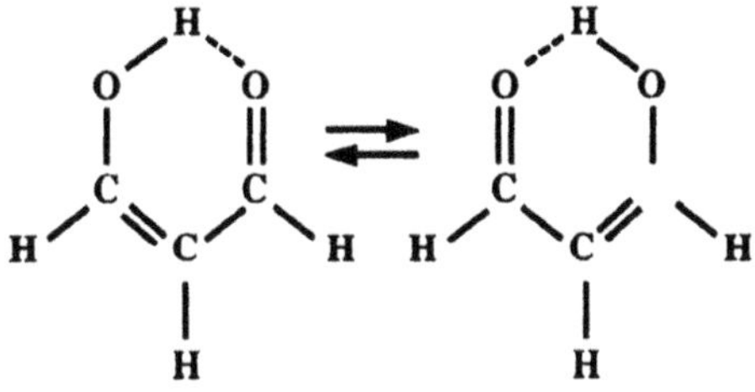

The splitting between the two states with in phase or out of phase oscillations of the hydrogen atom in the local minima is 21 cm^{-1} (Baughum *et al.* (1981). In a one-dimensional simulation of the process, not surprisingly, the splitting is related to the transmission probability of the previous section. A semi-classical analysis yields for the splitting (Connor (1969)):

$$\Delta E = \frac{\hbar\omega}{\pi}T^{\frac{1}{2}} \simeq \frac{\hbar\omega}{\pi}e^{-\theta} \tag{8}$$

where θ has been defined in equation (5). ω is the frequency in one of the wells. The last form is valid when $\theta \gg 1$ (small transmission probability and small splitting).

However this simple view: (a) finding a coordinate for nuclear motion from one local equilibrium (e.g. Structure I for malonaldehyde) to the other local equilibrium (Structure II); (b) calculating a one dimensional potential showing a double minimum; (c) evaluating the splitting for instance from Eq(8) is an oversimplification. In view of the sensitivity of the transmission probability versus the barrier characteristics and the mass of the particle which is transmitted, an agreement between the result of this simple program and an experimental observation can only be largely fortuitous. We use malonaldehyde to show how the splitting has been calculated with increasing degree of sophistication (but this does not always guarantee increased accuracy!).

It is now agreed that both the isomeric forms I and II and the transition state are planar, with for the latter a C_{2v} symmetry:

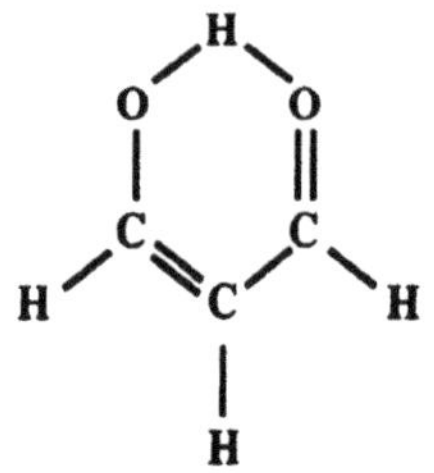

The first calculations of the potential barrier have been made with the semi- empirical Complete Neglect of Differential Overlap Method (CNDO) (Schuster (1969), Kato *et al.* (1977)). The CNDO method is sometimes not yielding the correct geometry. The barrier height defined as the difference between the energy of the geometrically optimized transition structure and one of the local structures is definitely too low (1 to 2 kcal/mole). With such a barrier height the vibrational levels of the double well potential are above the barrier. Ab initio methods give more satisfactory estimates of the barrier height, although there can be some variation with the type of method (choice of basis sets, extent of inclusion of electron correlation). Table 2 gives a non-exhaustive list of barrier heights calculate with ab-initio methods. All heights at the SCF level are similar (about 10 kcal/mole) and there is some reduction due to electron correlation effects which are more important in the transition state than in a stable configuration. The splittings are calculated with n-dimensional methods to be described now.

Table 2: Barrier heights (in kcal/mole) and splittings (in cm^{-1}) calculated by various groups applying ab initio methods to malonaldehyde at SCF or CI-level. The models use Reaction Path Hamiltonians (1-Dimensional treatment) or Reaction Surface Hamiltonians (n-Dimensional treatments with $n = 2$ or 3).

	Barrier heights (kcal/mole)		Splitting (cm^{-1})	Model
Reference	SCF	CI	ΔE	
Karlström *et al.* (1976)	11.5	10.0	–	
Bouma *et al.* (1978)	10.3	9.8	–	
Bicerano *et al.* (1983)	11.4	8.0	40	1-D
Frisch *et al.* (1985)	10.3			
Carrington and Miller (1986)	10.3		60	2-D
Shida *et al.* (1989)	10.6	6.3	9	3-D
Bosch *et al.* (1990)	7.81	–	10.6	2-D

3.1 1-DIMENSIONAL TREATMENTS

A simple 1-dimensional picture of the tunneling dynamics has been developed by Bicerano *et al* (1983). The semi-classical expression (2) for the splitting is adopted. The 1-dimensional barrier is approximated by an Eckart potential (Eckart (1930)). The barrier penetration integral θ (Eq. 5) is

$$\theta = \frac{2\pi}{\hbar\omega_i}\left(V_{\text{eff}} - \sqrt{E_0 V_{\text{eff}}}\right) \tag{9}$$

where V_{eff} is the effective barrier height. The difference between V_{eff} and the barrier height V_0 calculated from electronic energies is due to zero point vibrational energies which may differ in the reactant and the transition state. Thus V_{eff} is

$$V_{eff} = V_0 + \frac{1}{2}\sum_{k=1}^{F-1} \hbar \left(\omega_k^{\ddagger} - \omega_k\right) \tag{10}$$

F being the number of normal modes. The Fth mode is the reactant mode of zero point energy $E_0 = (1/2)\hbar\omega_F$ in the reactant state. This is the degree of freedom leading to the transition state. ω_i is the angular frequency deduced from the curvature at the top of the barrier (this is the frequency of an oscillator in the inverted barrier). Bicerano *et al.* (1983) identify the reactant mode with the O-H stretch. This approach requires only the knowledge of some local parameters. It has produced 40 cm^{-1} for the splitting, about twice the observed value.

Improved 1-Dimensional treatments can be developed along a line which is reminiscent of the Born-Oppenheimer separation for electron and nuclear motion. This leads to the so-called reaction path Hamiltonian (Marcus (1966), Miller *et al.* (1980), Basilevsky and Ryaboy (1982)). The first step is to find the Minimum Energy Path (MEP) leading from the transition geometry (a saddle point on the Potential Energy Surface (PES) to the isomeric forms (local minima). This is done with coordinates which allow to view the process as the motion of a particle of unit mass on the PES, by defining new cartesian coordinates with the rule

$$X_1 = \sqrt{M_1}x_1, \quad X_2 = \sqrt{M_1}y_1, \quad X_3 = \sqrt{M_1}z_1 \tag{11}$$

for nucleus 1, etc... The reaction coordinate S is the coordinate along the reaction path which generally shows some curvature (particularly high when the transfer takes place in a heavy-light-heavy combination). The remaining coordinates are taken to be displacements perpendicular to the reaction path. Because the transformation from Cartesian Coordinates to the new coordinates is not an orthogonal one, this leads to non-trivial changes in the form of the kinetic energy terms of the Hamiltonian (not to speak of additional complications due to the necessity to reduce the displacements to those which do not contribute to overall translations or rotations (Miller *et al.* (1980)). For instance for a collinear triatomic system, the Hamiltonian is found to take the form (Marcus (1966)):

$$H(S, P_S, Q, P) = \frac{1}{2}\frac{P_S^2}{[1 + QK(S)^2]} + V_0(S) + \frac{1}{2}P^2 + \frac{1}{2}\omega(S)^2Q^2 \tag{12}$$

The coordinates and momenta are S and P_S for the degree of freedom along the reaction path and Q and P for motion perpendicular to it. $V_0(S)$ is the energy along the MEP, while the energy in the Q direction is Taylor expanded and stopped at quadratic terms. $K(S)$ is the curvature of the reaction path. In order to reduce the treatment to one dimension, one can invoke the adiabatic approximation: if the harmonic motion transverse to the reaction path is faster than motion along S, a Hamiltonian for the slow motion can be obtained by averaging over fast motion for every possible state of the Q oscillator. When this oscillator is in the zero point level this separation gives the Hamiltonian (after ignoring the so-called diagonal corrections) (Miller *et al.* (1980)):

$$H(S, P_S) = \frac{P_S^2}{2\mu(S)} + V_0(S) + \frac{1}{2}\omega(S) = \frac{P_S^2}{2\mu(S)} + V_{eff}(S) \tag{13}$$

with

$$\mu(S) = \left[1 - \frac{K(S)^2}{\omega(S)}\right]^{3/2}$$

The transmission probability through the barrier, which is basic for expressing the splitting can then be obtained through Eq.(5) by evaluating

$$\theta = \int_{S_l}^{S_r} ds\sqrt{2\mu(S)}\sqrt{E - V_{eff}(S)} \tag{15}$$

where $S_{\substack{<\\>}}$ are the turning points for motion through the barrier and E the "translational energy" in the reactant mode (for instance at zero temperature the zero point energy in one of the local minima of $V_{eff}(S)$). This approach is based on the idea that the reaction path is the most efficient path for tunneling because the barrier is lower than along any other path. However this path does not necessarily maximize the tunneling probability. The transverse oscillator is exploring the valley even in its lower state (zero-point motion). It was shown by Marcus and Coltrin (1977) for the $H + H_2$ system that the path joining the turning points of the transverse motion on the concave side (this corresponds to stretching the 3-body system) yields a smaller phase integral θ, meaning an increase in tunneling probability. This is due to the fact that the increase in barrier height and thickness is more than compensated by the shortening of the path. The effective mass along this new path is different. This method for reducing a reaction to a one-dimensional model has been extensively developed by Truhlar and his associates (Skodje *et al.* (1981), Truhlar *et al.* (1982)) for chemically reacting systems and extended (Hancock and Truhlar (1989)) to the determination of tunneling splittings in isomerization processes.

3.2 N-DIMENSIONAL TREATMENTS

It is sometimes not possible to reduce to one degree of freedom the dynamics of hydrogen transfer. Malonaldehyde is again used as the example. Recent treatments of the isomerization process make use of either 2 (Carrington and Miller (1986) or 3 (Shida *et al.* (1989)) coordinates to describe the motion. In principle an optimum choice for the coordinates should result from some steepest descent algorithm. A simple approach is based on intuition. For malonaldehyde possible choices are:

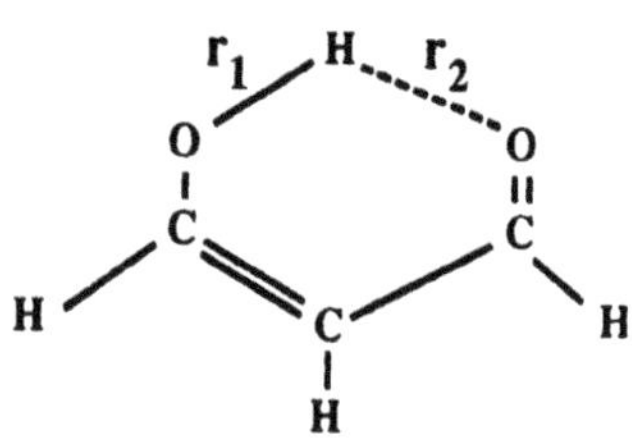

2 coordinates

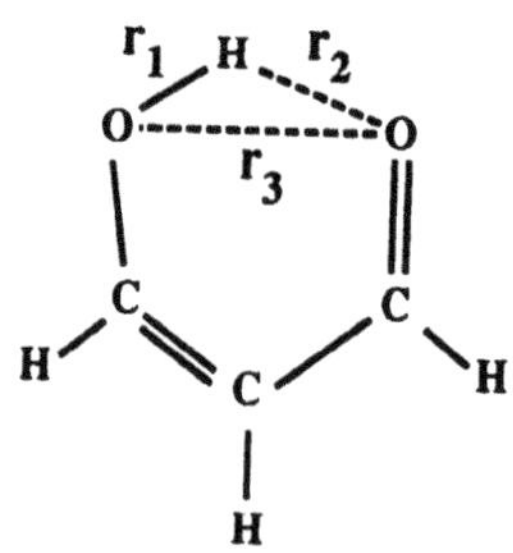

3 coordinates

Bosch *et al.* (1990) use the ab initio PES to fix the parameters of a 2-dimensional model Hamiltonian describing H-transfer (Makri and Miller (1987)).Such methods assume that motion in any direction perpendicular to $\underset{\sim}{r}$ space, $\underset{\sim}{r}$ representing collectively the reaction coordinates (2 or 3 in the present case) is harmonic. An adiabatic approximation similar to that described for the reaction path Hamiltonian gives for the effective Hamiltonian in $\underset{\sim}{r}$ space (Carrington and Miller (1986)):

$$H(\underset{\sim}{r}, \underset{\sim}{P_r}) = \frac{1}{2}\underset{\sim}{P_r} \cdot \underset{\approx}{G_{rr}}(r) \cdot \underset{\sim}{P_r} + V_0(\underset{\sim}{r})^{-1} + \frac{1}{2}\sum_k \hbar\omega_k(\underset{\sim}{r}) \tag{16}$$

this applying when all transverse oscillations are in their zero-point level. $V_0(\underset{\sim}{r})$ is the lowest potential energy for given $\underset{\sim}{r}$ and the $\omega_k(\underset{\sim}{r})$'s are the normal frequencies obtained from an harmonic approximation to the full potential in the neighbourhood of $\underset{\sim}{r}$. $G_{rr}(r)$ is analogous to the G matrix of a normal mode analysis (Wilson *et al.* (1980)). Once the reaction surface Hamiltonian has been obtained there remains to obtain its eigenvalues. This can be done through basis set expansions in every degree of freedom (in $\underset{\sim}{r}$ space). The splittings given in Table II have been obtained in this way.

Another method originating from nuclear physics has been recently applied to some molecular model Hamiltonians (Walet *et al.* (1989)). The procedure aims at reaching the maximum decoupling between the collective coordinates (describing the large amplitude motion) and the harmonic modes. Potential and kinetic energy coupling terms are simultaneously reduced to a minimum.

4. From Quantum Beats to Relaxation

In an isolated molecule, a potential surface with two equivalent wells induces a splitting which can be measured by spectroscopic methods, as we have seen in the case of malonaldehyde. It is also possible to envisage, through coherent excitation, the preparation of the molecule in a state associated with a local minimum. This would be followed by oscillations (quantum beats) of period $h/\Delta E$ (1.6 ps for malonaldehyde) between the two wells. For a molecule in the condensed phase (liquid, glass or crystal) we cannot ignore the role of the environment. When the molecule has been for some time in one of the wells, the solvent atoms or molecules are in a thermal equilibrium appropriate for this configuration. Hopping of the molecule toward the other well destroys this equilibrium and induces a solvent relaxation. Because of the large number of degrees of freedom, a dense set of final states is available and there is negligible probability for a recurrence back to the initial state. We wish in the following to recall that some simple models have the capacity to describe the transition from one regime to the other (from beats to irreversible decay).

Let us consider with Bixon (1982) only the lowest level in each of the two equivalent wells describing two isomers. For us these isomers can be either a supermolecule made of two molecules between which there is hydrogen transfer or two energetically equivalent forms of a molecule with an intramolecular hydrogen bond. The kets are denoted $|L\rangle$ and $|R\rangle$ (left and right). The eigenstates of the complete Hamiltonian are in this approximation:

$$|\phi_\pm\rangle = 2^{-\frac{1}{2}}\left[|L\rangle \pm |R\rangle\right] \tag{17}$$

of energies $E_\pm = \bar{E} \pm \varepsilon$, with the splitting ΔE equal to 2ε. The effect of the environment is taken into account by introducing a fluctuating time dependent perturbation which destroys the degeneracy of the left and right states and modifies the matrix elements responsible for the transfer. This perturbation is denoted $\mathcal{H}_1(t)$ of matrix elements

$$V_{LL}(t) = \langle L|\mathcal{H}_1(t)|L\rangle,\ V_{LR}(t) = \langle L|\mathcal{H}_1(t)|R\rangle,\ \text{etc.} \tag{18}$$

The effective 2×2 Hamiltonian in the $|L\rangle$, $|R\rangle$ basis is

$$H = \begin{pmatrix} V_{LL}(t) & \varepsilon + V_{LR}(t) \\ \varepsilon + V_{RL}(t) & V_{RR}(t) \end{pmatrix} \tag{19}$$

The problem will be simplified by ignoring the effect of the bath on the transfer matrix element and setting $V_{LL}(t)$ equal to zero since only the energy difference between the two states is relevant (see however Wertheimer and Silbey (1980)). The wave function at time t is:

$$|\psi(t)\rangle = a(t)|L\rangle + b(t)|R\rangle \tag{20}$$

The density matrix is

$$\rho = \begin{pmatrix} aa^* & ab^* \\ a^*b & bb^* \end{pmatrix} = \begin{pmatrix} \rho_{LL} & \rho_{LR} \\ \rho_{RL} & \rho_{RR} \end{pmatrix} \tag{21}$$

and its time evolution is given by

$$i\hbar\dot{\rho} = [H, \rho] \tag{22}$$

Instead of working with the elements of the density matrix it is customary (Feynman *et al.* (1957)) for a two-level system to build the three real combinations of the polarization vector:

$$\begin{aligned} r_1 &= ab^* + a^*b &&= \rho_{LR} + \rho_{RL} \\ r_2 &= i(ab^* - a^*b) &&= i(\rho_{LR} - \rho_{RL}) \\ r_3 &= aa^* - bb^* &&= \rho_{LL} - \rho_{RR} \end{aligned} \tag{23}$$

Supplemented with $r_4 = \rho_{LL} + \rho_{RR} = 1$, the knowledge of the r_i's is equivalent to that of the density matrix. The equation of motion (22) becomes in these variables:

$$\begin{aligned} \dot{r}_1 &= \hbar^{-1} V(t) r_2 && (24a) \\ \dot{r}_2 &= \hbar^{-1} V(t) r_1 - 2\hbar^{-1}\varepsilon r_3 && (24b) \\ \dot{r}_3 &= 2\hbar^{-1}\varepsilon r_2 && (24c) \end{aligned}$$

where $V(t)$ represents $V_{RR}(t)$. If we imagine the molecule to be prepared at time $t = 0$ in the left well we are interested primarily by r_3 which shows how the difference in populations in the two wells is changing. Equation (24a) can be formally integrated:

$$r_1(t) = \hbar^{-1} \int_0^t V(t') r_2(t') dt' \tag{25}$$

Equation (24b) becomes:

$$\dot{r}_2 = -\hbar^{-2} \int_0^t V(t) V(t') r_2(t') dt' - 2\hbar^{-1}\varepsilon r_3 \tag{26}$$

We assume now with Bixon (1982) that the random field represented by $V(t)$ has a very brief correlation time, with the property

$$\langle V(t) V(t') \rangle = 2\lambda\hbar^2 \delta(t - t') \tag{27}$$

the average being performed on a large number of realizations of the stochastic process. Since the molecular variable r_2 can be assumed to be changing on a time scale which is much larger than the solvent fluctuations, it is possible to assume a decorrelation:

$$\langle V(t) V(t') r_2(t') \rangle \ = \ \langle V(t) V(t') \rangle \langle r_2(t') \rangle \tag{28}$$

Using the two relations (27) and (28) we obtain when averaging equation (26):

$$\langle \dot{r}_2 \rangle \ = \ -\lambda \langle r_2 \rangle - 2\hbar^{-1}\varepsilon \langle r_3 \rangle \tag{29}$$

Averaging Eq.(24c) gives:

$$\langle \dot{r}_3 \rangle \ = \ 2\hbar^{-1}\varepsilon \langle r_2 \rangle \tag{30}$$

We can similarly proceed to a formal integration of Eq.(24b) to find from (24a) the time evolution of $\langle r_1 \rangle$. There is found:

$$\langle \dot{r}_1 \rangle \ = \ -\lambda \langle r_1 \rangle \tag{31}$$

Only equations (29) and (30) are needed. $\langle r_2 \rangle$ can be eliminated between these two equations by taking the time derivative of Eq.(30). There is found

$$\langle \ddot{r}_3 \rangle + \lambda \langle \dot{r}_3 \rangle + 4\hbar^{-2}\varepsilon^2 \langle r_3 \rangle \ = \ 0 \tag{32}$$

$\langle \ddot{r}_3 \rangle$ denotes the time derivative of $\langle \dot{r}_3 \rangle$. This is the equation of a damped harmonic oscillator. If λ is zero (no solvent perturbation), $\langle r_3 \rangle$ evolves periodically, with the period $h/2\varepsilon = h/\Delta E$. When $\lambda \neq 0$ three regimes have to be distinguished, depending on the value of λ. The initial conditions are $\langle r_3(0) \rangle = 1$ and $\langle \dot{r}_3(0) \rangle = 0$ (cf. Eq (30) with $\langle r_2(0) \rangle = 0$).
(a) $\delta < 0$ or $\lambda < 4\hbar^{-1}\epsilon$. The oscillator is underdamped (or weakly perturbed). Integration of Eq.(32) yields:

$$\langle r_3(t) \rangle \ = \ \exp[-\lambda t/2] \left\{ \cos\left(\delta'^{\frac{1}{2}} t/2\right) + \lambda \delta'^{-\frac{1}{2}} \sin\left(\delta'^{\frac{1}{2}} t/2\right) \right\} \tag{33}$$

with $\delta' = -\delta$.
(b) $\delta = 0$ or $\lambda = 4\hbar^{-1}\varepsilon$. The solution of Eq.(32) is

$$\langle r_3(t)\rangle \;=\; \exp\left[-\lambda t/2\right]\left[1+\lambda t/2\right] \tag{34}$$

This is almost a rate equation, with a rate which is proportional to the splitting.
(c) $\delta > 0$ or $\lambda > 4\hbar^{-1}\varepsilon$. The oscillator is overdamped (strong molecule-solvent interaction). The solution of Eq.(32) is

$$\begin{aligned}\langle r_3(t)\rangle \;=\; & 0.5\left[1+\delta^{-\frac{1}{2}}\lambda\right]\exp\left[-\frac{\lambda t}{2}+\delta^{\frac{1}{2}}t/2\right]+ \\ & +0.5\left[1-\delta^{-\frac{1}{2}}\lambda\right]\exp\left[-\frac{\lambda t}{2}-\delta^{\frac{1}{2}}t/2\right]\end{aligned} \tag{35}$$

If the interaction is very strong ($\lambda >> 4\varepsilon/\hbar$), $\delta^{\frac{1}{2}}$ can be replaced by $\lambda - 8\varepsilon^2/\lambda\hbar^2$. We have for $\langle r_3(t)\rangle$ the approximate solution:

$$\langle r_3(t)\rangle \approx \exp\left[-4\varepsilon^2 t/\lambda\hbar^2\right] - 4\varepsilon^2/\lambda^2\hbar^2\exp\left[-\lambda t\right] \tag{36}$$

The second term is small and disappears very quickly. As stressed by Bixon (1982), this result shows that the initial configuration of the molecule has a lifetime which can be considerably lengthened by the interaction with the environment. A rate $4\varepsilon^2/\lambda\hbar^2$ can be associated with the time evolution of the two populations. As t goes to infinity there is equalization of the two populations ($\lim_{t\to\infty}\langle r_3(t)\rangle = 0$), but this is true of all regimes (except of course if $\lambda = 0$). In fact all components of the averaged polarization vector go to zero when $t \to \infty$ since as shown by Equations (29–31) the steady state solutions are achieved when

$$\langle r_1\rangle \;=\; \langle r_2\rangle \;=\; \langle r_3\rangle \;=\; 0 \tag{37}$$

This is not in agreement with general thermodynamic arguments since this results also in equal populations in the $|\pm\rangle$ molecular states, although they have different energies. This can be corrected easily, as shown below. Before we go to this aspect of the relaxation of a two-level system, we show that it is possible to build an even simpler model (Lefebvre and Moiseyev (1990)) which, at zero temperature, gives equivalent results. In order to express the fact that there is a relaxation which follows the transition from one configuration to the other, we assume that there is interaction of the right state with a continuum of states. This model can be symbolized by the set of energy levels and couplings shown in Figure 4. The time-dependent wave-function can be written:

$$|\psi(t)\rangle = a(t)|L\rangle + b(t)|R\rangle + \int dE e^{-\frac{i}{\hbar}Et}C_E(t)|E\rangle \tag{38}$$

The amplitudes obey the equations:

$$\begin{aligned} i\hbar\dot{a}(t) &= \varepsilon b(t) && (39a)\\ i\hbar\dot{b}(t) &= \varepsilon a(t) + V\int dE C_E(t)e^{-\frac{i}{\hbar}Et} && (39b)\\ i\hbar\dot{C}_E(t)e^{-\frac{i}{\hbar}Et} &= Vb(t) && (39c)\end{aligned}$$

In order to eliminate the bath degrees of freedom we integrate formally (39c) with the initial condition $C_E(0) = 0$:

$$C_E(t) = \frac{V}{i\hbar}\int_0^t b(t')e^{\frac{i}{\hbar}Et'}dt' \tag{40}$$

Eq.(39b) becomes:

$$i\hbar\dot{b}(t) = \varepsilon a(t) + \frac{V^2}{i\hbar}\int_0^t b(t')dt'\int dE e^{\frac{i}{\hbar}E(t-t')} \tag{41}$$

This equation shows that the rate of change of b could depend on the past values of b (see Eq.(26) for a similar memory effect). This would be the case if we had introduced an energy dependent coupling instead of the constant coupling V. We use the identity

$$\int_{-\infty}^{+\infty} dE e^{\frac{i}{\hbar}E(t-t')} = 2\pi\hbar\delta(t-t') \tag{42}$$

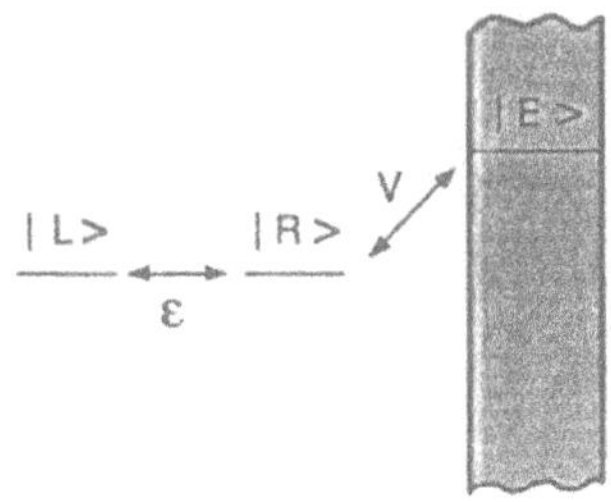

Figure 4
Two discrete states and a continum coupled in a chain-like manner to model the transfer of a particle between two equivalent wells and the relaxation taking place in the final well.

This assumes the continuum to be indefinite, which is a valid hypothesis if the molecular energies are not close to the threshold of the continuum. The integration in Eq.(41) is now straightforward and we obtain for the two amplitudes a and b the coupled equations:

$$i\hbar\dot{a} = \varepsilon b \qquad (43a)$$
$$i\hbar\dot{b} = \varepsilon a - i\Gamma b \qquad (43b)$$

with $\Gamma = \pi V^2$. We now look for the equation giving the time evolution of the quantities r_i (Eq. 23), with however the necessity to take also into account

$$r_4 = aa^* + bb^* \tag{44}$$

since the norm of the $\underset{\sim}{r}$ vector is no longer conserved. After some simple manipulations there is obtained:

$$\dot{r}_1 = -\hbar^{-1}\Gamma r_1 \qquad (45a)$$
$$\dot{r}_2 = -\hbar^{-1}\varepsilon r_3 - \hbar^{-1}\Gamma r_2 \qquad (45b)$$
$$\dot{r}_3 = 2\hbar^{-1}\varepsilon r_2 + \hbar^{-1}\Gamma(r_4 - r_3) \qquad (45c)$$
$$\dot{r}_4 = -\hbar^{-1}\Gamma(r_4 - r_3) \qquad (45d)$$

Deriving (45c) and using (45a) and (45b) there is found

$$\ddot{r}_3 + 2\hbar^{-1}\Gamma\dot{r}_3 + 4\hbar^{-2}\varepsilon^2 r_3 = 0 \tag{46}$$

This is strictly Eq.(32) if we identify $2\hbar^{-1}\Gamma$ with λ. If we solve Eq.(46) with the initial conditions $r_1(0) = r_2(0) = 0$, $r_3(0) = r_4(0) = 1$ we have also $\dot{r}_3(0) = 0$. Therefore Eq.(46) has the solutions which have been discussed previously. The two models give identical results for the time evolution of r_3 (in fact $\langle r_3 \rangle$ in the previous model) with however some difference in the physical interpretation, since in Bixon's model $\langle r_3 \rangle$ goes to zero because the two populations become equal, while in the second model both populations go to zero. We note that the introduction of an imaginary component into the energy of a damped state (Eq. 43b) has been very popular in the treatment of non-radiative transitions in polyatomic molecules (Jortner and Mukamel (1974), Avouris *et al.* (1977)). We will find later that this subject crosses our present problem more than once.

It is possible to extend Bixon's treatment to account for a non-zero temperature of the bath. Adapting some results given by Wertheimeir and Silbey (1980) (see also Silbey and Harris (1989)) the set of equations (29-31) is replaced by:

$$\langle \dot{r}_1 \rangle = (A-1)\lambda - (A+1)\lambda \langle r_1 \rangle \quad (47a)$$
$$\langle \dot{r}_2 \rangle = -2\hbar^{-1}\varepsilon \langle r_3 \rangle - (A+1)\lambda \langle r_2 \rangle \quad (47b)$$
$$\langle \dot{r}_3 \rangle = 2\hbar^{-1}\varepsilon \langle r_2 \rangle \quad (47c)$$

A is $exp\left[-2\varepsilon/k_BT\right]$ (k_B: Boltzmann constant). The damped oscillator obeys now the equation:

$$\langle \ddot{r}_3 \rangle + \lambda(1+A)\langle \dot{r}_3 \rangle + 4\hbar^{-2}\varepsilon^2 \langle r_3 \rangle = 0 \quad (48)$$

so that in the overdamped regime the rate constant becomes:

$$k = 4\varepsilon^2/\lambda(1+A)\hbar^2 \quad (49)$$

There is a further increase in the stability of the initial state as a result of a more efficient destruction of the resonance condition. We note that with these modifications the equilibrium populations obey Boltzmann law. Equilibrium means

$$\langle \dot{r}_1 \rangle = \langle \dot{r}_2 \rangle = \langle \dot{r}_3 \rangle = 0 \quad (50)$$

In order to see the consequences of these conditions we go to the eigenstate $|\pm\rangle$ representation. We have

$$r_1 = \rho_{++} - \rho_{--}; \qquad r_2 = i(\rho_{-+} - \rho_{+-}); \qquad r_3 = \rho_{+-} + \rho_{-+} \quad (51)$$

The consequences of Eq.(50) are $\rho_{-+} = \rho_{+-} = 0$ (no polarization in this representation) and:

$$\rho_{++}/\rho_{--} = A = \exp\left[-2\varepsilon/k_BT\right] \quad (52)$$

The literature contains other treatments of isomerization with more elaborate descriptions of the molecule-environment interaction and which lead also to the equation of a damped oscillator. We quote a few of them. Harris and Stodolsky (1981) examine the density matrix of a chiral molecule which undergoes collisions. This is also a situation where two molecular states are relevant (the ground states of the two optical isomers). The optical activity is measured by the component $\langle r_3 \rangle$ of the polarization vector. The collisions are assumed to interrupt the oscillations which would take place in the isolated molecule. The equations for the time evolution of two of the components of the polarization vector are also of the form of Eqs. (29) and (30), λ being now a parameter depending on the density of colliding atoms, their velocity and the effectiveness of a collision into dampening the off-diagonal elements ρ_{LR} and ρ_{RL} of the density matrix. The overdamping of the oscillator means that the collisions tend to preserve the optical activity. Nikitin (1974) has also studied the effect of collisions on a tunneling process. The model allows for random modulations of both splitting and coupling. The parameter playing the role of λ (Eq. 32) is a function of the strength of the perturbation and depends on the collision frequency and the average duration of a collision. The treatment results in an equation for a damped oscillator with the three regimes described above.

We have also to mention that in a very exhaustive treatment of a two-level system coupled to a bath of harmonic oscillators Leggett and collaborators (1987) obtain for particular assumptions on the bath parameters (through the choice of the so-called spectral densities, see next Section) either damped oscillations or exponential relaxation. An important result of this treatment which cannot be found with the methods mentioned so far is that it is possible to block completely the system in the initial state. We have obtained only a life-time going to infinity as the ratio ε/λ goes to zero.

Finally, there is an interesting simple physical picture of the stabilization of the initial state as the coupling to the medium is increased (Simonius (1978), Harris and Stodolski (1981), Leggett et al. (1987), Silbey and Harris (1989)). In the absence of interactions (no collisions in the gas phase), Eqs. (29) and (30) are simply

$$\langle \dot{r}_2 \rangle = -2\hbar^{-1}\varepsilon\langle r_3 \rangle \ ; \quad \langle \dot{r}_3 \rangle = 2\hbar^{-1}\varepsilon\langle r_2 \rangle \tag{53}$$

$\langle r_2(t) \rangle$ different from zero means that there is some coherence being built up ($\langle r_2 \rangle$ is a combination of the off-diagonal elements of the density matrix). The effect of a collision can be compared to that of a measurement process which reduces the wavepacket. After a first collision arising at time τ, the system is in either one of the two possible states with $\langle r_2(\tau) \rangle = 0$ and $\langle r_3(\tau) \rangle = \cos(2\hbar^{-1}\varepsilon\tau)$. With these initial conditions the system starts again to oscillate between the two states (or the two wells) until there is another collision. After N such interruptions $\langle r_3(t) \rangle$ is:

$$\langle r_3(t) \rangle = \left[\cos\left(\frac{2\varepsilon}{\hbar}\tau\right)\right]^N \simeq \left[1 - \frac{4\varepsilon^2\tau^2}{\hbar^2}\right]^N \simeq 1 - \frac{4\varepsilon^2\tau^2 N}{\hbar^2} \approx \exp\left[-\frac{4\varepsilon^2\tau t}{\hbar^2}\right] \tag{54}$$

with $t = N\tau$. Defining the collision frequency by $\lambda = \tau^{-1}$, the relaxation rate is $k = 4\varepsilon^2/\lambda\hbar^2$, in agreement with Eq.(36). Collision can be here interpreted in a broad sense as meaning any interaction with a bath.

We will see in the next section that there is another mechanism explaining the lengthening of the lifetime of the molecule in the initial well. From the point of view of the treatment given in the present section, this amounts to a reinterpretation of the parameter ε: the hopping of the hydrogen atom is accompanied by a readjustment of one or more of the vibrational modes describing the environmental atoms. The coupling parameter ε of the 'naked' hydrogen transfer mechanism is multiplied by a Franck-Condon amplitude between displaced oscillators with a value which can be much lower than unity.

Quantum beats following the preparation of the molecule in one of the wells could in principle be observed in the $10^{-12} - 10^{-14}$ s time scale for a cold molecule in collision-free conditions (Barbara *et al.* (1989)). This situation can be achieved in a seeded supersonic jet. Time resolved experiments (Barbara *et al.* (1989)) have been so far yielding the rate of appearance of a different isomer after a molecule with an intramolecular hydrogen-bond has been raised to an excited state.

5. Calculation of Relaxation Rates

The conditions are assumed to be those of a pure relaxation process (no possibility of recurrences). This section will use a model which is a simplified version of those found recurrently in various contexts: absorption spectra of trapped electrons in a crystal (Huang and Rhys (1950), Lax (1952), Kubo and Toyozawa (1955), Perlin (1964)), electron transfer (Marcus (1956), Levich (1966), Dogonadze *et al.* (1968), Ulstrup and Jortner (1975), Cukier (1988)), atom transfer (Trakhtenberg *et al.* (1981), (1982),Cukier and Morillo (1989), Skinner and Tromsdorff (1988), Borgis *et al.* (1989)) and radiationless transitions (Englman and Jortner (1970), Freed and Jortner (1970), Fain (1979)). Such a simple picture will serve to describe the various effects accounted for in many of the available expressions for the transfer rate of hydrogen atoms.

A hydrogen atom (or a proton) can occupy two sites with equivalent or unequivalent energies, say E_1 and E_2. There is a coupling ε between the site states leading to a splitting 2ε when $E_1 = E_2$. The tunneling problem for the hydrogen atom when the entities bound to it are motionless is considered as being solved through the knowledge of these parameters (even though they may be very difficult to estimate with ab initio methods). This two-state system is now coupled to a harmonic oscillator which could represent the intra or intermolecular motion of the two entities carrying the hydrogen atoms (Siebrand *et al.* (1984), Borgis *et al.* (1989)) or possibly the solvent (Cukier and Morillo (1989)). We assume a linear coupling between oscillator and atom. The Hamiltonian can be written:

$$\mathcal{H} = E_1|L\rangle\langle L| + E_2|R\rangle\langle R| + \varepsilon\{|L\rangle\langle R| + |R\rangle\langle L|\} + \frac{P^2}{2m} + \frac{1}{2}m\omega^2Q^2 + gQ|L\rangle\langle L| - gQ|R\rangle\langle R| \quad (55)$$

$\Delta E = E_1 - E_2$ measures the asymmetry of the double well potential of the transferred atom. Only the lowest state in each well is considered, as in the previous section. If we ignore the coupling ε, the solutions are

$$|L;n\rangle \quad \text{or} \quad |R;\bar{n}\rangle \quad (56)$$

with $|n\rangle$ and $|\bar{n}\rangle$ being the eigenstates of shifted harmonic oscillators with the Hamiltonians

$$h = \frac{P^2}{2m} + \frac{1}{2}m\omega^2\left(Q + \frac{g}{m\omega^2}\right)^2 - \frac{1}{2}\frac{g^2}{m\omega^2} \quad (57a)$$

$$\bar{h} = \frac{P^2}{2m} + \frac{1}{2}m\omega^2\left(Q - \frac{g}{m\omega^2}\right)^2 - \frac{1}{2}\frac{g^2}{m\omega^2} \quad (57b)$$

The coupling between the atom and the oscillator produces shifts in opposite directions of the equilibrium position of the oscillator, depending on the position of the atom in the left or right site. The energies of the two classes of states are

$$E_n = E_1 + \left(n + \frac{1}{2}\right)\hbar\omega - \frac{1}{2}\frac{g^2}{m\omega^2} \quad (58a)$$

$$E_{\bar{n}} = E_2 + \left(\bar{n} + \frac{1}{2}\right)\hbar\omega - \frac{1}{2}\frac{g^2}{m\omega^2} \quad (58b)$$

In the following the uniform shift of the energy $-\frac{1}{2}\frac{g^2}{m\omega^2}$ is ignored. The initial states of the combined system H atom plus oscillator are identified with the states of kets $|L;\ n\rangle$ (however see Leggett *et al.* (1987)) for a criticism of this approach which cannot pretend to be rigorous). Figure 5 depicts the two potentials defining the zeroth order states $|L;\ n\rangle$ and $|R;\bar{n}\rangle$ with the assumption $E_1 - E_2 = \Delta E > 0$. There are two possibilities: the crossing point of the potentials supporting the displaced oscillator states is either between the minima (a model usually met in the context of H transfer (Cukier and Morillo (1989), Borgis *et al.* (1989)) or beyond the minimum of the right potential (a model used frequently for electron transfer (Cukier (1988), Jortner (1976)). In the Figure the coordinate shift is denoted ΔQ for convenience. Other relevant parameters in Figure 5 are E_S and E_A.

We have

$$E_S = \frac{1}{2}k\Delta Q^2 \quad (59)$$

This is the potential energy of the oscillator when the hydrogen has jumped to the right site while the oscillator is still in the equilibrium position appropriate for the left site. It is called the relaxation energy. E_A is an activation energy with the expression

$$E_A = \frac{(\Delta E - E_S)^2}{4E_S} \quad (60)$$

This is the energy needed by a particle at the minimum of potential V_1 to pass classically onto the potential V_2.

With the Hamiltonian (55) the matrix element between a state belonging to V_1 with a state of the same energy belonging to V_2 is:

$$\langle L;\ n|\mathcal{H}|R;\ \bar{n}\rangle = \varepsilon\langle n|\bar{n}\rangle \quad (61)$$

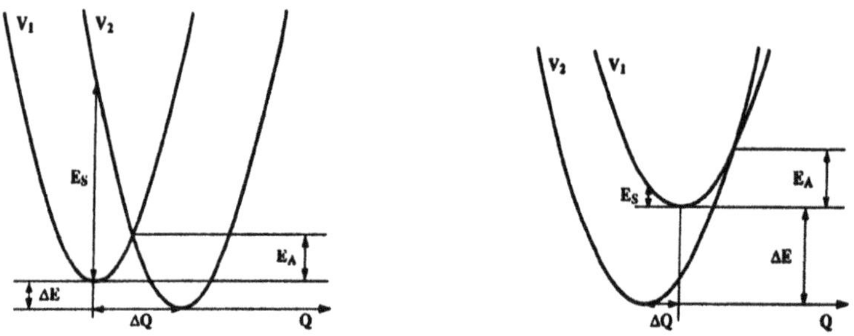

Figure 5
The parameters of the two basic models to couple a two state system to a harmonic oscillator. E_S is the oscillator relaxation energy. E_A is an activation energy. There is initially a thermal equilibrium among the states supported by potential V_1. The diagrams are appropriate for an exothermic reaction.

Conservation of energy requires that $(\bar{n} - n)\,\hbar\omega = \Delta E$. We have assumed that V_1 is higher than V_2 so that with ΔE positive by convention $\bar{n}$ is larger than n. The endothermic case where V_1 is below V_2 is treated below. The expression for the rate k is, according to the Golden Rule:

$$k = \frac{2\pi}{\hbar}\varepsilon^2 \left[1 - \exp\left(-\beta\hbar\omega\right)\right] \sum_{n=0}^{\infty} \exp\left[-\beta n\hbar\omega\right] |\langle n|\bar{n}\rangle|^2 (\hbar\omega)^{-1} \tag{62}$$

β is $(k_B T)^{-1}$. The factor $[1 - \exp(-\beta\hbar\omega)]$ comes from the inverse of the partition function (the zero point energy is a factor in both numerator and denominator). The last factor is the density of states in the final manifold of states. Many ingenious methods have been devised to evaluate such expressions (Huang and Rhys (1950), Lax (1952), Kubo and Toyozawa (1955), Perlin (1964)). These methods are even capable of handling more than one oscillator. For the simple case at hand we will make use of the relation between the Franck-Condon Factors $|\langle n|\bar{n}\rangle|^2$ and Laguerre polynomials and use a remarkable identity which appears to be tailored for such a calculation (Ulstrup and Jortner (1975), Trakhtenberg *et al.* (1981), 1982), Jortner and Ulstrup (1979)) . The Franck-Condon factor $|\langle n|\bar{n}\rangle|^2$ with $\bar{n} > n$ is (Trakhtenberg *et al.* (1981), Fulton and Gouterman (1961)):

$$|\langle n|\bar{n}\rangle|^2 = \exp[-d^2] d^{2(\bar{n}-n)} \left(\frac{n!}{\bar{n}!}\right) \left[L_n^{\bar{n}-n}\left(d^2\right)\right]^2 \tag{63}$$

with

$$d^2 = \frac{m\omega}{2\hbar}(\Delta Q)^2 \tag{64}$$

$L_n^{\bar{n}-n}$ is an associated Laguerre polynomial. We have the relation (Gradshteyn and Ryzhyk (1980):

$$\sum_{n=0}^{\infty} n! \frac{L_n^p(x) L_n^p(y) z^n}{(n+p)!} = \frac{(xyz)^{-p/2}}{1-z} \exp\left\{-\frac{z(x+y)}{1-z}\right\} I_p\left(\frac{2\sqrt{xyz}}{1-z}\right) \tag{65}$$

I_p is a modified Bessel function of integer order (Abramovitz and Stegun (1973)). Comparison of Eq.(65) with Eqs.(62) and (63) leads to the following identification:

$$z = \exp(-\beta\hbar\omega);\; x = y = d^2;\; p = \bar{n} - n = \Delta n \tag{66}$$

and the rate constant becomes:

$$k = \frac{2\pi}{\hbar^2\omega}\varepsilon^2 \exp\left[\frac{\beta\Delta n\hbar\omega}{2}\right] \times$$
$$\exp\left[-d^2 coth\frac{\beta\hbar\omega}{2}\right] I_{\Delta n}\left[\frac{d^2}{sh\frac{\beta\hbar\omega}{2}}\right] \tag{67}$$

This expression can be made physically transparent in the high temperature and low temperature limits defined by $\beta\hbar\omega << 1$ and $\beta\hbar\omega >> 1$. The high temperature limit is valid at room temperature if the oscillator is the slow heavy - heavy motion (100 to 200 cm^{-1} for the quantum is a reasonable estimate). We assume also that $\Delta E/\hbar\omega$ and $d^2 = E_S/\hbar\omega$ are large so that there are the asymptotic expansions:

$$coth\frac{\beta\hbar\omega}{2} \sim \frac{2}{\beta\hbar\omega} + \frac{\beta\hbar\omega}{6} \; ; \qquad \frac{1}{sh\left(\frac{\beta\hbar\omega}{2}\right)} \sim \frac{2}{\beta\hbar\omega} - \frac{\beta\hbar\omega}{12} \tag{68}$$

For large ν and z but with $\nu^2/z << 1$, the Bessel function can be approximated as (Abramovitz and Stegun (1967)):

$$I_\nu(z) \sim (2\pi z)^{-\frac{1}{2}} \exp\left[z - \frac{\nu^2}{2z}\right] \tag{69}$$

ν is here $\Delta E/\hbar\omega$ and the condition $\nu^2/z << 1$ is equivalent to $(\Delta E)^2 << E_S \cdot k_B T$. The rate constant becomes:

$$k = \frac{\varepsilon^2}{\hbar}\left(\frac{\beta\pi}{E_S}\right)^{\frac{1}{2}} \exp\left\{\beta\left(\frac{\Delta E}{2} - \frac{E_S}{4} - \frac{(\Delta E)^2}{4E_S}\right\}\right. \tag{70}$$

with the replacement of $\Delta n\hbar\omega$ by ΔE. But this is also

$$k = \frac{\varepsilon^2}{\hbar}\left(\frac{\beta\pi}{E_S}\right)^{\frac{1}{2}} \exp[-\beta E_A] \tag{71}$$

with E_A defined by Eq. (60). This form given for the rate goes back to Marcus [61] who established it for an electron transfer process. E_A (cf. Figure 5) is the activation energy. The rate is proportional to the probability to reach the energy at the crossing point of the two potentials. The relation (71) is an Arrhenius law. In this high temperature regime there is no need to invoke tunneling to explain the relaxation of the oscillator (which could be the medium) when there is hopping of the hydrogen atom. Formula (71) gives the solution to one of the problems described in the introduction: observation of a temperature dependence of the rate and therefore of an activation energy does not require a consideration of the excited states of the hydrogen atom in the two wells. Excitation of the heavy-heavy motion is a much more plausible mechanism (Dogonadze *et al.* (1968)).

In the low temperature case with $\beta\hbar\omega >> 1$ we have

$$coth\frac{\beta\hbar\omega}{2} \to 1 \; ; \qquad sh\frac{\beta\hbar\omega}{2} \to \infty \tag{72}$$

so that the useful formula for the Bessel function is (Abramovitz and Stegun (1967))

$$I_\nu(z) \xrightarrow[z\to 0]{} \left(\frac{z}{2}\right)^\nu / \nu! \tag{73}$$

with

$$z = 2d^2 \exp\left[-\frac{\beta\hbar\omega}{2}\right] \quad \text{and} \quad \nu = \frac{\Delta E}{\hbar\omega} \tag{74}$$

There is no temperature dependence of the rate which becomes:

$$k = \frac{2\pi}{\hbar^2\omega}\varepsilon^2 \exp\left[-\frac{E_S}{\hbar\omega}\right]\left[\frac{E_S}{\hbar\omega}\right]^{\frac{\Delta E}{\hbar\omega}} \Big/ \left(\frac{\Delta E}{\hbar\omega}\right)! \tag{75}$$

This formula has a very simple interpretation. Only the zero-point level of the initial manifold is populated. The rate is also:

$$k = \frac{2\pi}{\hbar}\varepsilon^2 |\langle 0|\bar{n}\rangle|^2 (\hbar\omega)^{-1} \tag{76}$$

since the Franck-Condon factor is:

$$|\langle 0|\bar{n}\rangle|^2 = \exp[-d^2] d^{2\bar{n}}/\bar{n}! \tag{77}$$

with $d^2 = \frac{E_S}{\hbar\omega}$ and $\bar{n} = \Delta E/\hbar\omega$. The relaxation is entirely due to tunneling of the oscillator when the zero-point level of the initial state is below the crossing point. Another possible form for k is obtained by using Stirling formula:

$$\bar{n}! \simeq \sqrt{2\pi\bar{n}}\ \bar{n}^{\bar{n}} e^{-\bar{n}} \tag{78}$$

for the factorial. The alternative form for k is:

$$k = \frac{\sqrt{2\pi}\varepsilon^2}{\hbar\sqrt{\hbar\omega\Delta E}} \exp\left[\frac{\Delta E - E_S}{\hbar\omega}\right]\left[\frac{E_S}{\Delta E}\right]^{\frac{\Delta E}{\hbar\omega}} \tag{79}$$

We examine now briefly the endothermic case where the potential V_2 is higher than the potential V_1. We use again the notation ΔE to represent the energy mismatch, with ΔE positive. The formula for k is

$$\begin{aligned} k = \quad & \frac{2\pi}{\hbar^2\omega}\varepsilon^2 e^{-d^2}\left(1 - e^{-\beta\hbar\omega}\right) e^{-\beta\frac{\Delta n\hbar\omega}{2}} \\ & \times \sum_{\bar{n}=0}^{\infty} e^{-\beta\bar{n}\hbar\omega}\left(\frac{\bar{n}!}{n!}\right)\left[L_{\bar{n}}^{n-\bar{n}}\left(d^2\right)\right]^2 \end{aligned} \tag{80}$$

with now $\Delta n = n - \bar{n} = \Delta E/\hbar\omega$. The formula takes into account the fact that there is a rate only for levels with $n \geq \Delta E/\hbar\omega$. With the help of Eq.(65), k is given the form:

$$k = \frac{2\pi}{\hbar^2\omega}\varepsilon^2 \exp\left[-\frac{\beta\Delta n\hbar\omega}{2}\right] \exp\left[-d^2 coth\frac{\beta\hbar\omega}{2}\right] I_{\Delta n}\left(\frac{d^2}{sh\frac{\beta\hbar\omega}{2}}\right) \tag{81}$$

Comparing Eq.(81) with Eq.(67) shows that a sign change has occurred in the argument of the first exponential. The high temperature limit leads for the rate to:

$$k = \frac{\varepsilon^2}{\hbar}\left(\frac{\beta\pi}{E_S}\right)^{\frac{1}{2}} \exp\left\{-\beta\frac{(\Delta E + E_S)}{4E_S}\right\} \tag{82}$$

This is again an activated process where the energy needed allows to reach the crossing point. The low temperature limit of k is:

$$k = \frac{\sqrt{2\pi}\varepsilon^2}{\hbar\sqrt{\hbar\omega\Delta E}} \exp\left[\frac{\Delta E - E_S}{\hbar\omega}\right]\left(\frac{E_S}{\Delta E}\right)^{\frac{\Delta E}{\hbar\omega}} e^{-\beta\Delta E} \tag{83}$$

This expression has a simple meaning. The population in the level of V_1 facing the zero-point level of V_2 is proportional to $\exp[-\beta\Delta E]$ and the rate depends on the Franck-Condon factor between these two states. When the temperature goes to zero, the rate vanishes: there is no level of V_2 facing the zero-point level of V_1.

The formula (71) is probably the most useful to describe the basic features of hydrogen transfer when an additional degree of freedom (inter or intramolecular, or bath) is introduced. There are many other models which have been advanced to improve this expression. 1) The intersite matrix element ε has been considered as independent of either molecular or medium degrees of freedom. The necessity of introducing at least a dependence versus heavy-heavy motion (the heavy entities carrying the hydrogen atom) is now recognized in many treatments of the subject (Trakhtenberg *et al.* (1981),(1982), Borgis *et al.* (1989), Borgis and Hynes (1991), Suarez and Silbey (1991)). This changes the factor ε^2 of Eq.(71) into a quantity $\langle\varepsilon^2(Q)\rangle$, where the thermal average is on the heavy-heavy motion. In essence this means that the Condon approximation is not valid: the intersite matrix element (and therefore the splitting) varies considerably with the heavy-heavy distance (tunneling is assisted by reducing the distance over which the atom is transferred). This correction to formula (71) can therefore enhance considerably the rate. This is in contrast to electron transfer where the Condon approximation is more justified. 2) The oscillator which is present in the scheme leading to the rate expression given in this section is generally associated loosely with the medium. It is becoming now clear that two classes of degrees of freedom should be considered: the molecular (or intermolecular) degrees of freedom and the bath modes. Borgis and Hynes (1989)] have introduced both the heavy-heavy oscillator and an additional coordinate for the bath. Suarez and Silbey (1991) introduce the molecular coordinate and a bath of harmonic oscillators. These improvements of the basic model are difficult to test since they introduce many additional parameters in the rate. Such models with two classes of oscillators have already been explored by Jortner (1976) in the context of electron transfer. Since the medium oscillators and the molecular oscillators can be of quite different frequencies (10 to 100 cm^{-1} for a molecular solid, 700 to 3000 cm^{-1} for molecular motion), a regime which is easy to realize is that with a high temperature for the medium modes and a low temperature for the molecular modes. 3) Consideration of a 2 level system interacting with a bath of harmonic oscillators can also be very informative, although probably insufficient to give a complete account of the hydrogen transfer mechanism. We have already quoted the extensive work of Leggett et al. (1987). Silbey and Harris (1989) have developed an interesting variational approach based on the fact that the bath oscillators will not respond all in the same way to the hopping motion of the hydrogen atom. High frequency oscillators are able to follow the tunneling particle, while the low frequency ones should be less affected. If g_i denotes the coupling constant for an oscillator of frequency ω_i (cf the Hamiltonian in Eq.(55)), with the oscillators displaced by $\pm\frac{g_i}{m\omega_i^2}$, the matrix element responsible for the transfer at zero temperature is (assuming $\Delta E = 0$):

$$\langle L; \prod_i \langle 0_i|H|R; \prod_j |\bar{0}_j\rangle \;=\; \varepsilon \exp\left[-\sum_i \frac{g_i^2}{m\omega_i^3}\right] \;=\; \varepsilon_{\text{eff}} \tag{84}$$

$\langle 0_i|$ and $|\bar{0}_j\rangle$ represent the zero-point states of the oscillators when the hydrogen atom is in the left or the right well. The matrix element is reduced: the transfer is less effective because the hydrogen and the oscillator must move together. This effect is present in Eq.(62) for the rate valid for a single oscillator model. For a multioscillator treatment, there is an overestimation of the effect since low-frequency oscillators should not participate. Silbey and Harris ((1984),(1989)) introduce a parameter f_i per oscillator which is to be determined variationally to give its propensity to be displaced. There is a very suggestive relation for f_i:

$$f_i \;=\; g_i\left(1+\frac{2\varepsilon_{\text{eff}}}{\omega_i}\right) \tag{85}$$

For a high frequency oscillator $f_i \rightarrow g_i$ while for low frequency

$$f_i \;=\; g_i\left(\frac{\omega_i}{2\varepsilon_{\text{eff}}}\right) \tag{86}$$

The treatment of a system interacting with a bath of harmonic oscillators requires in addition some law to explain the way the couplings and the frequencies are distributed. This information comes through the so-called spectral density:

$$J(\omega) = \frac{g^2(\omega)}{\omega}\rho(\omega) \tag{87}$$

with $g(\omega)$ giving the coupling with oscillator of frequency ω and $\rho(\omega)$ being the density of oscillator frequencies. The choice of $J(\omega)$ can have profound effects on the dynamics (Leggett *et al.* (1987)). 4) Finally, there is to mention the recent use (Borgis and Hynes (1991), Azzouz and Borgis ((1993)) of molecular dynamical simulations to examine the correlation functions present in the general expression for the rate. This is an efficient method for assessing the importance of the various mechanisms.

REFERENCES

Abramovitz, M. and Stegun, I. A. (1967) Handbook of Mathematical Functions, Dover Publications, New york.

Avouris, P., Gelbart, W. M. and El-Sayed M. A. (1977) "Non-Radiative Electronic Relaxation under Collision-Free Conditions", Chem. Rev. **77** , 793-833.

Azzouz, H. and Borgis, D. (1993) "A Quantum-Dynamics Study of Proton-Transfer Reactions along Asymmetrical H Bonds in Solution", J. Chem. Phys. **98** , 7361-7374.

Barbara, P. F., Walsh, P. K. and Brus, L. E. (1989) "Picosecond Kinetic and Vibrationaly Resolved Spectroscopic Studies of Intramolecular Excited-State Hydrogen Atom Transfer", J. Phys. Chem. **93** , 29-34.

Basilevsky, M. V. and Ryaboy, V. M. (1982) "Quantum Dynamics of Linear Triatomic Molecules" in Advances in Quantum Chemistry, **15** , 1-83.

Baughcum, S. L., Duerst, R. W., Rowe, W. F., Smith, Z. and Wilson, E. B. (1981) "Microwave Spectroscopic Study of Malonaldehyde (3-Hydroxy-2-Propenal). Structure, Dipole Moment, and Tunneling", J. Am. Chem. Soc. **103** , 6296-6303.

Bicerano, J., Schaefer III, H. F. and Miller, W. H. (1983) "Structure and Tunneling Dynamics of Malonaldehyde. A Theoretical Study", J. Am. Chem. Soc. **105** , 2550-2553.

Bixon, M. (1982) "Medium Stabilization of Localized Born-Oppenheimer States", Chem. Phys. **70** , 199-206.

Borgis, D. C., Lee, S. and Hynes, J. T. (1989) "A Dynamical Theory of Non-Adiabatic Proton and Hydrogen Transfer Reaction Rates in Solution", Chem. Phys. Letters **162** , 19-26.

Borgis, D. C. and Hynes, J. T. (1991) "Molecular-Dynamics Simulation for a Model Nonadiabatic Proton Transfer Reaction in Solution" J. Chem. Phys. **94** , 3619-3628.

Bosch, E., Moreno, M., Lluch, J. M. and Bertran, J. (1990) "Bidimensional Tunneling Dynamics of Malonaldehyde and Hydrogenoxalate Anion. A Comparative Study", J. Chem. Phys. **93** , 5685-5692.

Bouma, W., Vincent, M. A. and Radom, L. (1978) "*Ab initio* Molecular Orbital Studies of Sigmatropic Rearrangements", Int. J. of Quantum Chem. **14** , 767-777.

Carrington, T. and Miller, W. H. (1986) J. Chem. Phys. **84** , 4364-4370.

Connor, J. N. L. (1969) "On the Semiclassical Approximation for Double Well Potentials", Chem. Phys. Letters **4** , 419-420.

Cukier, R. I. (1988) "Non-Arrhenius Rate Constants for Non-Adiabatic Electron Tranfers: The Role of Quantum and Solvent Dynamics", J. Chem. Phys. **88** , 5594-5605.

Cukier, R. I. and Morillo, M. (1989) "Solvent Effects on Proton-Tranfer Reactions", J. Chem. Phys. **91** , 857-863.

Dogonadze, R. R., Kuznetzov, A. M. and Levich, V. G. (1968) "Theory of Hydrogen-Ion Discharges on Metals: Case of High Overvoltages", Electrochim. Acta **13** , 1025-1044.

Eckart, C. (1930) "The Penetration of a Potential by Electrons" Phys. Rev. **35** , 1303-1309.

Englman, R. and Jortner, J. (1970) "The Energy Gap Law for Radiationless Transitions in Large Molecules", Mol. Phys. **18** , 145-164.d

Fain, B. (1979) "Adiabatic Rate Processes in Condensed Media", Chem. Phys. Letters **67** , 267-272.

Feynman, R. P., Vernon, F. L. and Hellwarth, R. W. (1957) "Geometrical Representation of the Schrödinger Equation for Solving Maser Problems", J. Appl. Phys. **28** , 49-52.

Freed, K. and Jortner, J. (1970) "Multiphoton Processes in the Non-Radiative Decay of Large Molecules", J. Chem. Phys **52** , 6272-6291.

Frisch, M. J., Scheiner, A. C. and Schaefer III, H. F. (1985) "The Malonaldehyde Equilibrium Geometry: A Major Structural Shift Due to the Effects of Electron Correlation", J. Chem. Phys. **82** , 4194-4198.

Fulton, R. L. and Gouterman, M. (1961) "Vibronic Coupling. I.Mathematical Treatment for Two Electronic States", J. Chem. Phys. **35** , 1059-1071.

Garrett, B. C. and Truhlar, D. G. (1979) "Semiclassical Tunneling Calculations", J. Phys. Chem. **83** , 2921-2926.

Gradshteyn, S. I. and Ryzhik, I. M. (1980) Tables of Integrals, Series, and Products, Academic Press, New York.

Hancock, G. C. and Truhlar, D. G. (1989) "Reaction-Path Analysis of the effect of Monomer Excitation Splitting of the Hydrogen Fluoride Dimer", J. Chem. Phys. **90** , 3498-3505.

Harris, R. A. and Stodolsky, L. (1981) "On the Time Dependence of Optical Activity", J. Chem. Phys. **74** , 2145-2155.

Harris, R. A. and Silbey, R. (1985) "Variational Calculation of the Tunneling System Interacting with a Heat Bath. II. Dynamics of an Asymmetric Tunneling System", J. Chem. Phys. **83** , 1069-1074.

Herzberg, G. (1967) Molecular Spectra and Molecular Structure. III. Electronic Spectra and Electronic Structure of Polyatomic Molecules, Van Nostrand Company, Princeton.

Huang, K. and Rhys, A. (1950) "Theory of Light Absorbtion and Non-Radiative Transitions in F-Centres", Proc. Roy. Soc. **A204** , 406-423.

Hudson, R. L., Shiotani, M. and Williams F. (1977) "Hydrogen Atom Abstraction by Methyl Radicals in Methanol Glasses at 15-100 K: Evidence for a Limiting Rate Constant Below 40 K by Quantum-Mechanical Tunneling" Chem. Phys. Letters **48** , 193-196.

Isaacson, A. D. and Morokuma, K. (1975) "Molecular Studies of Hydrogen Bonds. VIII. Malonaldehyde and Symmetric Hydrogen Bonding in Neutral Species" J. Am. Chem. Soc. **97** , 4453-4457.

Jortner, J. and Mukamel, S. (1974) "Preparation and Decay of Excited Molecular States", in R. Daudel and B. Pullman(eds),The world of Quantum Chemistry, Reidel Publishing Company, Dordrecht, pp. 145-209.

Jortner, J. (1976) "Temperature Dependent Activation Energy for Electron Transfer between Biological Molecules", J. Chem. phys. **64** , 4860-4867.

Jortner, J. and Ulstrup, J. (1979) "Tunneling in Low-Temperature Atom-Transfer Processes", Chem. Phys. Letters **63** , 236-239.

Karlström, G., Jönsson, B., Roos, B. (1976) "Correlation Effects on Barriers to Proton Transfer in Intramolecular Hydrogen Bonds. The Enol Tautomer of Malondialdehyde Studied by ab Initio SCF-CI Calculations", J. Am. Chem. Soc. **98** , 6851-6854.

Kato, S., Kato, H. and Fukui, K. (1976) "Hydrogen-Bonded Proton in Malonaldehyde", J. Am. Chem. Soc. **99** , 684-691.

Korst, N. N. and Nikitin, E. E. (1965) "Relaxation in a Double Well Potential", Theor. Exp. Chem. **1** , 11-21, (in russian).

Kubo, R. and Toyozawa, Y. (1955) "Application of the Method of Generating Function to Radiative and Non-Radiative Transitions of a Trapped Electron in a Crystal", Prog. of Theor. Phys. **13** , 160-182.

Lax, M. (1952) "The Franck-Condon Principle and its Application to Crystals", J. Chem. Phys **20** , 1752-1760.

Le Roy, R. J., Murai, H. and Williams, F. (1980) "Tunneling Model for Hydrogen Abstraction Reactions in Low-Temperature Solids. Applications to Reactions in Alcohol Glasses and Acetonitrile Crystals", J. Phys. Chem. **102** , 2325-2334.

Lefebvre, R. and Taylor, H. (1989) "Resonance and Transmission", J. of Mol. Struct.(Theochem) **199** , 327-336.

Lefebvre, R. and Moiseyev, N. (1990a) "Artificial Resonance Procedure for the Determination of Quantum Mechanical Rate Constant in the Tunneling Regime", J. Chem. Phys. **93** , 7173-7178.

Lefebvre, R. and Moiseyev, N. (1990b) "Automerization of Cyclobutadiene", J. Am. Chem. Soc. **112** , 5052-5054.

Leggett, A. J., Chakravarty, S., Dordey, A. T., Fisher, M. P. A., Garg, A. and Zwerger, W. (1987) "Dynamics of the Dissipative Two-State System", Rev. Mod. Phys. **59** , 1-85.

Levich, V. G. (1966) "Present State of the Theory of Oxidation-Reduction in Solution (Bulk and Electrode Reactions)", in Advances in Electrochemistry and Electrochemical Engineering, P. Delahay (ed.), **4** , 249-371.

Makri, N. and Miller, W. H. (1987) "Basis Set Methods for Describing the Quantum Mechanics of a "System" Interacting with a Harmonic Bath", J. Chem. Phys. **86** , 1451-1457.

Marcus, R. A. (1956) "On the Theory of Oxidation-Reduction Reactions Involving Electron Transfer.I", J. Chem. Phys. **24** , 966-978.

Marcus, R. A. (1966) "On the Analytical Mechanics of Chemical Reactions. Quantum Mechanics of Linear Collisions", J. Chem. Phys. **45** , 4493-4504.

Marcus, R. A. and Coltrin, M. E. (1977) "A New Tunneling Path for Reactions such as $H + H_2 \rightarrow H_2 + H$", J. Chem. Phys. **67** , 2609-2613.

McCurdy, C. W. and Garrett, B. C. (1986) "Quantum Mechanical Microcanonical Rate Constants from Direct Calculations of the Green's Function for Reactive Scattering", J. Chem. Phys. **84** , 2630-2642.

McKinnon, W. R. and Hurd, C. M. (1983) "Tunneling and the Temperature Dependence of Hydrogen Transfer Reactions", J. Phys. Chem., **87** , 1283-1285.

Merzbacher, E. (1967) Quantum Mechanics, Wiley and Sons, New York.

Meyer, R. and Ernst, R. R. (1987) "Hydrogen Transfer in Double Minimum Potential: Kinetic Properties derived from Quantum Dynamics", J. Chem. Phys., **86** ,784-801.

Meyer, R. and Ernst, R. R. (1990) "Transitions Induced in a Double Minimum System by Interaction with a Quantum Mechanical Heat Bath", J. Chem. Phys., **93** , 5518-5532.

Miller, W. H., Handy, N. C. and Adams, J. E. (1980) "Reaction Path Hamiltonian for Polyatomic Molecules", J. Chem. Phys. **72** , 99-112.

Nikitin, E. (1967) Theory of Elementary Atomic and Molecular Processes in Gases, Clarendon Press, Oxford.

Perlin, Yu. I. (1964) "Modern Methods in the Theory of Many-Phonon Processes", Sov. Phys. Uspekhi **6** , 542-565.

Schuster, P. (1969) "LCAO-MO Calculations on the Enol Form of Acetylacetone and its Metal Complexes (LCAO-MO-Studies on Molecular Structure III)", Chem. Phys. Letters **3** , 433-436.

Shida, N., Barbara, P. F. and Almlöf (1989) "A theoretical Study of Multidimensional Nuclear Tunneling in Malonaldehyde", J. Chem. Phys. **91** , 4061-4072.

Siebrand, W., Wildman, T. A. and Zgierski, M. Z. (1984) "Golden Rule Treatment of Hydrogen-Transfer Reactions.1./Principles", J. Am. Chem. Soc. **106** , 4083-4089. "2. Applications" *ibid.* 4089-4096.

Silbey, R. and Harris, R. A. (1984) "Variational Calculation of the Dynamics of a Two Level System Interacting with a Bath", J. Chem. Phys. **80** , 2615-2617.

Silbey, R. and Harris, R. A. (1989) "Tunneling of Molecules in Low-Temperature Media: An Elementary Description", J. Phys. Chem. **93** , 7062-7071.

Simonius, M. (1978) "Spontaneous Symmetry Breaking and Blocking of Metastable States", Phys. Rev. Letters **40** , 980-983.

Skinner, J. L. and Tromsdorff H. P. (1988) "Proton Transfer in Benzoic Acid Crystals: A Chemical Spin-Boson Problem. Theoretical Analysis of Nuclear Magnetic Resonance, Neutron Scattering, and Optical Experiments", J. Chem. Phys. **89** , 897-907.

Skodje, R. T., Truhlar, D. G. and Garrett, B. C. (1981) "A General Small-Curvature Approximation for Transition-State-Theory Transmission Coefficients", J. Phys. Chem. **85** , 3019-3023.

Suarez, A. and Silbey, R. (1991) "Hydrogen Tunneling in Condensed Media", J. Chem. Phys. **94** , 4809-4816.

Trakhtenberg, L. I., Klochikhin, V. L. and Pshezhetsky, S. Ya. (1981) "Tunneling of a Hydrogen Atom in Low Temperature Processes", Chem. Phys. **59** , 191-198.

Trakhtenberg, L. I., Klochikhin, V. L. and Pshezhetsky, S. Ya. (1982) "Theory of Tunnel Transitions of Atoms in Solids", Chem. Phys. **69** , 121-134.

Truhlar, D. G., Isaacson, A. D. Skodje, R. T. and Garrett, B. C. (1982) "Incorporation of Quantum Effects in Generalized-Transition-State Theory", J. Phys. Chem. **86** , 2252-2261.

Ulstrup, J. and Jortner, J. (1975) "The Effect of Intramolecular Quantum Modes on Free Energy Relationships for Electron Transfer Reactions", J. Chem. Phys., **63** , 4358-4368.

Walet, N. R., Klein, A. and Do Dang, G. (1989) "Reaction Paths and Generalized Valleay Approximation", J. Chem. Phys. **91** , 2848-2858.

Wertheimer, R. and Silbey, R. (1980) "On Excitation Transfer and Relaxation Models in Low-Temperature Systems", Chem. Phys. Letters **75** , 242-248.

Wilson, E. B., Decius, J. C. and Cross, P. C. (1980) Molecular Vibrations, Dover Publications, New York.

BUILDING A BRIDGE BETWEEN AB INITIO AND SEMIEMPIRICAL THEORIES OF MOLECULAR ELECTRONIC STRUCTURE

Karl F. Freed
James Franck Institute and Department of Chemistry
The University of Chicago, Chicago, Illinois 60637 USA

1 Introduction

The phenomenal increase in computer capabilities, new theoretical developments, and advances in algorithms have sparked a likewise phenomenal increase in the capabilities of ab initio electronic structure methods for accurately describing molecular electronic structure. Thus, ab initio electronic structure computer "packages" are now available to the general scientific community, and their usage is in a rapid stage of growth. While there is a concomitant increase in the sizes of molecules that may be accurately treated with ab initio methods, interesting molecular systems always exist that far exceed the capabilities of the most advanced ab initio electronic structure approaches. Consequently, many approximate and semiempirical electronic structure methods are and must always continue to be necessary for describing the electronic properties of these large systems.

The early development of theories for chemical bonding focused upon semiempirical approaches[1,2,3] for simple pi-electron systems. Most modern semiempirical approaches are direct generalizations of the earlier successful pi-electron methods. One generic example is Hückel theory for the electronic structure of pi-electron systems. The pi-electron methods form a central position in the history and the development of semiempirical methods[1,2]. Deficiencies with the predictions of the one-electron Hückel pi-electron methods led to the introduction of Pariser-Parr-Pople (PPP) theory[1,4], which explicitly contains the electron-electron repulsion. Further developments involve the inclusion of "all" valence electrons[5]. The structures of these newer (and current) semiempirical methods are direct generalizations of the earlier successful pi-electron methods.

The advent of digital computers has provided the tools necessary for performing ab initio computations for the smallest of molecules, and these earliest computations already raised serious questions concerning the

J. L. Calais and E. S. Kryachko (eds.), Structure and Dynamics of Atoms and Molecules: Conceptual Trends, 25–67.

fundamental assumptions of semiempirical methods, questions which are continually reinforced by comparisons with accurate ab initio computations and which persist even today for the most advanced semiempirical approaches. A review of these questions provides insights and challenges associated with relating ab initio and semiempirical electronic structure approaches. First of all, the most sophisticated semiempirical methods employ a minimal basis set of orbitals, whereas the lowest level of ab initio computations requires at least a split valence orbital basis[6], twice the valence basis size of that in the all-valence-electron semiempirical approaches. Highly accurate, correlated ab initio computations[7], on the other hand, require very extensive basis sets with several levels of polarization functions[8]. This glaring difference alone leads many ab initio theorists to question whether the semiempirical approaches are merely sophisticated curve fitting methods, bereft of any true theoretical foundations. Thus, this comparison raises the first and most fundamental question posed by semiempirical methods, namely, whether it is theoretically meaningful to transform the full molecular Schrödinger equation with a formally infinite basis set into a valence only problem involving a minimal valence basis[9,10]. (Some frontier orbital methods reduce the valence space further to one or a few highest occupied and lowest unoccupied molecular orbitals).

While all electronic structure methods strive for accuracies of one kcal/mol, there are strong differences in how semiempirical and ab initio approaches achieve this goal. The ab initio methods can generate this chemical accuracy only by using large basis sets and high levels of electron correlation, thereby limiting treatment to "smaller" molecules because of the great computational expense. (However, the definition of "small" in this context is constantly growing with time!) Semiempirical methods likewise strive[11] for accuracies of fractions of a kcal/mol for bond and low lying electronic excitation energies of rather large and complex molecules. Their achieved accuracies are about 4kcal/mol for hydrocarbons and less for more complicated molecules[12]. Nevertheless, the semiempirical approaches accomplish this admirable feat for molecules which are quite large and for which the most advanced, correlated, ab initio methods would be prohibitive. Thus, the accomplishments of semiempirical electronic structure methods pose the following fundamental question: How can the semiempirical methods achieve such accuracy with just a minimal valence basis set? The general rationale argues[3] that the valence orbital integrals over one and two-electron interactions in the semiempirical methods (called the semiempirical parameters below) have been adjusted to incorporate correlation effects, thereby obviating the need for using an extended basis and also producing the quoted high accuracies. By assuming that semiempirical parameters implicitly include correlation contributions, many additional questions arise as follows[1,9]: Is the process theoretically justified? If correlation is introduced into semiempirical parameters using a self-consistent field formulation and then if the semiempirical method uses configuration interaction (CI) in the computation of energies, has there been double counting[1,9,10] of correlation contributions? Early semiempirical approaches are parametrized either to compute bond energies or electronic excitation energies, etc., but now advanced methods are attempting to obtain geometries, bond, excitation, ionization, etc., energies from a single semiempirical scheme[13]. CI is increasingly being used,[14,15] and the double counting question then looms stronger.

These questions are heightened further by the general philosophy behind the current construction of semiempirical methods, a philosophy which is best exemplified by Sadlej[16]: "Let us, therefore, assume that our wavefunctions arise from non-empirical calculations. The resulting assertions and conclusions will then be extended to corresponding semiempirical variants of these methods. How far this procedure is valid can only be justified a posteriori by comparing the results with those arising from non-empirical calculations or with experimental data. Nevertheless, it is commonly accepted that useful theorems valid for non-empirical functions are also valid for semiempirical versions of the same method. Unfavorable theorems and conclusions are eliminated by an appropriate choice of empirical parameters." Thus, the all-valence-electron methods are generally parametrized[17] by considering the form of Hamiltonian matrix elements within a minimal valence orbital basis set[5], but this is done with an underlying hope[3] that "correlation contributions" from the complete set of omitted orbitals may somehow be incorporated into the semiempirical parameters.

If it were possible to derive exact expressions for correlation contributions to the individual integrals, then a number of the assumptions regarding the structure and transferability of semiempirical parameters could be tested directly[1], thereby hopefully reducing the laborious trial and error process currently required in deriving new parameter sets[17]. Lacking the theoretical expressions for the correlation contributions to individual integrals, semiempirical parameterizations have proceeded under the assumption that it is merely necessary to alter the form of theoretical integrals to incorporate correlation corrections. This bold assumption requires testing based on rigorous quantum mechanical theories enabling the ab initio computation of all semiempirical parameters[1]. It is, of course, likely that these sophisticated tests would suggest revision of current practices. Nevertheless, even given the strong assumption of merely modifying theoretical integrals, it has been necessary to proceed by a laborious trial and error parameterization procedure that lengthens as the semiempirical method becomes more sophisticated[14,15,17]. Thus, a derivation of computationally usable expressions for these individual correlation contributions would enable addressing such a basic question as to whether semiempirical methods are neglecting significant parameters. For instance, the zero differential overlap (ZDO) approximation argues that certain repulsion integrals become small in a Löwdin orbital basis and are therefore negligible. However, the theoretical nearest-neighbor, two-center, pi-electron exchange integral for carbon atoms is about 0.3eV and is quite significant for methods that strive towards accuracies of less than a kcal/mol. Does the inclusion of correlation make this parameter small enough to be truly negligible, or does correlation enhance its value to the extent that its retention is warranted in semiempirical methods? Similar arguments apply to other ZDO approximation neglected interactions. The question has numerous applications as Hirsch[18] has argued that a two-electron contribution to resonance integrals is necessary for understanding the electronic structure of high temperature superconductors, while Iwata and Freed[19] have shown how the same interaction contributes, in part, to the breakdown of the paring theorem in alternant hydrocarbons[20], leading, for example, to a dissymmetry between the spectra of cations and anions.

Several fundamental questions center on whether other parameters exhibit correlation contributions with different dependences on bond lengths, etc., than assumed in semiempirical models because the prior parameterizations have proceeded without the benefit of rigorous, systematic, ab initio theoretical guidance concerning the structure of correlation contributions to individual parameters. Some semiempirical practitioners already appreciate[14] the role of theory in formulating improved semiempirical approaches[21]: "... parameterization based only on experimental results can lead to strange physical behavior in regions where no or a small number of experimental data are available. Ab initio calculations may help fix those problems." Thus, resolving some of the questions posed above could have several beneficial consequences. The first is that ab initio theoretical descriptions of parameter dependences must be richer and more accurate than the more intuitive ones currently in use. Additional parameters may be necessary, and alternative variations with bond lengths may be uncovered.

Ab initio theory may aid in further extending the capabilities of current semiempirical methods. These extensions include alleviating present difficulties in parameterizing semiempirical methods with extended valence spaces. Completely satisfactory parameterizations are still lacking for transition metal systems, especially if the desire is to obtain semiempirical methods that cover the whole range of formal charge states, high and low spin, and bond as well as electronic excitation energies. Extensions would be welcomed to computations of the parameters in semiempirical treatments of solids, especially high temperature superconductors[22], conducting polymers, and nonlinear optical materials[23]. Furthermore, semiempirical methods are notoriously poor in representing dipole moments of small molecules, while they are extensively used for describing nonlinear optical properties of large molecules[24]. The same semiempirical methods are parametrized to describe the energetics of small molecules[12,14,15,17,25,26,27] before using them for larger systems, so the difficulties with treating dipole moments of small molecules signals the presence in semiempirical methods of serious systematic errors whose resolution with theoretical guidance[28] must have important practical ramifications.

While theory might contribute to many possible improvements of semiempirical methods, a more thorough understanding of the latter may enhance purely ab initio capabilities. Semiempirical methods are predicated on the use of localized, transferable parameters, and the successes of semiempirical methods suggest that correlation can indeed be represented in a localized fashion, a form in which many ab initio theories have sought to express correlation for more efficient computation in large molecular systems[29]. A more thorough understanding of how correlation enters into individual semiempirical parameters may therefore aid with the development of localized ab initio correlation treatments.

The first and most fundamental question in devising practical theoretical connections between ab initio and semiempirical approaches lies in explaining whether and why the full Schrödinger equation can be transformed to a minimal valence shell problem with perhaps some different structure than that customarily assumed by semiempirical methods[9,10,30]. Thus, section II considers the establishment of this formal bridge, followed by section III in which we outline recent ab initio computations based upon the resultant theoretical formulation, computations which assess the possible accuracy of calculating the correlation contributions to individual parameters. This section indicates the requisite high degree of correlation, the large basis sets, and hence the huge scale computations, which have only recently become possible to the degree necessary for treating interesting molecular systems. Section IV describes recent advances in calculating the correlation contributions to individual parameters.

2. Theoretical bridge between ab initio and semiempirical methods

We confine attention to those semiempirical methods that are characterized by using a set of valence orbitals and a semiempirical effective valence shell Hamiltonian $H_M{}^v$, which contains the empirical parameters. Thus, our considerations do not include density functional[31] or many-body Green's function approaches[32], which necessitate the use of different theoretical formulations. Early attempts at a theoretical justification of semiempirical methods[1] are flawed because most are based on overly optimistic beliefs concerning the type of ab initio computation that is necessary to achieve experimental accuracy. Many more recent attempts focus narrowly on oversimplified models as discussed below. Thus, we begin by assessing what are the components of semiempirical methods that may be tested theoretically.

An optimal semiempirical method enables accurate computation of bond energies, activation barriers to chemical reactions, electronic excitation energies, ionization potentials, etc., with *a single set of parameters for all properties and electronic states of interest.* Consequently, this optimal many-state approach cannot be understood with widely used theoretical "justifications" of semiempirical models that are based on the properties of only a single electronic state (usually the ground state). These single state arguments generally begin with the expression for the self-consistent field (SCF) energy and then append pairwise additive correlation energies. The many-state $H_M{}^v$ must be understood with a more sophisticated theory involving many electronic states of a given molecular system.

2.1. DERIVATION OF THE EXACT EFFECTIVE VALENCE SHELL ONLY EQUATIONS

The many-state analog of $H_M{}^v$ may simply be determined by analyzing the logical (i.e., purely mathematical) content of this $H_M{}^v$. Given a particular $H_M{}^v$ along with its set of valence orbitals, it is, in principle, possible to solve for all its eigenfunctions $\psi_i{}^v$ and eigenvalues E_i by expanding the semiempirical wavefunction $\psi_i{}^v$ in terms of the valence functions $|p\rangle$, where $|p\rangle$ is, for instance, a determinantal function containing electrons only in the valence orbitals (with a filled core implicit)[9,10,30]. Thus, we have the semiempirical Schrödinger equation,

$$H_M^v \psi_i^v = E_i \psi_i^v, \tag{2.1}$$

where ψ_i^v is given by the full valence shell CI expansion.

$$\psi_i^v = \sum_p C_{pi} |p\rangle. \tag{2.2}$$

Combining equations (2.1) and (2.2) and using standard matrix notation converts these into

$$H_M^v C_i = E_i C_i. \tag{2.3}$$

When applying the semiempirical methods to large molecular systems, it is often desirable to approximate the $\psi_i{}^v$ using less than a full valence shell CI computation, but the introduction of this numerical approximation represents a secondary question that may be addressed subsequently. (It is well understood that the choice of semiempirical parameters depends somewhat on the amount of CI that is used in conjunction with the method.) We, therefore, seek to determine whether the full electronic Schrödinger equation,

$$H\psi_i = E_i \psi_i, \tag{2.4}$$

can be expressed equivalently in a valence only representation of the form in Eq. (2.3), but with an exact effective valence shell Hamiltonian matrix H^v replacing the semiempirical model $H_M{}^v$ in Eq. (2.3) and with E_i in Eq. (2.3) identical to the exact energy in (2.4).

The formal derivation of this exact effective valence shell Hamiltonian is quite straightforward[9,10,30,33,34]. Let $|p\rangle$ also designate the zeroth order valence space wavefunction with the filled core orbitals explicitly included. Now collect the full Hamiltonian H matrix elements $\langle p|H|p'\rangle$ between the valence configurations $|p\rangle$ and $|p'\rangle$ into the block matrix H_{PP}, where the subscript P designates the valence space. An exact ab initio computation requires the introduction of a set (in principle, infinite) of complimentary configurations $\{|q\rangle\}$ which contain occupied excited orbitals, vacant core orbitals, or both. The Hamiltonian matrix $\langle q|H|q'\rangle$ in the "excited" (or "correlating") space is likewise collected into the block matrix H_{QQ}. Introducing the off-diagonal blocks between P and Q-spaces as H_{PQ} and H_{QP} and the P and Q-space blocks of the CI column vectors C_i^P and C_i^Q, respectively, equations (2.4) with the full CI analog of Eq. (2.2) are transformed into the block matrix equations,

$$\begin{pmatrix} H_{PP} & H_{PQ} \\ H_{QP} & H_{QQ} \end{pmatrix} \begin{pmatrix} C_P \\ C_Q \end{pmatrix} = E \begin{pmatrix} C_P \\ C_Q \end{pmatrix}, \qquad (2.5)$$

where the state index i has been dropped for convenience. Multiplying out the block matrix equations from Eq. (2.5) yields the pair of matrix equations

$$H_{PP}C_P + H_{PQ}C_Q = EC_P, \qquad (2.6a)$$

$$H_{QP}C_P + H_{QQ}C_Q = EC_Q. \qquad (2.6b)$$

Formally solving equation (2.6b) as

$$C_Q = (E1_Q - H_{QQ})^{-1}H_{QP}C_P \qquad (2.6c)$$

enables the substitution of Eq. (2.6c) into (2.6a) to give the exact effective valence shell Schrödinger equation as

$$H^vC_P = EC_P, \qquad (2.7)$$

where the exact effective valence shell Hamiltonian is obtained as

$$H^v = H_{PP} + H_{PQ}(E1_Q - H_{QQ})^{-1}H_{QP}. \qquad (2.8)$$

Equation (2.8) is a matrix solely in the space of valence functions. However, the correlation corrections enter into Eq. (2.8) from the remainder of the infinite (complete) basis set

Equation (2.7) has the identical formal structure as the semiempirical full CI equations (2.3) when it is recognized[9,10,30] that the two basis sets involve formally identical sets of valence functions $|p\rangle$. Thus, the full semiempirical matrix Schrödinger equation (2.3) represents an attempt to mimic the exact matrix equation (2.7). Consequently, the above simple derivation *proves that it is meaningful to reduce the full electronic Schrödinger equation to a valence orbital only equation, provided the Hamiltonian in this valence space is taken as the exact effective valence shell Hamiltonian H^v of (2.8).* The valence orbital basis may be a minimal basis, or it may contain fewer frontier orbitals. The exact H^v is, however, different for these two choices of valence spaces.

Although equations (2.7) and (2.8) justify the first and most fundamental minimal basis tenet of semiempirical methods, there is no guarantee yet that the exact H^v of (2.8) has almost the precise structure postulated by some semiempirical methods. First of all, the explicit dependence of Eq.

(2.8) on the state energy E differs from the assumed semiempirical forms. The next subsection, however, introduces expansions which remove this E-dependence to produce computationally viable expressions for H^v that are independent of E and therefore identical for all states of interest. Thus, we now turn to an examination of some properties of and some computational procedures for H^v.

2.2. PROPERTIES OF AND COMPUTATIONAL METHODS FOR H^v

The first $\mathbf{H_{PP}}$ contribution to H^v in Eq. (2.8) is just the Hamiltonian matrix in the space of valence functions, while the second term on the right hand side of Eq. (2.8) represents the "correlation contribution" because it involves excitations into a complete set of excited orbitals, excitations out of the core orbitals, and all possible combinations of both. The matrix elements in $\mathbf{H_{PP}}$ contain the "theoretical" or "bare" integrals, while the correlation contribution exactly adjusts all these integrals for correlation as implicitly assumed by the semiempirical methods. Many ab initio theorists mistakenly believe that the semiempirical methods approximate $\mathbf{H_{PP}}$ alone, but the formal derivation of Eq. (2.8) proves that it is possible to incorporate all correlation contributions into the valence only problem of Eq. (2.7).

Unfortunately, the exact computation of H^v from Eq. (2.8) is impossible because the inverse matrix is infinite, and the use of large basis sets renders the matrix much too huge to be inverted for large systems. Thus, some approximations are necessary beyond the well-studied ones of using finite basis sets. One convenient approach to computing the H^v of Eq. (2.8) involves the introduction of many-state (van Vleck type[35,36,37,38]) Rayleigh-Schrödinger perturbation theory. As now described, these formulations produce perturbative expressions for an approximate H^v which no longer depends on the individual states, e.g., as does Eq. (2.8) by virtue of the presence in the denominator of the unknown state energy E.

For convenience, we present the perturbation theory formulas for the case in which the valence space is taken to be exactly degenerate. The application of perturbation theory requires the decomposition of the full Hamiltonian H into its zeroth order part H_0 and the perturbation V,

$$H = H_0 + V, \tag{2.9}$$

where the valence space determinantal functions $|p\rangle$ and the secondary space determinantal functions $|q\rangle$ are eigenfunctions of H_0,

$$H_0|p\rangle = E_p{}^0|p\rangle, \quad H_0|q\rangle = E_q{}^0|q\rangle. \tag{2.10}$$

The condition of complete degeneracy is that $E_p{}^0 = E_P{}^0$ for all $|p\rangle$ in the valence space. The exact energies E in the denominator or Eq. (2.8) must likewise be expanded about the zeroth order energies $E_P{}^0$, and a wide variety of formalisms exist for accomplishing this expansion[39,40,41]. (The details are somewhat more involved for the non-degenerate case.)

Degeneracy simplifies the multireference perturbation expression for the Hermitian H^V into

$$\langle p|H^V|p'\rangle = \langle p|H|p'\rangle + \sum_q \frac{\langle p|V|q\rangle\langle q|V|p'\rangle}{E^0{}_p - E^0{}_q}$$

$$+ \sum_{q,q'} \frac{\langle p|V|q\rangle\langle q|V|q'\rangle\langle q'|V|p'\rangle}{(E^0{}_p - E^0{}_q)(E^0{}_p - E^0{}_{q'})} + \sum_{q,p''} \frac{\langle p|V|q\rangle\langle q|V|p''\rangle\langle q|V|p''\rangle}{2(E^0{}_p - E^0{}_q)^2}$$

$$+ \text{h.c.} + \cdots, \tag{2.11}$$

where h.c. designates the Hermitian conjugate of the preceding term.

The individual third order contributions to Eq. (2.11) [containing the sums over q,q' and q,p'', respectively] have portions incorrectly scaling as the square of the number of electrons in the system. Brandow[39], however, has shown that upon use of a "complete" valence space (see the definition below), the improper terms cancel identically between these two third order terms and the combination of these two third order contributions is indeed size-extensive, scaling properly with the number of electrons in the system[42]. Brandow's linked-cluster theorem is proven to all orders in multireference configuration perturbation theory[39], and Brandow presents a convenient diagrammatic method[34,39] for reducing the original perturbation expressions, which contain N-body matrix elements of V as in Eq. (2.11), to ones in terms of the matrix elements of the one and two-electron interactions in the Hamiltonian within a complete set of core {c}, valence {v} and excited {e} orbitals. The complete valence space requires that the set of valence functions $\{|p\rangle\}$ represents what is often called a "complete active space" in which the core orbitals are all doubly occupied and the remaining electrons occupy only valence orbitals. However, this complete active space is just that required to make H^V and the semiempirical $H_M{}^V$ act on identical spaces[9,10,30].

The expansion in Eq. (2.11) exhibits each N-electron matrix element $\langle p|H|p'\rangle$ as being modified by the correlation contributions which first enter at second order in V. Simple arguments enable showing how this, in turn, implies that the individual matrix elements of the one and two-body portions of H^V contain correlation contributions, thereby formally justifying the alternation of bare H matrix elements for correlation. The actual computation of these matrix elements from Eq. (2.11) is rather involved and is facilitated by use of many-body theory techniques[34,39]. Since the latter would take us too far afield, we briefly provide a motivational sketch of a possible (but more tedious) procedure that avoids introducing many-body theory.

The first tedious step involves substituting N-electron Slater determinants for the functions $|p\rangle$ and $|q\rangle$ into Eq. (2.11) in order to reduce the sums-over-states expressions into ones involving sums over core, valence, and excited orbitals and their matrix elements of the one and two-body interactions in the original Hamiltonian H. When the system contains no valence electrons, Eq. (2.11) provides the nondegenerate perturbation expression for the correlated core energy E_C. Next, we choose the N=1 one-valence-electron states and substitute these N=1 states into the sums-over-states perturbation equations. Then, represent each matrix element in the matrix $\langle p|H^v|p'\rangle$, on the one hand, as an expansion in terms of sums over orbitals. On the other hand, use of the operator representation

$$H^v = E_C + \sum_{i=1}^{N_v} H_i^v,$$

with H_M^v an effective one-body operator, enables expressing the matrix elements $\langle p|H^v|p'\rangle$ as matrix elements of H_i^v in the valence orbital basis. Equating the two matrices provides explicit perturbative formulas for computing the valence shell matrix elements $\langle v|H_i^v|v'\rangle$, where $|v\rangle$ and $|v'\rangle$ are valence orbitals.

Passing on to the N=2 two-valence-electron states, the matrix elements from the right hand side of Eq. (2.11) in the sum over orbitals form is then equated to the matrix elements constructed from the left hand side using the operator representation

$$H^v = E_C + \sum_{i=1}^{N_v} H_i^v + \frac{1}{2!}\sum_{i\neq j=1}^{N_v} H_{ij}^v$$

where H_{ij}^v is the effective two-body operator and the effective one-body operator is that already obtained from the N=1 states. This produces perturbation expansions for the two-electron matrix elements $\langle v_1v_2|H_{ij}^v|v_3v_4\rangle$, with valence orbitals $|v_1\rangle$ through $|v_4\rangle$.

Let us now analyze in detail one correction term arising from the second order portion of Eq. (2.11) in order to provide further understanding of some general structure of H^v. Consider the contributions from excited configurations $|q\rangle$ containing a single $v \rightarrow e$ valence to excited orbital excitation. After evaluation of the N-electron matrix elements, these single excitations yield contributions with the numerator,

$$\langle v_1v_2|\frac{1}{r_{12}}|v_3e\rangle\langle v_4e|\frac{1}{r_{12}}|v_5v_6\rangle, \qquad (2.12)$$

where $1/r_{12}$ is the electron-electron interaction in atomic units, v_1 through v_6 designate valence orbitals, while e is an excited orbital. The contribution in Eq. (2.12) may, in principle, have all six valence orbitals $v_1,\cdots,v_6$ all different, thereby providing a matrix element of H^v that is *off-diagonal in three spin-orbital indices.* Such a term can only be represented as part of a three-body effective interaction H_{123}^v. Similar analysis demonstrates that the third order terms in Eq. (2.11) yield four-body effective operators H_{1234}^v, etc. The true H^v, therefore, has the general structure of[9,10,30,43]

$$H^V = E_C + \sum_{i=1}^{N_V} H_i^V + \frac{1}{2!}\sum_{i\neq j=1}^{N_V} H_{ij}^V + \frac{1}{3!}\sum_{i\neq j\neq k=1}^{N_V} H_{ijk}^V + \cdots, \qquad (2.13)$$

where N_V is the number of valence electrons, E_C is the fully correlated core energy, and H_i^V, H_{ij}^V, H_{ijk}^V, are the one, two, and three-body effective valence shell operators that act only on the space of the valence shell orbitals. The correlation contributions to Eqs. (2.8) and (2.11) imply that, for instance, the individual matrix elements of the effective operator H_{12}^V may be written as the sum of the bare (often called theoretical) matrix element $\langle v_1v_2|r_{12}^{-1}|v_3v_4\rangle$ plus the correlation portion,

$$\langle v_1v_2|H_{12}^V|v_3v_4\rangle = \langle v_1v_2|\frac{1}{r_{12}}|v_3v_4\rangle + \langle v_1v_2|\Delta_{12}^V|v_3v_4\rangle. \qquad (2.14)$$

The process of extracting the correlation contributions like Δ_{12}^V from Eq. (2.11) is greatly facilitated by use of many-body techniques, such as those of Iwata and Freed[43] or more generally of Brandow's diagrammatic methods[34,39,40].

The many-body operators in Eq. (2.13) are part of the price paid in order that the H^V of Eqs. (2.8) and (2.11) be exact. The existence in H^V of these non-classical many-body operators is but one indication that the true H^V differs from the forms commonly assumed by semiempirical methods since the latter do not explicitly contain objects such as the three-electron interactions with valence shell matrix elements $\langle v_1v_3v_3|H_{123}^V|v_4v_5v_6\rangle$. Consequently, insight into the structure of the true H^V can only emerge from ab initio computations using Eq. (2.11) since no other ab initio formulation is currently available that can compute individual matrix elements of the correlation contributions to H_1^V, H_{12}^V, H_{123}^V, $\cdots$. However, before using ab initio methods to compute these correlation corrections, it remains to be determined whether this ab initio effective valence shell Hamiltonian formulation is of suitable ab initio accuracy to warrant drawing conclusions concerning the true H^V from such computations.

3. The ab initio H^V method

There is rather wide spread usage of ordinary non-degenerate Rayleigh-Schrödinger perturbation theory, which is termed either many-body perturbation theory or MP theory in the literature. (The "many-body" label is usually applied when the method employs techniques of second quantization and Feynman diagrams, while the MP designation also implies a more restrictive choice of zeroth order orbitals and orbital energies than that desirable for multireference configuration studies.) These non-degenerate methods require a zeroth order description that involves a single determinantal wave function, and most common schemes proceed in an order-by-order fashion, e.g., MPi refers to computations through ith order. Applications are generally restricted to $i \leq 4$, and the necessary computer codes are contained in commercially available electronic structure computer packages.

Deficiencies of this non-degenerate perturbation approach are quite obvious when considering developments with the variationally based configuration interaction (CI) methods, which likewise began with treatments based on a single reference configuration. Both single reference CI and perturbation methods encounter difficulties in treating inherently open-shell systems[7,44], such as those arising with the ground states of some radicals, many excited electronic states, transition states for chemical reactions, etc. Thus, just as CI methods have been extended[7], it is quite natural that the perturbation theory formulations likewise be extended to methods involving many reference configurations.

The available multireference configuration CI computations provide some guidance as to requirements of the multiconfigurational perturbation approaches as follows: The multireference CI calculations appear to work quite well, provided that the reference space (the P-space of Sect. 2) contains all of the dominant configurations and provided that the single and double excitations out of these configurations are included in the secondary space (the Q-space of Sect. 2). However, these types of CI methods are formally not size-extensive and, therefore, do not scale properly with the number of electrons in the molecule. Davidson corrections[45] are generally appended to correct for lack of size-extensivity, producing what has been the most viable ab initio method for computing global potential energy surfaces covering wide ranges of geometries. However, when there are several important configurations, the Davidson corrections are no so obviously determined. Multiconfigurational perturbation methods, on the other hand, are inherently size-extensive provided they employ a complete reference space as described in Sect. 2. Experience with multireference configuration CI methods further suggests the need in the analogous perturbation approaches for using sufficient numbers of reference configurations to incorporate all the important configurations for the states of interest.

The use of large reference spaces in perturbation methods, however, would appear to introduce serious difficulties. For instance, the first multireference configuration perturbation calculations for molecules are those of Kaldor[46] for excited states of the H_2 and BH molecules. Kaldor only considers the apparently simplest cases in which there are two electrons occupying two nearly degenerate spatial orbitals φ and ψ, and he applies the Brandow[39] formulation of quasidegenerate many-body perturbation theory to this system. The (complete active space) Brandow theory requires the use of a four (spatial) configuration reference space for the two active electrons, a reference space containing the configurations $\varphi\psi$, $\psi\varphi$, $\varphi\varphi$, and $\psi\psi$. However, Kaldor finds it necessary to restrict the computations to include only the pair of nearly degenerate $\varphi\psi$ and $\psi\varphi$ configurations because the remaining $\varphi\varphi$ and $\psi\psi$ configurations are very different in energy from the quasidegenerate pair $\varphi\psi$ and $\psi\varphi$. Thus, the inclusion of the $\varphi\varphi$ and $\psi\psi$ configurations in the reference space would severely violate the quasidegenerate constraints on the Brandow complete reference space formulation of multiconfigurational perturbation theory and would therefore produce erroneous results.

Kaldor's calculations demonstrate the precarious nature of complete reference space perturbation computations for the simplest possible case of a two-orbital, two-electron valence space. However, the situation must become even less quasidegenerate and consequently more problematic as the number of spatial valence orbitals (and the number of valence electrons) is increased further to be consistent with the corresponding demands of multireference configuration CI calculations. In fact, it is this inherent difficulty with complete reference space computations that led Kaldor and Hose[47] to develop incomplete reference space formulations of many-body perturbation theory in which the reference space only contains the important quasidegenerate configurations, suggesting that these methods are useful solely in rather limited situations of strict or near degeneracy with no other important reference configurations of very differing zeroth order energies.

Various model problems have been studied to ascertain the convergence properties of quasidegenerate perturbation theory. Simple three-level model computations[48,49] indicate that these perturbation expansions *may in some instances* converge to the wrong eigenvalue. This *may occur* when there is an eigenvalue of the matrix H_{QQ} that lies close to a desired reference (P) space eigenvalue. The perturbation expansion *may converge* to the offending Q-space state, called an intruder state, instead of the desired one in the reference space. Unfortunately, the *possible occurrence of intruder state problems* has led to the wide spread belief that these problems always occur[50], despite many useful examples to the contrary[33,51]. Nevertheless, the above considerations indicate the fundamental difficulties associated with developing an accurate ab initio formulation of complete active space multiconfigurational perturbation theory, especially with the large valence spaces that are required for comparisons with semiempirical methods. The simplest examples are provided by atoms[52,53,54] whose study suffices for attempting to understand the correct choice of one-center integrals for semiempirical methods.[55,56,57] First row atoms have a valence space which contains the 2s and 2p orbitals. However, the 2s-2p energy separation grows considerable in crossing the first row, making it highly unlikely that these systems satisfy the quasidegeneracy criterion required for use of the Brandow formulation or variants thereof. Matters become much worse when studying transition metal atoms[54] or diatomic molecules involving a pair of first row atoms[58,59,60]. The latter case has eight spatial valence orbitals, leading to zeroth order valence space configurations with an energy span that may exceed the ionization potentials and that is, therefore, by no stretch of the imagination quasidegenerate!

2.1. DESCRIPTION OF THE AB INITIO H^v METHOD

Because an understanding of semiempirical methods requires the use of a complete active space formulation of multiconfigurational reference perturbation theory, the corresponding ab initio method must treat rather large, manifestly not quasidegenerate valence spaces. This formidable problem has been handled by simply using a zeroth order Hamiltonian H_0 that forces the zeroth order valence space states to be exactly degenerate, leaving the considerable spread in original valence space energies as a large perturbation that enters into the theory only beginning in third order[34]. When the valence space contains four or more electrons, the third order quasidegenerate many-body perturbation theory requires the evaluation of several hundred diagrams and represents an enormous and complicated computational task. These third order corrections are necessary in order to assure that the large perturbation, emerging from the actual spread of zeroth order valence space energies, does not cause any practical difficulties with the theory.

Just as an optimal semiempirical method produces all energies using a single minimal basis and a single $H_M{}^v$ with the same atomic orbital basis "integrals" for all states, *the diagonalization of the effective valence shell Hamiltonian H^v, in principle, generates all valence (P-space) state energies simultaneously from a single computation.* The orbitals in the H^v method obviously involve appropriate linear combinations of the original basis functions, but somehow the perturbation contributions in Eq. (2.11) are correcting the effective valence shell Hamiltonian H^v for the differences in orbitals that are optimal for individual states as in SCF or MCSCF computations.

There is therefore a sharp contrast between traditional wavefunction methods, which employ bare Hamiltonian matrix elements and different orbitals for each state, and the H^v method which has correlation modified effective matrix elements and the same orbitals for all states. This sharp contrast, resembles one that is well known for the different representations of the time-dependent Schrödinger equation. A discussion of this analogy further aids in understanding the operation of the H^v method and therefore of semiempirical electronic structure methods. Traditional SCF, CI, etc., wavefunction methods optimize a set of orbitals for each state separately and then use these orbitals to compute correlation corrections. This whole process is carried out with the original, bare Hamiltonian H, much as the Schrödinger picture of time-dependent quantum mechanics has the states change with time and the operators remain the same. On the other hand, the H^v method corresponds more closely to the Heisenberg picture of time-dependent quantum mechanics in which the states remain the same, while the operators change with time. This situation corresponds to the H^v method where the orbitals are the same for all states, but the H^v operator changes to describe the different orbital polarizations, etc., between the different valence states.

Semiempirical methods implicitly assume that this single-orbital, single-$H_M{}^v$ procedure is well defined. The ability of the ab initio H^v method to generate many states simultaneously from the diagonalization of a single H^v operator would provide justification for this bold semiempirical assumption as normal intuition, based upon traditional ab initio computations, suggests that this should not be possible. The ability to use a common set of orbitals in the ab initio H^v method, on the other hand, would be beneficial in computations of off-diagonal electronic matrix elements (transition dipoles[28], spin-orbit and derivative couplings, etc.) since the use of different and therefore nonorthogonal orbitals in traditional wavefunction methods complicates the computation of these transition matrix elements.

The effective valence shell Hamiltonian method is completely specified once choices have been made[61] of a) normalization, b) order of truncation of the perturbation expansion, c) the valence space, d) the orbitals (core, valence, and excited), and e) the zeroth order Hamiltonian H_0. We choose to work only with the Hermitian form of H^v and to retain all contributions through third order. The Hermitian form is natural for comparison with semiempirical $H_M{}^v$, while the use of large valence spaces with more than two valence electrons precludes treating higher than third order. Perhaps, there exist useful partial summations to all orders of certain contributions[34], and further analysis of this possibility would be desirable. Hence, the only real freedom in the H^v method is associated with choosing c) through e) above.

In performing comparisons of the ab initio H^v method with semiempirical approaches, a natural choice of valence space is the one taken for the corresponding semiempirical method. Semiempirical computations for large molecules involve the use of very large valence spaces, which are currently impractical for H^v computations. However, highly correlated ab initio methods are most suitable for describing smaller molecules. Furthermore, a sufficient test of the assumptions inherent in semiempirical methods requires only the treatment of smaller molecules because the semiempirical methods assume that only one and two-center integrals (or parameters) are present. Thus, the simplest tests of one-center integrals necessitates computations for atoms[52-59], while the analysis of semiempirical two-center integrals may be accomplished based on ab initio H^v computations for diatomic molecules. More interesting tests of the transferability assumption of semiempirical methods, on the other hand, require consideration of small polyatomic molecules[62,63]. Our desire is to apply the H^v method to molecules of similar sizes to those treated by the other most advanced correlated electronic structure methods. Many of these cases lead to rather large valence spaces of the all-valence-electron variety, thereby raising questions concerning the optimal choice of valence spaces. However, the initial H^v studies consider systems for which the valence space may readily be taken as identical to that of the corresponding semiempirical approach. The orbitals, however, must somehow be optimized for maximal ab initio accuracy and, therefore, must differ in precise form from the (actually often partially unspecified) orbitals in the semiempirical methods.

The leading, first order approximation in the H^v method is provided by the full valence shell CI computation produced by diagonalizing the matrix $\mathbf{H_{PP}}$ of Eq. (2.8). Thus, we minimize the perturbation corrections in equation (2.11) and thereby facilitate the convergence of the perturbation expansion by having $\mathbf{H_{PP}}$ provide an optimal description for all the valence states since all of their energies, in principle, emerge from a single H^v. Consequently, in principle, the valence space, and hence the core and valence orbitals, might be chosen from natural orbitals or from some systematic averaged complete active space self-consistent field (CASSCF) formulation for all valence states. However, it is rather unrealistic to expect the H^v method to provide a meaningful description of the high lying states in large valence spaces, and this suggests limiting the orbital optimization to a group of lower lying states, which, in general are the states of interest in both ab initio and semiempirical applications. The CASSSCF orbitals for a single state contain some low occupancy valence orbitals that are useful for correlating that state but that are rather poor for describing other excited states in which this orbital is occupied. Lacking the ability to perform state averaged CASSCF computations with states of differing symmetries, the H^v computations have primarily chosen valence orbitals from a sequence of SCF computations for several low lying states as providing a reasonable compromise designed to "optimize" the description afforded by $\mathbf{H}_{\mathbf{PP}}$ for the low lying states of interest. Other choices involving CASSCF orbitals are currently being tested.[64]

We illustrate this choice of valence orbitals by the example of the pi-electron states of trans-butadiene[65]. The semiempirical pi-electron Hamiltonians describe the pi-electron states of this molecule with a basis of four localized atomic-like p_π orbitals[1], and, consequently, the ab initio H^v method naturally should begin with a set of four "optimized" pi-electron molecular orbitals. The core sigma and valence pi-orbitals have been obtained from the following sequence of single state SCF computations[65]:

$$\begin{aligned}
&(1)\ (\text{core})(1\pi_u)^2(1\pi_g)^2, && {}^1A_g,\\
&(2)\ [(\text{core})(1\pi_u)^2(1\pi_g)^1](2\pi_u)^1, && {}^3B_u,\\
&(3)\ [(\text{core})(1\pi_u)^2(1\pi_g)^1(2\pi_u)^0](2\pi_g)^1, && {}^3A_g,\\
&(4)\ [(\text{core})(1\pi_u)^2(1\pi_g)^2(2\pi_u)^0(2\pi_g)^0], && {}^1A_g, \quad (3.1)
\end{aligned}$$

where (core) designates a filled sigma core. Step (1) is just a SCF computation for the trans-butadiene ground state, and this calculation serves to define the core and the $1\pi_u$ and $1\pi_g$ valence orbitals. The next step (2) is a ^3B_u state SCF computation in which the core and $1\pi_u$ and $1\pi_g$ valence orbitals are frozen to be the orbitals generated from step (1), while only the $2\pi_u$ orbital is varied. Frozen orbitals are placed inside square brackets, and only those orbitals outside the square brackets are permitted to vary in each step. Because step (2) involves variation of only the $2\pi_u$ orbital, this step corresponds to the computation of an improved virtual orbital (IVO), which is "optimal" for the ^3B_u state given the already obtained ground state orbitals. Although the ^3B_u state is an electronically excited state, the IVO computation is equivalent to that for a simple ground state. Step (3) continues this practice in computing an IVO for the $2\pi_g$ orbital, while step (4) freezes all core and excited orbitals to obtain the virtual orbitals from a ground ^1A_g state computation. Because step (4) employs a ground state Fock operator, the virtual orbitals are determined in the field of the N core and valence electrons for reasons discussed below in conjunction with a description of the choice of zeroth order Hamiltonian H_0. The sequence of SCF computations produces orbitals that are better suited to the ground and lowest excited states. Thus, several other choices of orbitals have been tested and are currently under investigation[64], but the above scheme works about the best (of several tried[65]) for describing all the low lying pi-electron states simultaneously. Nevertheless, some computations in progress for larger conjugated systems find that the highest lying IVO's from this sequential scheme develops too much Rydberg character. Thus, there is a still need for further improvement of the above scheme for choosing valence orbitals.

The only remaining ingredient requiring specification in the H^v method is the zeroth order Hamiltonian H_0. The most common choices for perturbation methods involve taking H_0 to be a sum of one-electron operators h_1, which are individually diagonal among the set of core $\{|c\rangle\}$, valence $\{|v\rangle\}$, and excited $\{|e\rangle\}$ orbitals. The Moller-Plesset partitioning, for instance, considers h_1 to be a ground state Fock operator, while MCSCF theories provide several alternative multiconfigurational Fock operators. However, the most general form for this one-electron h_1 may be written in the form,

$$h_1 = \sum_c \varepsilon_c |c\rangle\langle c| + \sum_v \varepsilon_v |v\rangle\langle v| + \sum_c \varepsilon_e |e\rangle\langle e|, \qquad (3.2)$$

where the diagonal one-electron "orbital energies" $\{\varepsilon_c, \varepsilon_v, \varepsilon_e\}$ are, *in principle*, arbitrary. The first order theory is obtained from the valence block H_{PP} of the full Hamiltonian and, therefore, is independent of the choice of h_1. Thus, h_1 should be chosen as somehow optimizing the performance of the perturbation expansion. Hence, there is no reason to expect that an "optimal' approach is necessarily provided by the traditional choices for Moller-Plesset single configuration nondegenerate perturbation theory. By analogy with nondegenerate perturbation methods, a natural first choice involves defining the orbital energies as those emerging from the SCF, MCSCF, or CASSCF equations used to compute the corresponding orbitals. The sequential SCF approach, illustrated by Eq. (3.1), employs different Fock operators to obtain the valence orbitals that are unoccupied in the ground state SCF wavefunction.

Although the H^v method admits of very general forms for the zeroth order h_1 in Eq. (3.2), a poor choice of orbital energies can doom to failure the perturbation expansion, even through third order. It is possible, however, to understand the stringent restrictions[34] on the orbitals and orbital energies by examining the energy denominators appearing in the sums-over-orbitals representation of the perturbation expansion for H^v. Some of the second order contributions to matrix elements of the two and three-electron effective operators $H_{12}{}^v$ and $H_{123}{}^v$ contain energy denominators of the form[34,52]

$$\text{denominator} = \varepsilon_v - \varepsilon_e + \varepsilon_{v'} - \varepsilon_{v''}, \qquad (3.3)$$

where v, v', and v" refer to valence shell spin-orbitals, and e is an excited orbital. (Similar denominators appear with $\varepsilon_v - \varepsilon_e$ replaced by $\varepsilon_c - \varepsilon_v$, where ε_c is a core orbital energy.) The contribution $\varepsilon_v - \varepsilon_e$ is always positive (as is $\varepsilon_c - \varepsilon_v$), so long as we make the natural choice of all valence orbital energies lower than those in the excited space (and higher than those in the core). However, when the valence space is not degenerate, the other contribution $\varepsilon_{v'} - \varepsilon_{v''}$ may be of either sign. Thus, if the minimum valence to excited orbital excitation energy $\min|\varepsilon_v - \varepsilon_e|$ is equal to the spread in valence orbital energies $\max|\varepsilon_{v'} - \varepsilon_{v''}|$, the denominator in Eq. (3.3) may diverge[34], and the perturbation expansion becomes ill-defined already in second order. Experience with second order H^v computations suggests the quasidegenerate condition[34]

$$\min\{|\varepsilon_v - \varepsilon_e|, |\varepsilon_c - \varepsilon_v|\} > 2\max|\varepsilon_{v'} - \varepsilon_{v''}| \tag{3.4}$$

to assure proper behavior of the second order treatment. However, we have already explained the necessity for using large valence spaces to produce accurate ab initio computations and, of course, for comparison with semiempirical methods. Large valence spaces yield large spreads in valence orbital energies (for any reasonable choice of orbital energies) and thereby rapidly violate the quasidegenerate conditions in Eq. (3.4). The situation becomes even more problematic in third order where some denominators contain two contributions of the form $\varepsilon_{v'} - \varepsilon_{v''}$. Computational experience[34] shows that these terms are well behaved provided that the factor of 2 on the right hand side of Eq. (3.4) is replaced by a factor of 3. The more stringent quasidegenerate constraint of Eq. (3.4) with the factor of 3 is met by first row atoms with (four orbital) valence spaces composed of their 2s and 2p orbitals, but already fails for the five orbital valence space description of first row diatomic hydrides[66]. (Some illustrative examples are given below.) Several model analyses of quasidegenerate perturbation methods have obtained divergent behavior because the orbital energies employed severely violate the quasidegenerate constraint[49].

The above examples illustrate the predicament that the quasidegeneracy constraint appears to preclude the use of large valence spaces. In order to avoid the appearance of small energy denominators in the H^v perturbation expansions for large, non-degenerate valence spaces, the valence orbital energies in H_1 of Eq. (3.2) may be shifted closer together without sacrificing ab initio rigor. While this may at first sight appear to be artificial in comparison with nondegenerate perturbation methods which naturally employ, for instance, SCF orbital energies in H_0, the completely general definition of h_1 in Eq. (3.2) admits, in principle, of *any* choice for the orbital energies. Of course, a poor choice may lead to the divergence of the perturbation expansion in Eq. (2.11). Thus, an analysis of the perturbation expansion provides the only means for assessing the relative merits for various definitions of h_1. Reducing the valence orbital energy spread to satisfy Eq. (3.4) with a factor of 3, replacing the factor of 2 on the right hand side of this equation, serves to eliminate the numerator problems, but then it introduces into the perturbation V (and, hence, into the denominators of the perturbation expansion) correction terms containing the shift in the orbital energies. These correction terms first contribute in third order, and, therefore, we have chosen to regularly perform third order H^v computations to assure that the shifts introduce none of the above denominator problems and that indeed the changes in state energies between second and third orders are considerably smaller than those between first and second orders. The latter is a reasonable definition of practical convergence of the H^v method through third order, so long as there is good agreement with benchmark full CI computations[62,64,67] and with experiment when large basis sets are used[65].

The above described denominator problems are completely eliminated by forcing the valence space to be completely degenerate in zeroth order. This is accomplished by introducing into h_1 of Eq. (3.2) a single averaged valence orbital energy $\bar{\varepsilon}_v$ to replace the separate and generally different ε_v. Such a method is perhaps more honestly called "forced degenerate perturbation theory". When H^v computations are designed to provide all low lying excitation energies simultaneously or to compare with semiempirical methods, the average valence orbital energy $\bar{\varepsilon}_v$ is taken as a equal weighted average of all valence orbital energies, but other choices are currently under investigation[64]. Extensive computations and comparisons with full CI calculations indicate that this procedure works well in third order, while there tends to be some overcorrecting in second order compared to third order and full CI for some model problems. Perhaps, other orbital energy averaging schemes could be developed to improve the practical convergence of H^v computations. (Some of these averaging methods correspond to infinite order diagram summations[34]). The ideal situation would be to have very accurate results from second order H^v calculations since the computational labor required is rather minimal.

3.2. ILLUSTRATIONS OF RECENT AB INITIO H^V CALCULATIONS

The optimal way of testing new ab initio electronic structure methods is through comparisons with benchmark full CI computations because differences with full CI cannot be ascribed to inadequacies of the basis set as they may when comparisons are made with experiment. Some comparisons between full CI and the H^v method have been made for the CH_2 molecule[62], for the insertion reaction of Be into H_2 at various geometries[67], and most recently for the spectroscopic constants of the nitrogen molecule ground state[64]. These nontrivial comparisons attest to the capabilities of the H^v method, but the examples we present now are interesting cases in which large basis set H^v calculations for difficult systems are compared both to experiment and to other highly accurate correlated ab initio computations.

The first example involves the computation of the pi-electron vertical excitation energies for the trans-butadiene molecule[65] which presents an interesting challenge in many respects. The low-lying pi-electron spectrum of trans-butadiene and other small conjugated molecules contains both valence-like and Rydberg states, as well as some states of mixed Rydberg-valence parentage. The accurate description of such a complicated spectrum is quite non-trivial because correlation in these three classes of states is both quantitatively and qualitatively different. Consequently, the ab initio method must adequately describe the differential correlation between these states of varying spatial character, a task whose difficulty is compounded by the highly open shell nature of some electronically excited pi-electron states.

We begin discussion of the trans-butadiene H^v computations with a valence space resembling that of semiempirical pi-electron theory in containing four p_π molecular orbitals. The primitive basis set contains 126 Gaussian functions and is constructed from a Huzinaga (4s3p/2s) basis augmented by (d/p) polarization functions and two diffuse sets of p-functions on all the carbon centers[65]. This basis set is of sufficient quality and size to warrant direct comparisons with experiment. The orbitals and orbital energies are defined as described already by the sequence of SCF computations in Eq. (3.1).

This four orbital H^v computation is also useful in subsequently developing methods for testing the assumptions of pi-electron theories. Since semiempirical pi-electron methods with localized, transferable basis sets are designed, in principle, to provide only the valence-like states, it is only reasonable to expect the four-orbital valence space trans-butadiene H^v computations to represent these valence-like states. Before turning to these computations, we briefly note some significant differences between the calculations now discussed and those necessary for testing semiempirical theories. The optimized pi-electron molecular orbitals of Eq. (3.1) depart considerably from the localized, transferable basis of atomic p_π valence orbitals implicit in semiempirical pi-electron theories. Nevertheless, the success of the ab initio H^v computations with these orbitals would demonstrate the practical validity inherent in the semiempirical assumption for using a minimal valence space, an assumption whose *formal validity* is already justified by our derivation of the effective valence shell Hamiltonian.

The four-orbital valence space trans-butadiene H^v computations[65] present a reasonably good description of the low-lying pi-electron valence and do not suffer from intruder state problems, even though many possible low-lying Rydberg states penetrate into the valence-like spectrum. The lowest two triplet 1^3B_u and 1^3A_g states excitation energies departing from experiment by only 0.13eV each. A larger error of 0.52eV appears for the 1^1B_u state which contains an admixture of some Rydberg character into a primarily valence-like state. The nearby Rydberg excited states are totally absent from the four-valence-orbital third order H^v calculations, despite the fact that they represent potential intruder states whose presence is widely believed[50] as dooming to failure any multireference configuration perturbation formalism. However, the couplings between Rydberg and valence configurations are sufficiently small that they neither pose intruder state "problems" or permit entry of Rydberg states into the third order H^v calculations, even though we actually would like to describe these Rydberg states. Thus, these H^v computations once again produce a practical verification for the semiempirical assumption of using a minimal valence space.

A more thorough ab initio description of the trans-butadiene pi-electron spectrum requires treating both the low-lying Rydberg and valence-like states. Thus, H^v computations have also been performed using six pi-electron valence orbitals[65], four of valence-like character and two Rydberg pi-orbitals. The computed excitation energies compare well with both experiment and other highly correlated ab initio methods. The core and four valence-like orbitals are taken from the SCF sequence in Eq. (3.10), but step (4) of this sequence is replaced by the steps

$$(4)\quad [(\text{core})(1\pi_u)^2(1\pi_g)^1(2\pi_u)^0(2\pi_g)^0](3\pi_u)^1, \qquad {}^3B_u,$$

$$(5)\quad [(\text{core})(1\pi_u)^2(1\pi_g)^1(2\pi_u)^0(2\pi_g)^0(3\pi_u)^0](3\pi_g)^1, \quad {}^3A_g,$$

$$(6)\quad [(\text{core})(1\pi_u)^2(1\pi_g)^2(2\pi_u)^0(2\pi_g)^0(3\pi_u)^0(3\pi_g)^0], \quad {}^1A_g, \qquad (3.5)$$

where the Rydberg molecular orbitals (one for each possible symmetry) are obtained as IVO's in steps (4) and (5) and the excited orbitals are computed in step (6). Table I presents the third order H^V computations of the trans-butadiene pi-electron excitation energies[65] along with experimental values and the results of subsequent CASPT2 computations by Roos and coworkers[68]. The CASPT2 method is also a perturbative approach, but it involves diagonalization and then perturbation as opposed to the perturb then diagonalize H^V method. More specifically, the CASPT2 trans-butadiene computations begin with a complete active space SCF computation with six active pi-orbitals. Then, single reference state second order perturbation computations are performed with each of the CASSCF wavefunctions separately. The H^V computations exhibit an average deviation from experiment of only 0.10eV, whereas the deviation for the CASPT2 method is the slightly larger 0.16eV. Table I also indicates the spatial character of the states (V denotes valence and R Rydberg) as well as the dimension of the primary space for each electronic state symmetry. The transition to the six-orbital valence space improves the H^V excitation energies of the valence-like 1^3B_u and 1^3A_g states by 0.11 and 0.10eV, respectively, but there is a considerable 0.43eV improvement in the energy of the 1^1B_u state,

Table I: Third order H^V computations for trans-butadiene.

State	H^V	Experiment	CASPT2	Dimension	Type
1^3B_u	3.20eV	3.22eV	3.20eV	54	Valence
1^3A_g	4.88	4.91	4.89	57	Valence
1^1B_u	6.01	5.92	6.23	48	.8V+.2R
2^1A_g	6.20	?	6.27	57	Valence
2^1B_u	6.91	7.07	6.70	48	Rydberg
3^1A_g	7.17	7.4	7.47	57	Rydberg
1^2B_g	8.72	9.07		35	Rydberg
1^2A_u	10.98	11.4		35	Rydberg

which is of mixed Rydberg-valence character. In addition, the H^V computations now describe the 2^1B_u and 3^1A_g Rydberg states so well that the computations confirm the tentative assignment of the former and provide the assignment of the latter[65]. Thus, the six-valence orbital H^V calculations represent the most accurate to date for the trans-butadiene pi-electron vertical excitation energies. Table I also presents the $1\pi_u$ and

$1\pi_g$ vertical ionization potentials[69], which emerge from the *same* H^v computation as the listed excitation energies. The only additional computational labor is associated with the almost trivial diagonalization of the H^v matrix for the ion states.

Analogous H^v calculations have been completed for cyclo-butadiene[70] and others are currently in progress for benzene and hexatriene[71]. No experimental data are available for the cyclo-butadiene pi-electron states, so our computations for this molecule are designed to provide accurate predictions for guiding attempts to observe the elusive ultraviolet spectrum of this interesting molecule. The benzene H^v calculations use a 156 basis function correlation consistent basis, augmented with two sets of diffuse p-orbitals on each carbon. Preliminary calculations are available with a valence space of six valence-like orbitals. The average deviation from experimental pi-electron vertical excitation energies to eight valence-like states is only 0.28eV, only slightly worse than the corresponding 0.26eV from the CASPT2 method with a similar active space of six valence orbitals[72].

The pi-electron H^v computations are important in several respects. Firstly, they demonstrate the high ab initio accuracy of the H^v method in computations with large basis sets and for highly nontrivial systems. Secondly, they provide numerical justification for the implicit semiempirical assumption that accurate electronic energies can emerge from an effective Hamiltonian containing a small number of valence orbitals. The accuracy of ab initio H^v calculations for pi-electron systems indicates that the H^v method can be applied to test common semiempirical assumptions concerning the description of pi-electron systems. Since these pi-electron systems represent the birthplace of modern semiempirical methods where many of the fundamental principles of semiempirical approaches have been developed, the pi-electron systems also provide a fertile testing grounds for other more specific assumptions of semiempirical methods that are present in the all-valence-electron methods. Before turning to these tests in the next section, however, we consider another illustration of recent H^v computations.

One important application of electronic structure theories is to the computation of molecular potential energy surfaces and pathways of chemical reactions. The overwhelming majority of the most accurate potential surface computations, and especially those for open-shell systems, such as excited and transition states, rely on multireference configuration CI computations with appended Davidson corrections[7]. These CI treatments are the method of choice because they provide the best uniformity of correlation treatment as a function of molecular geometry. Multireference coupled cluster methods, on the other hand, cannot yet treat the high asymptotic degeneracy necessary for describing the dissociation of the nitrogen molecule triple bond[73]. Thus, we turn to the interesting problem of whether the size-extensive H^v method is capable of providing global potential energy surfaces. Prior computations for many potential energy curves of the diatomic molecules CH[66], CH^+[66], Li_2[74], LiH[75], O_2[59] O_2^+[59], and S_2[60], recent ones for N_2[64], and prior computations for triatomic hydrides[63,67] attest to the ability of the H^v method to accurately generate a series of valence state potential curves simultaneously from a single computation of the third order H^v, provided the valence space is of the all-valence-electron type (i.e., eight valence orbitals for O_2, N_2, and S_2, but five for CH and LiH, etc.) While the eight valence orbital spaces for O_2, N_2, and S_2 already severely strain the quasidegeneracy constraints of the H^v method, as the size of the molecule grows further, the all-valence-electron valence space becomes for too large for use of the H^v method, and restricted valence spaces are necessary. Thus, we consider whether the H^v method can treat interesting global potential energy surfaces with less than the all-valence-electron valence space. The system chosen for this study is the methyl mercaptan (CH_3SH) molecule[76,77], which is an important component in atmospheric pollution. The all-valence-electron valence space for CH_3SH contains twelve spatial orbitals and is probably at the current limits of feasibility for the H^v method. However, a reduced valence space is clearly highly desirable. Our interests lie primarily in the ground and lowest two excited A'' states which are pertinent to the photodissociation and polarized emission studies of Butler and coworkers[76]. The first excited A'' state of CH_3SH is a valence-like state, but the second A'' state is Rydberg, thereby making the computation of adequate potential surfaces highly non-trivial. Hence, our experience with conjugated hydrocarbons suggests the necessity for also having Rydberg orbitals in the valence space in order to properly treat the Rydberg state. This requirement again emphasizes the question of how well the H^v method fares with less than the Rydberg-extended all-electron valence space.

The H^v computations for the CH_3SH potential surfaces[77] use a 4-31G** (with 62 functions and diffuse s and p functions on carbon and sulfur) in order to permit comparisons with previous ab initio computations using the same basis. H^v calculations with a larger 102 function basis have also been performed at selected geometries to assess the basis set errors[77]. CASSCF calculations for the ground and low-lying excited states indicate that the $1^1A''$ excited state wavefunction is obtained from the dominant ground state configuration by a 3a'' $\rightarrow$ 11a' excitation, while the $2^1A''$ CASSCF wavefunction is primarily a 3a'' $\rightarrow$ 12a' excitation. This convenient situation suggests that the H^v computation might proceed with just a three orbital valence space containing the ground state occupied 3a'' and unoccupied 11a' and 12a' orbitals. However, these considerations of zeroth order wavefunctions do not assess the quasidegeneracy properties of the resultant valence space. The H^v method requires a reasonable energy gap both between the core and valence orbitals and between the valence and excited orbitals, as well as not too large a spread in valence orbital energies.

Figures 1 and 2 display the CH_3SH orbital energies as functions of the SH and CS distances[77], respectively, for several of the orbitals. Notice in both Figs. 1 and 2 that the 3a'' orbital energy crosses the ground state occupied 10a' orbital energy, suggesting that criteria of quasidegeneracy might require inclusion of the 10a' orbital in the H^v valence space to provide a valence space that is adequate over the whole (two-dimensional) range of geometries. The 10a' and 11a' orbital energies likewise cross at larger SH distances and at large CS distances. This crossing of orbital energies for orbitals of the same spatial symmetry is permissible because the 10a' and 11a' orbitals are eigenfunctions of different Fock operators, namely, the 10a' orbital is obtained from the ground state SCF wavefunction, while the 11a' orbital is an IVO from a $1^1A''$ state computation. In a similar vein, the 11a' and 13a' orbitals become nearly degenerate for a CS distance of 3au, while these two orbitals are rather close in energy over the whole range of SH distances, suggesting the need

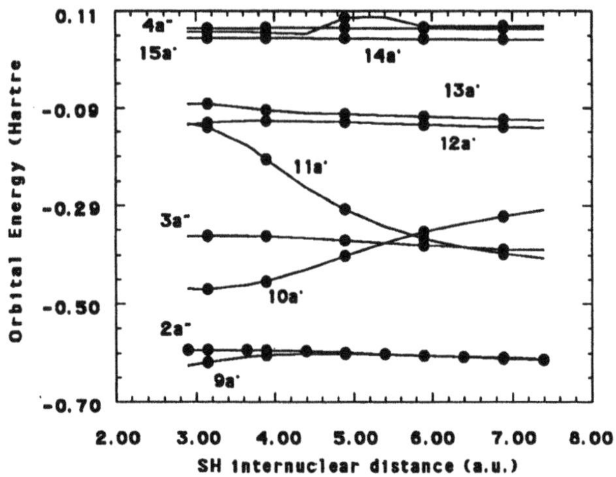

Figure 1. Selected CH_3SH orbital energies as functions of the SH distance

for including the 13a' orbital in the H^V valence space. This leads to a five orbital valence space containing the ground state occupied 10a' and 3a" orbitals as well as the ground state unoccupied (using IVO's) 11a', 12a', and 13a' orbitals. Thus, the 9a' and 2a" orbitals are the highest orbitals in the H^V core, while the H^V excited space has the lowest orbitals as the 14a', 15a', and 4a" orbitals. Notice that a reasonable gap exists between the core and valence orbital energies as well as between the valence and excited orbital energies. The spread in valence orbital energies, however, is sufficiently large that the H^V method must be applied to this five valence orbital space with averaged valence orbital energies. Thus, third order computations are necessary to include contributions from the difference between the actual orbital energies and the averaged values used in the zeroth order Hamiltonian.

Known vertical excitation energies, bond energies, and ionization potentials serve as tests of the accuracy of the H^V computations. The vertical excitation energies to the $1^1A''$ and $2^1A''$ state are obtained with the 62 function production basis as 5.72 and 6.37eV, respectively, compared to the experimental values of 5.45 and 6.07eV. Extending the basis to 102 functions by uncontracting the 4-31G** basis and adding two sets of s and p diffuse functions on carbon and sulfur and two sets of diffuse functions on the hydrogen atoms leads to an improvement of the excitation energies to 5.50 and 5.92eV, respectively[77]. The 10a' and 3a" orbital ionization potentials are computed with the 62 function basis as 12.21 and 9.69eV, respectively, compared to the experimental 12.08 and 9.44. Increasing the basis size to 102 functions yields the negligible changes to 12.22 and 9.66eV, respectively[77]. The absence of basis functions with higher angular momenta precludes the computation of highly accurate bond energies. The

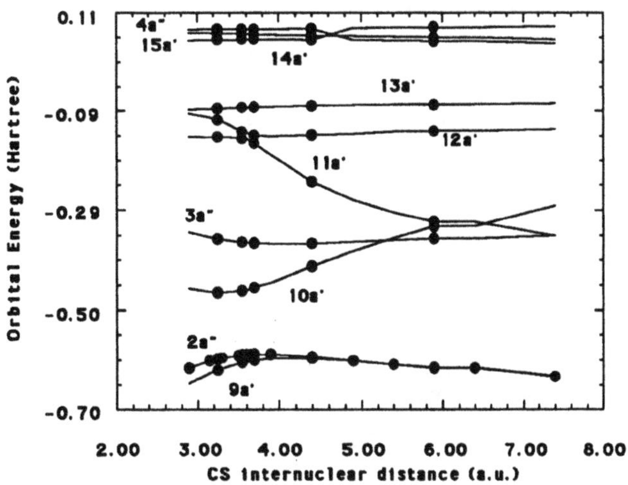

Figure 2. Selected CH_3SH orbital energies as functions of the CS distance

62 (102) function computations predict the CS and SH bond energies as 65.6 (62.4) and 82.4kcal/mole (81.5), while the experimental values are 72.6 and 86.1kcal/mole, respectively, yielding an accuracy comparable with that expected from the types of basis sets used.

Figures 3 and 4 depict cuts of the potential energy surfaces along the SH and CS distances, respectively. The potential surfaces correspond rather closely to the analogous surfaces for methanol, as they should. The full two-dimensional surfaces are being used in computations of the dynamics of photodissociation of CH_3SH.

4. Ab initio computations of correlated semiempirical-like integrals

Both the formal theory of section 2 and the ab initio H^v computations, described in section 3 and elsewhere, already explain certain fundamental aspects of semiempirical methods. First of all, the formal theory complely

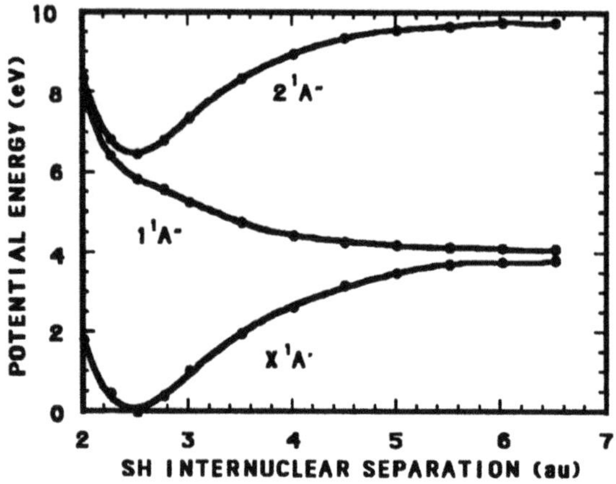

Figure 3. Cuts in CH_3SH potential energy surfaces along SH distance.

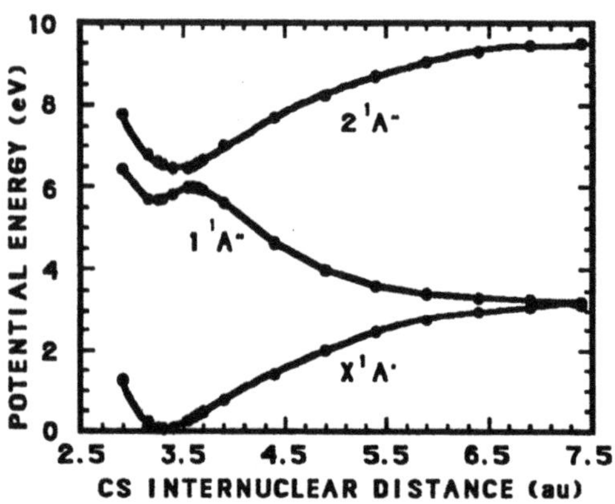

Figure 4. Cuts in CH_3SH potential energy surfaces along CS distance.

justifies the semiempirical assumption that it is, in principle, meaningful to represent the exact solutions of the molecular Schrödinger equation in terms of a valence electron only problem. This exact formal H^V theory also indicates that the simplest representation requires a complete valence shell CI calculation, although further approximations may be possible to reduce the computational labor from the full valence shell CI with a concomitant adjustment of the resultant approximate H^V matrix elements to compensate for the necessary approximations. The ab initio H^V

computations, on the other hand, demonstrate the practical feasibility of using a minimal or smaller valence shell basis and performing state-of-the-art correlated computations based on the same H^v formulation.

The formal theory and computations, however, also indicate the presence of several differences between the exact H^v and the semiempirical model Hamiltonians $H_M{}^v$ in current usage. Foremost among these differences is the presence in the exact H^v of three, four, ··· electron effective interactions, which have no counterpart in semiempirical models. Since, as discussed below, the matrix elements of the three-body effective H^v interactions (either called H^v integrals or true parameters below) are rather large in some systems, the existence of these considerable non-classical many-body interactions already signals important departures between the true H^v and semiempirical model $H_M{}^v$. Thus, ab initio H^v computations are necessary to probe the structure of the exact H^v and to test the fundamental assumptions inherent in current model $H_M{}^v$. Our ultimate goal is the generation of systematic ab initio information that can be used to improve and extend semiempirical electronic structure methods.

Perhaps the next most fundamental assumptions of semiempirical methods, beyond that inherent in the use of a minimal valence space, are those associated with the zero differential overlap approximation (ZDOA) and the assumption of transferability for the matrix elements of $H_M{}^v$ within a localized atomic orbital basis. The ZDOA effectively enables elimination of all three and four-center repulsion integrals in $H_M{}^v$ and even of some two and one-center repulsion integrals, depending on the degree of neglect of differential overlap. The ZDOA leads to considerable simplification of $H_M{}^v$ over that from the valence shell matrix elements of the full Hamiltonian H, and this simplification is partially responsible for the ease and speed in applications of semiempirical methods. No less fundamental is the transferability assumption that the semiempirical $H_M{}^v$ may be expressed in terms of a localized, transferable, molecule independent basis set with one-center and two-center matrix elements in this basis that depend, respectively, only on the atom and pair of atoms involved, independent of the remaining molecular environment.

Because semiempirical methods derive their one-center matrix elements from atomic spectral data, the ab initio H^v test of the customary semiempirical one-center matrix elements merely requires performing H^v calculations for atoms with the H^v valence space chosen to correspond with that used in the semiempirical method[52-57]. More interesting questions arise in the computation of two-center integrals[63,78,79,80]. First of all, a test of the transferability hypothesis requires use of a molecular valence space basis that is constructed from bases appropriate to the constituent atoms[78]. Thus, the molecular valence space must be constrained to be the union of the atomic valence spaces, with the molecular orbitals represented as linear combinations of the atomic valence orbitals. These "constrained" valence orbitals are naturally expected to provide a poorer initial description of H_{PP} and therefore to require more Q-space correlation contributions for correcting this less than optimal starting point. Thus, the constrained orbital ab initio H^v computations are anticipated to display somewhat slower convergence than the "full" H^v calculations in which valence orbitals are variationally determined as described in section 3. In order that the H^v method be capable of providing insights into the transferability assumption and into forms for the true two-center parameters, we require that the constrained H^v computations be capable of providing a reasonably accurate description of the valence state energies. This, in turn, demands that the H^v method adequately correct for differences between the full and constrained valence orbitals. As already explained in section 3, the full ab initio H^v computations employ a single set of (variationally determined "full") valence orbitals, and the H^v method appears to describe perturbatively the influences associated with differences in optimal orbitals between different valence states. Thus, we may hope that the constrained H^v calculations likewise rectify some deficiencies of the constrained valence orbitals and thereby provide useful valence space energies and, more importantly, two-center true parameters. For brevity, the term constraining errors is used below to indicate the errors in computed valence state energies arising from use of constrained valence orbitals.

If the constraining errors are indeed small, H^v computations for diatomic molecules suffice to determine the two-center true parameters and therefore complete the test of semiempirical assumptions concerning the bond length dependence of two-center integrals. However, calculations for diatomic molecules are inadequate for performing a thorough test of the transferability approximation. While the diatomic molecule examples[78-80] do enable considering whether the one-center integrals remain the same in diatomics as they are in atoms[81], H^v computations for polyatomic molecules must provide the ultimate test of transferability of two-center integrals and the lack of dependence of the one and two-center true parameters (from atoms and diatomic molecules) on the molecular environment[63]. Thus, a quite extensive series of constrained ab initio H^v computations are necessary for probing the transferability and ZDOA assumptions of semiempirical methods. Until the recent explosion in available computational resources and the significant recent advancement in H^v method algorithms and computer codes, our progress with these fundamental questions has been limited to the consideration of atoms, a few diatomic molecules, one triatomic molecule, and trans-butadiene. However, with the recent advances in code performance and computational facilities, we have made enormous progress through a huge number of H^v calculations for several highly non-trivial conjugated pi-electron molecules[65,70,71].

This section continues with some examples from the H^v computations for atomic systems and the insights they provide into assumptions of semiempirical methods. For brevity, we omit a summary of the diatomic molecule H^v calculations of true two-center parameters[78-80] and whether the one-center diatomic parameters are indeed the same as in the atom. Several general features of the diatomic molecule calculations are, however, also exhibited by the more extensive H^v treatments of the pi-electron systems, and the interested reader is referred to the original papers on the diatomic molecule computations[78-80] for detailed tests of the bond length dependence of various two-center integrals. This section then closes with a brief description of the constrained H^v computations[82] for the ab initio effective valence shell dipole operator[83,84,85], which provides dipole moments and transition dipole moments of all valence states. Constrained calculations of the effective dipole operator yield important insights into severe limitations of current semiempirical methods for molecular dipole properties.

4.1. COMPUTATIONS OF ONE-CENTER H^V PARAMETERS FOR ATOMIC SYSTEMS

A series of H^v computations have been performed for the atomic systems of C through F[52], Si through P[53], and Ti through Cr[54], with the all-electron valence spaces of 2s2p, 3s3p, and 4s3d, respectively. We illustrate the basic conclusions by reference to the first row atoms and second order H^v calculations using unaveraged valence orbital energies and SCF orbitals for either the neutral atom or one of its ions. The particular SCF orbitals are taken as those from the ground state of C (neutral), N (+1 ion), O (+2 ion), F (+3 ion) as providing a compromise for describing all charge states[52]. The computations yield valence state excitation energies and valence orbital ionization potentials and electron affinity for the neutral atoms and all their ions simultaneously from a single computation of the H^v for the atom. These energies are obtained[52] with average deviations from experiment of 0.32 (22), 0.31 (29), 0.38 (40), and 0.33eV (43) for the C, N, O, and F systems, respectively, where the numbers in parentheses indicate the numbers of energies involved in the comparison. Improved accuracy is currently possible with third order computations and larger basis sets than the 4s3p2d Slater bases used previously. (For instance, the average error in the oxygen computations is reduced by to 0.26eV in third order[86] and should be diminished further upon expansion of the basis set.) Nevertheless, these 1980 computations are of sufficient accuracy to illustrate the salient physical points.

The most important feature is the comparison of computed ab initio H^v one-center integrals with those of standard semiempirical methods as exemplified by MINDO/3[87]. The following comparison considers the subset of H^V integrals that may also be uniquely represented as sums and differences of experimental energies for various neutral and ion states of the system since this also enables a comparison of H^V integrals directly with experiment[56]. For instance, the ionization potentials of the $(1s)^2 2s$ and $(1s^2)2p$ states provide the exact experimental values of the one-electron H^v integrals $\langle s|s\rangle \equiv \langle 2s|H_1{}^v|2s\rangle$ and $\langle x|x\rangle \equiv \langle 2p_x|H_1{}^v|2p_x\rangle$, respectively. Likewise, the two-valence-electron state energy differences $\{E[^1D(p^2)] \pm E[^3P(p^2)]\}/2$ yield (upper sign) the one-center repulsion integral $\langle xy|xy\rangle \equiv \langle 2p_x 2p_y|H_{12}{}^v|2p_x 2p_y\rangle$ and (lower sign) the one-center exchange integral $\langle xy|yx\rangle$, respectively. Thus, Table II compares ab initio[52] H^v one-center, one-electron and selected one-center, two-electron integrals with experiment[88] and with MINDO/3 values[87].

Notice that the H^v integrals are in good agreement with experiment. The differences arise from incomplete inclusion of correlation contributions by virtue of the use of a second order perturbation truncation and a limited basis set. The MINDO/3 integrals, on the other hand, are vastly different from the H^V or experimental values! The MINDO/3 one-center, one-electron

Table II. Comparison of H^V, experiment, and MINDO/3 integrals.

	C			N			O			F		
Integral	H^V	Expt	M/3	H^V	Expt	M/3	H^V	Expt	M/3	H^V	Expt	M/3
$-\langle s\|s\rangle$	64.70	64.50	51.79	98.24	97.90	66.06	138.5	138.1	91.73	185.4	185.2	129.9
$-\langle x\|x\rangle$	57.33	56.49	39.18	88.32	87.90	56.40	126.1	126.1	78.80	171.0	171.2	105.8
$\langle sx\|sx\rangle$	19.13	18.20	11.47	23.51	22.69	12.66	27.78	27.16	14.48	32.13	31.63	17.25
$\langle sx\|xs\rangle$	3.41	3.10	2.43	4.20	3.93	3.14	4.97	4.72	3.94	5.75	5.56	4.83
$\langle xy\|xy\rangle$	19.36	18.17	9.84	23.81	23.02	11.59	28.20	27.86	12.98	32.86	32.66	14.91
$\langle xy\|yx\rangle$	0.68	0.52	0.62	0.95	0.82	0.70	1.21	1.10	0.77	1.50	1.41	0.90

integrals are 30-55eV more positive than H^V or experiment, while the MINDO/3 one-center, two-electron repulsion integrals are considerably more positive (by 8-17eV) than those from H^V or experiment. The H^V and experimental repulsion integrals $\langle sx|sx\rangle$ and $\langle xy|xy\rangle$ are, in fact, considerably larger than the neutral atom "bare," or theoretical SCF, integrals by several electron volts. Thus, the ab initio true empirical H^V departs severely from the assumptions of standard semiempirical methods. Consequently, the semiempirical methods are either grossly in error or a more fundamental explanation exists.

The second order H^V, on the other hand, also contains three-electron interactions, which are formally absent from semiempirical methods. Several of the three-electron H^V integrals may likewise be expressed exactly in terms of differences of ion energy levels[56]. Thus, these experimental three-electron integrals are known to accuracies of roughly a cm^{-1}. Table III presents a comparison of the experimental three-electron H^V integrals with those computed from the second order H^V computations[52] using the convenient notation in which, for instance $\langle sxy|sxy\rangle \equiv \langle 2s2p_x2p_y|H_{123}^V|[2s2p_x2p_y-2p_y2p_x2s]\rangle$. (The tabulated three-body H^V integrals contain a direct and exchange contribution because the existence of only two possible spins for the electron implies the physical impossibility of separating the direct and exchange contributions. However, it is possible mathematically to define a set of spin-independent three-body integrals[89], but this separation is unnecessary here.)

Table III. Comparison of some three-body H^V integrals with experiment.

	C		N		O		F	
Integral	H^V	Expt	H^V	Expt	H^V	Expt	H^V	Expt
$\langle sxy\|sxy\rangle$	-2.69	-1.87	-2.45	-1.95	-2.23	-1.97	-2.13	-1.90
$\langle sxy\|syx\rangle$	0.02	-0.12	0.05	0.06	0.07	0.08	0.07	0.06
$\langle sxy\|xsy\rangle$	-0.44	0.06	-0.38	-0.27	-0.33	-0.21	-0.29	-0.23
$\langle xyz\|xyz\rangle$	-3.59	-2.97	-3.32	-2.85	-3.02	-2.87	-2.94	-2.79
$\langle xyz\|xzy\rangle$		-0.17	-0.15	-0.15	-0.13	-0.12	-0.13	-0.13

Notice the huge magnitudes, for instance, of $\langle xyz|xyz\rangle$, whose experimental values of roughly 3eV are fairly independent of atomic number (within a row of the periodic table) and are in reasonable agreement with the second order H^v computations. With hindsight, this large magnitude is not so surprising. The lowest order contributions to the three-body H^v interactions involve single orbital excitations [see Eq. (2.12)] and represent orbital polarization contributions which correct for differences in the orbitals that are "optimal" for different valence states. The other large three-body integral $\langle sxy|sxy\rangle$ in Table III is likewise quite independent of atomic number and is in decent agreement with the H^v calculations. Several other computed three-body integrals are in the range of 2-4eV for the first row atoms carbon through fluorine. The presence of these large three-body parameters contrasts sharply with semiempirical methods which do not contain explicit three-body interactions.

Without the formal H^v theory, it would have been impossible to deduce the presence of large three-body integrals. Some reflection suggests that the semiempirical methods must somehow be averaging[55] these large three-electron integrals into their one and two-body integrals similar to the general spirit in which a one-electron Fock or MCSCF Fock operator contains an average of two-body interactions. A particular illustrative (but not necessarily optimal) averaging scheme has been applied to the carbon atom ab initio H^v parameters[55] and produces averaged integrals close in magnitude to MINDO/3 values. For instance, the averaged H^v (and corresponding MINDO/3) integrals $\langle s|s\rangle$, $\langle x|x\rangle$, $\langle ss|ss\rangle$, and $\langle sx|sx\rangle$ are -50.28 (-51.79), -40.83 (-39.18), 13.59 (12.23), and 10.49eV (11.47), respectively. While not perfect agreement, the correspondence is enormously closer than that between the unaveraged H^v parameters and MINDO/3 integrals. This similarity between these two sets of integrals justifies the above assertion that semiempirical methods are effectively averaging the influence of the three and higher-body H^v integrals into the semiempirical parameters. (Third order H^v computations indicate that four-body integrals are roughly an order of magnitude smaller than the three-body parameters[86].)

It would appear that the semiempirical methods are attempting to have their cake and eat it. These methods implicitly average the large three-body H^v interactions into their one and two-electron parameters, and they expect that these averaged parameters are transferable between different molecules. The following analysis of this implicit averaging provides some general insights into the ranges of validity for conventional types of semiempirical methods: Just as portions of one-electron Fock operators are somewhat transferable if local environments are very nearly identical, a particular averaging scheme for three-electron integrals may be reasonable for systems with similar (but a wider range of) electronic structures. Organic molecules have carbon, oxygen, nitrogen, etc., within fairly narrow ranges of effective charges but with a range of hybridizations about the central carbon atom. Thus, it is possible that a suitable three-body averaging scheme exists and applies well to this range of local environments. This possibility accords with the development of fairly accurate all-valence electron semiempirical methods for these systems.

Transition metal atoms, on the other hand, occur in chemical compounds with a very wide range of formal charges and with high and low spin states. Third order H^v computations for Ti through Cr[54] obtain a large number of three-body integrals lying in the range of 2-3eV. We believe it rather unlikely that there exists a three-body averaging scheme for the transition metal which is adequate with all possible formal charges and spin couplings. This conjecture is in conformity with the absence of a semiempirical method for transition metals in which there is a common set of parameters for all charge and spin states of a given transition metal atom.

Calculations have been made of the bond length dependence of two-center, one and two-body H^v integrals for diatomic molecules[78-80]. Some of these bond length dependences have identical forms as those invoked by semiempirical methods, but only after correlation contributions are included. Other two-center integrals have a slightly different form from the traditional semiempirical parameterizations, while a few integrals depart considerably in their bond length dependence. However, the all-valence electron H^v treatments produce large three-body parameters, which should really be averaged into the one and two-body integrals for proper comparison of the averaged H^v with their semiempirical counterparts. Thus, additional work is of interest on developing "optimal" three-body averaging schemes. We now turn to tests of the fundamental semiempirical assumptions of transferability and zero differential overlap approximation that are common to all semiempirical methods.

4.2. AB INITIO COMPUTATION OF PARAMETERS IN PI-ELECTRON METHODS

Because of the high accuracy afforded by full ab initio H^v computations for pi-electron systems, it is quite natural to apply the more approximate constrained H^v computations to pi-electron systems in order to test and refine the assumptions of semiempirical pi-electron methods, such as that of Pariser-Parr-Pople (PPP) theory. The pi-electron systems provide an ideal system for beginning tests of the transferability and zero differential overlap approximations, which are common to the all-valence electron semiempirical methods. Thus, understanding and advancements made with the pi-electron systems have immediate ramifications for the more complicated all-valence electron systems. Pi-electron methods are still widely used[22,24,90] for studying the properties of conducting and nonlinear polymers[23] and for treating the electronic structure of complex biological molecules. Consequently, improvements in pi-electron methods have direct applicability to these interesting systems. In addition, virtually all electronic structure theories of solids are semiempirical in nature. For example, the widely used Hubbard-type models for solids have the same formal structure as PPP theory. Thus, fundamental understanding gleaned from studying pi-electron systems suggests possible improvements with semiempirical models for the electronic structure of solids.

Some ideal characteristics of the pi-electron systems for performing these tests include the fact that there are only a small number of pi-orbitals, and the symmetries are generally high. Consequently, the number of ab initio

H^v parameters for analysis is kept to a minimum. Another significant feature of the pi-electron H^v emerges because the three-body parameters are considerably smaller than in the all-electron example of atomic one-center integrals described in subsection 4.1. The largest three-body H^v parameters are generally the diagonal one-center integrals, but the pi-electron Hamiltonians have just a single pi-electron orbital per main atom center (carbon atoms in the examples below.) Thus, the largest three-electron integrals for the pi-electron systems are the two-center, diagonal interactions, which are considerably smaller than the approximate 4eV magnitudes found for first row atoms in the all-valence electron H^v. Old, smaller basis set trans-butadiene H^v computations[91] with a non-transferable set of pi-orbitals yield the largest three-body integrals as 0.74eV. Hence, the test of semiempirical assumptions for pi-electron systems presents the least complications from the necessity for averaging the three-body parameters into the one and two-electron integrals.

Full, ab initio H^v computations, such as those described in section 3, use a set of variationally determined orbitals. The construction of these orbitals represents an attempt to "optimize" the first order description H_{PP} that is provided by the core and valence orbitals. However, these "optimal" variational valence orbitals exhibit a strong molecule dependence. The semiempirical methods, on the other hand, assume the existence of a universal, transferable, molecule independent basis set of valence orbitals which depend solely on the constituent atoms and which are to be used for all valence states of both the neutral and ion states for all molecular systems. Thus, the comparison of ab initio H^v calculations with semiempirical methods requires that we perform more approximate H^v computations in which the constrained valence orbitals are constructed from such a transferable basis set. These constrained valence orbitals provide less than an optimal first order description from H_{PP}, and consequently, they must also slow the convergence of the H^v perturbation expansion. Thus, it is necessary first to test the ability of the H^v method in describing the pi-electron spectrum with this constrained valence space. While higher order computations with constrained orbitals might be of very high accuracy, our computations are limited to third order, and therefore we must consider the "practical" convergence of the third order constrained H^v calculations for the pi-electron systems.

The construction of constrained valence orbitals is quite straightforward for the pi-electron systems. The transferable, localized basis contributes one p_π-like valence orbital $p_{\pi(i)}$ on the ith carbon atom center, and these p_π basis functions are naturally nonorthogonal. Molecular orbitals φ_α^{mo} are then formed as linear combinations of these individual p_π orbitals[92,93],

$$\varphi_\alpha^{mo} = \sum_{i=1}^{N_v} C_{i,\alpha} p_{\pi(i)}, \tag{4.1}$$

where symmetry in many systems dictates all or a portion of the expansion coefficients $\{C_{i,\alpha}\}$. Since the core orbitals are not specified in semiempirical methods, our constrained orbital H^v computations employ a variationally optimized set of core and excited orbitals[93], determined according to the same sequential set of SCF calculations as in Eq. (3.5), except for the constraint in Eq. (4.1) upon the construction of the valence orbitals. Constrained H^v computations for pi-electron systems have so far considered only the simplest choice for the transferable pi-orbitals $p_{\pi(i)}$, namely, carbon atom SCF p_π orbitals[92,93], although other possibilities should be tested in the future. Notice that the resultant molecular orbitals, therefore, omit important polarization components, for instance, those that move some pi-electron density onto the hydrogens as is present in optimal (called "full") H^v and other ab initio calculations. The H^v computations described in section 3 indicate the ability of the H^v method to compensate for the state dependence of valence orbitals appearing in traditional wavefunction methods, but the constraints in Eq. (4.1) on the valence orbitals are quite a bit more severe.

Differences (called constraining errors), of course, appear between "full" and constrained H^v computations for the pi-electron systems. The constrained H^v calculations poorly represent excitation energies to valence states of primary Rydberg character[83], but this is as anticipated since the traditional pi-electron Hamiltonian is represented in terms of valence-like pi-orbitals and *should not* generate a satisfactory description of the low-lying Rydberg states. Hence, the analysis of constraining errors is limited to states of only valence-like parentage. The possibility exists that higher order H^v computations could converge well to the excitation energies (but not the spatial extent) of states with mixed Rydberg-valence parentage, such as the 1^1B_u state in trans-butadiene which has approximately 20% Rydberg character. However, the third order constrained H^v calculations appear unable to represent such mixed character states well. (For instance, the constraining error for the 1^1B_u state is a considerable -0.69eV.) Thus, constrained H^v computations are useful in testing basic assumptions of semiempirical pi-electron methods if the constraining errors are tolerable for the low-lying valence-like pi-electron states.

Table IV summarizes the constraining errors obtained for trans-butadiene with a ccPVDZ basis augmented with two Rydberg carbon p functions[93]. The constraining errors (difference between full and constrained calculation) for the ground state energy, excitation energies to three low-lying valence-like states, and the two lowest ionization potentials in Table IV are an average of 0.23eV. The three-body errors in Table IV designate the difference between full H^v computations with and without the three- and four-body interactions. The average three-body error in Table IV is 0.21eV. The combined error designates the difference between full H^v computations and constrained calculations with omitted three- and four-body interactions. The combined average error of 0.16eV indicates that the constraining and three-body errors are non-additive, but, more importantly, the small magnitude of the combined error demonstrates that a reasonably good approximation is provided by the approximate constrained H^v computations with only one and two-body effective interactions. If, on the other hand, the combined error is determined instead with respect to the experimental energies (available in Table I), then the combined error shrinks to a mere 0.08eV. Similar computations for several other conjugated hydrocarbons yield combined errors with respect to experiment that are comparable to those from the full H^v calculations. Hence, we have the remarkable occurrence that the approximate, constrained calculations with neglected three- and four-body interactions yield highly accurate descriptions of the valence-like states of all conjugated hydrocarbons using a single universal basis set of p_π valence orbitals for all these systems! Consequently, we can certainly use the constrained H^v one- and two-body interactions to test the basic approximations invoked by semiempirical methods. (Additional universal valence shell basis set computations are desirable for larger systems to determine if the combined error with respect to experiment remains so small.

Table IV. Constraining errors for low-lying pi-electron states of trans-butadiene.

State	Constraining Error	3-body Error	Combined Error	Type
X^1A_g	0.13	-0.19	-0.24	Valence
1^3A_g	-0.15	-0.17	-0.19	Valence
1^3B_u	-0.16	-0.16	-0.04	Valence
2^1A_g	-0.47	-0.48	-0.17	Valence
1^2B_g	-0.22	0.09	0.20	Valence
1^2A_u	-0.26	0.17	0.12	Valence

While it may be possible to reduce the constraining and three-body errors by investigating methods for "optimizing" the localized, transferable, atomic p_π basis and for averaging the three-body parameters into the one- and two-body parameters, respectively, the errors with the current calculations are sufficiently tolerable for using these constrained computations to assess for pi-electron systems the validity of the ZDOA and the transferability assumptions which are common to all semiempirical methods. We, therefore, begin by examining the repulsion integrals neglected by the ZDOA[94].

Because Löwdin orthogonalization is generally invoked as providing the justification for the ZDOA, the ZDOA neglected two-electron constrained H^v parameters are presented in the Löwdin carbon atom p_π basis using the conventional notation $\langle ij|kl\rangle \equiv \langle p_{\pi(i)}(1)p_{\pi(j)}(2)|H_{12}{}^v|p_{\pi(k)}(1)p_{\pi(l)}(2)\rangle$. Theoretical (or "bare") integrals are denoted with a subscript zero. The largest bare integrals among the ZDOA neglected ones are the nearest neighbor two-center exchange $\langle 12|21\rangle_0$, the two-electron hybrid integral $\langle 11|22\rangle_0$, and the two-electron contribution $\langle 12|11\rangle_0$ to the resonance integral. The latter two-bare integrals are identical but generally have different correlation contributions. The illustrations provided below consider examples drawn from ethylene, trans-butadiene, cyclobutadiene, benzene, and preliminary computations for trans-trans-hexatriene. In order to test parameter transferability, comparisons of H^v parameters are made using computations in which the geometries are taken as similar as possible. Thus, for instance, the ethylene geometry is distorted slightly to have the same double bond length and HCH angle as in trans-butadiene[93,94]. (The distorted geometry ethylene H^v parameters are rather close to those evaluated at the ethylene ground state geometry.) Likewise, computations for square cyclobutadiene use the same CC bond length as either trans-butadiene of benzene.

The inclusion of correlation contributions is important in providing the true justification of the ZDOA. The bare $\langle 12|21\rangle_0 = \langle 11|22\rangle_0$ for distorted ethylene are 0.126eV, but upon introducing correlation, the H^v parameters are reduced to 0.001 and -0.007eV, respectively[94]. These truly negligible values are likewise obtained for all other pi-electron systems studied to date. For example, they become 0.021 and -0.013eV for trans-butadiene. However, correlation converts the $\langle 12|11\rangle$ parameter from the bare value of -0.277 to the rather non-negligible 0.281eV. Again, constrained H^v computations for other linear pi-electron systems yield significant $\langle 12|11\rangle$, e.g., 0.235eV in trans-butadiene; however, the somewhat smaller -0.134 and -0.051eV are obtained for the cyclic systems of (square) cyclo-butadiene and benzene (both with the benzene CC bond length), respectively. More computations are necessary to ascertain whether this difference between linear and cyclic systems is significant, but the available results suggest that an improved pi-electron theory should consider inclusion of the two-electron contribution $\langle 12|11\rangle$ to the nearest neighbor resonance integral.

The non-negligible two-electron parameter $\langle 12|11\rangle$ is quite interesting. Iwata and Freed[19] have shown that this parameter (along with the three-body analogs such as $\langle 112|111\rangle$) contribute to a breakdown of the pairing theorem for alternant hydrocarbons, leading, for instance to the observed asymmetry between spectra for anions and cations. More recently, Hirsh has argued for the necessity of retaining the analogous nearest neighbor $\langle 12|11\rangle$ integral in theoretical description for the electronic structure of high temperature superconductors[18]. Constrained H^v computations should be performed to evaluate this (and other) correlated integrals for the copper oxide systems.

As the atomic centers become more removed, both the bare ZDOA neglected repulsion integrals and their correlation contributions diminish considerably. Thus, the trans-butadiene bare Löwdin basis $\langle 13|31\rangle$ and $\langle 14|41\rangle$ exchange integrals of 0.011 and 0.001eV, respectively, are transformed into the correlated third order H^v integrals of 0.003 and 0.0001eV[94]. While correlation again considerably reduces these integrals, their bare values are already quite negligible. Similar trends are obtained for the two-center hybrid integrals $\langle 11|33\rangle$ and $\langle 11|44\rangle$, while the H^v two-electron resonance integral $\langle 11|13\rangle$ of trans-butadiene is roughly -0.1eV and may therefore warrant retention in an improved pi-electron method.

The bare counterparts of the one-electron, one-center integrals $\alpha_i = \langle i|i\rangle \equiv \langle p_{\pi(i)}(1)|H_1{}^v|p_{\pi(i)}(1)\rangle$ are understood to depend on local geometry since electrons on the ith atomic center experience fields due to the surrounding atoms. Thus, the H^v α_1 differs from α_2 for trans-butadiene (by roughly 4eV) as well as for other inequivalent atomic centers, and the only potential transferability is associated with the correlation contributions to α which range between 1-3eV. Trans-butadiene and distorted ethylene have the correlation portion of α_1 transferable (i.e., equal) to within a remarkable 0.012eV out of the total correlation contribution of 1.44eV. However, this is a rather unusual case in which the geometries are very similar. The less similar cases of benzene at its equilibrium geometry and square cyclobutadiene (with the same CC bond length as in benzene) yield correlation contributions to α that differ by a substantial 0.6eV[71]. Thus, a rather systematic study is necessary to elucidate the nontrivial but significant geometry dependence of correlation contributions to the H^v α parameters.

Fortunately, the correlation contributions to the α parameter exhibit the worst transferability. The other one-electron pi-electron integrals are the resonance integrals $\beta_{ij} \equiv \langle p\pi(i)(1)|H_1{}^v|p\pi(j)(1)\rangle$ and are best analyzed again in a Löwdin orthogonalized basis set. Again using as close as possible to a common geometry, the H^v β_{12} for ethylene, trans-butadiene, and trans-trans-hexatriene (preliminary) are -3.43, -3.34, and -3.15eV, respectively, while the single bond β_{23} for the latter two molecules are -2.57 and -2.37eV, respectively[94]. The ZDOA neglected β_{13} and β_{14} in trans-butadiene are 0.24 and 0.06eV, respectively, thereby justifying the neglect of β_{14} but *not* of β_{13}. Computations of β_{13} for cyclobutadiene yield 0.24eV, while those for benzene (at its equilibrium geometry) obtain 0.29eV, further supporting the necessity for retention of this customarily neglected pi-electron integral.

The one-center, two-electron repulsion integrals $\gamma_{ii} \equiv \langle p\pi(i)(1)p\pi(i)(2)|H_{12}{}^v|p\pi(i)(1)p\pi(i)(2)\rangle$ contain the largest correlation contributions (roughly 4eV), but display comparable transferability to the β_{ij} and far superior transferability to the α_i. The bare integrals $\gamma_{ii}{}^0$ are exactly transferable when evaluated with the non-orthogonal localized pi-orbitals, so our discussion of repulsion integrals is conveniently presented using this basis. The transferability of γ_{11} between ethylene, trans-butadiene, and trans-trans hexatriene is quite remarkable, with values of 11.78, 11.83, and 11.83eV, respectively. The computed γ_{22} differs slightly from γ_{11} because correlating electrons on these two atomic centers experience different environments. The difference $|\gamma_{11}-\gamma_{22}|$ in trans-butadiene is the rather small 0.02eV but grows slightly to 0.14eV in trans-trans-hexatriene. The double bond γ_{12} H^v integrals are 7.47, 7.70, and 7.55eV, respectively, for ethylene, trans-butadiene, and trans-trans-hexatriene, while the single bond γ_{23} for the latter two molecules are 7.34 and 7.22eV, respectively. Since it is doubtful that our constrained, third order H^v calculations for the γ_{ij} are accurate to better than 0.1eV, these results provide strong support for the transferability of the *correlated* H^v repulsion integrals.

The constrained H^v computations for pi-electron systems provide deep insights into the transferability and ZDOA assumptions that are common to all semiempirical methods. Firstly, the computations focus on the correlated parameters and not on the irrelevant theoretical integrals customarily analyzed in developing semiempirical parameterization schemes. Correlation is found to diminish the magnitudes of all of the ZDOA discarded parameters to a truly negligible magnitude, except for the two-electron contribution to resonance integrals and next nearest neighbor resonance integrals, which are computed with values in the range of 0.2-0.3eV (depending on system and parameter) and which therefore merit retention in improved pi-electron methods. Rather good transferabilities are found for resonance and repulsion integrals, with the correlation contributions to the one-center, one-electron α parameters displaying the greatest sensitivity to local environment. Further extensive H^v computations are necessary to map out such interesting behaviors as the bond length and environment dependence of the true H^v parameters, as well as to introduce improvements from optimizing the atomic p_π orbitals and from averaging the influence of neglected three-body and other neglected small H^v parameters. Nevertheless, the constrained H^v computations to date provide excellent justifications for some, *but not all*, assumptions of semiempirical pi-electron methods, and, more importantly, these computations provide significant information into how correlation contributes to individual parameters. The latter insight is buttressed by a computational approach that enables evaluation of the individual parameters to determine a set of approximations and parameters for developing improved semiempirical methods.

4.3. AB INITIO COMPUTATIONS OF THE EFFECTIVE DIPOLE OPERATOR: COMPARISONS WITH SEMIEMPIRICAL METHODS

Just as section 2 demonstrates how it is possible to construct an effective valence shell Hamiltonian that acts only on the space spanned by a set of valence orbitals, a straightforward generalization produces effective operators for properties other than the energy. For instance, if A is a Hermitian operator corresponding to an observable and if ψ_i^v is an eigenfunction of H^v corresponding the exact eigenfunction ψ_i of the full Hamiltonian H, we may likewise derive an effective operator A^v such that

$$\langle\psi_i|A|\psi_j\rangle = \langle\psi_i^v|A^v|\psi_j^v\rangle. \tag{4.2}$$

Equation (4.2) implies that the exact expectation values (for $i=j$) and exact transition matrix elements (for $i\neq j$) of the full operator A may alternatively be evaluated as matrix elements of the valence shell operator A^v using the eigenfunctions of the effective valence shell Hamiltonian. (Both the exact and H^v eigenfunctions are taken as normalized to unity.[61]) Consequently, semiempirical treatments of properties other than energies are implicitly attempting to approximate the structure of the true A^v and its valence space matrix elements.

Numerical applications to date of the effective operator method have been limited to the computation of the effective valence shell dipole operator μ^v whose evaluation provides dipole moments, transition dipole moments, oscillator strengths, radiative lifetimes, etc.[83-85] The computations consider the leading order correlation corrections and formally resemble[83] a portion of the second order H^v calculations. Recent work has finally provided the rather non-trivial derivation of the next order terms.[95] Particular ab initio applications of the μ^v method consider first row diatomic hydrides. Calculated dipole properties for these systems compare rather favorable to those from experiment and from highly correlated ab initio methods with basis sets of similar quality. We omit details of the computed ab initio dipole properties and instead focus on comparisons with semiempirical methods.

The leading order correlated effective valence shell dipole operator μ^v contains a core contribution μ_C as well as both one- and two-body operators. The μ^v computations demonstrate that the two-body matrix elements are non-negligible and must be retained for ab initio accuracy. Hence, semiempirical dipole theories should again implicitly average these two-body dipole contributions into the retained one-body dipole operators. Constrained μ^v calculations for the CH molecule (and for CH^+) agree well with full computations over a range of internuclear separations that includes the ground state equilibrium internuclear separation. Thus, we compare the constrained one-electron μ^v matrix elements with their semiempirical counterparts.

Taking the carbon atom at the origin of the coordinate system, the z-axis along the molecular bond, and the internuclear separation as 2.1au, several ab initio constrained μ^v matrix elements qualitatively resemble those in semiempirical methods.[96] For example, the carbon 2s and $2p\pi$ diagonal matrix elements of the z-component z^v are 0.016 and 0.004au, respectively, in good agreement with the semiempirical treatment of these integrals as vanishing. Likewise, the ab initio off-diagonal $\langle 2s|z^v|2p\sigma\rangle$ matrix element on carbon is 0.934, while the corresponding semiempirical integral is the similar 0.888au. There are, however, some major differences: For instance, the two-center integral $\langle 1s_H|z^v|2p\sigma\rangle$ is obtained from the μ^v computations as 0.203au, but the semiempirical treatment (with two-center integrals retained) yields a considerably different 1.028au. Similarly, the semiempirical vanishing $\langle 2p\sigma|z^v|2p\sigma\rangle$ contrasts sharply with the ab initio computed 0.771au.

To understand the origins of this sharp discrepancy between the ab initio and semiempirical dipole matrix elements, it is useful to recall the customary point charge model for computing the semiempirical matrix elements. Each atom is considered as a point charge with the charge densities computed from the population densities that follow from the semiempirical wavefunction. The diagonal dipole matrix elements are obtained from the atomic charge densities, while the off-diagonal matrix elements are computed using single Slater functions. Thus, the point charge model *represents the use of an approximation to the theoretical dipole matrix elements.* As noted above and in the literature[3], an important component in the success of semiempirical methods for electronic energies lies in the empirical "adjustment" of valence shell matrix elements to include correlation contributions. However, these correlation contributions are completely omitted from the semiempirical dipole matrix elements!

This observation already raises serious concern. Semiempirical dipole moments are notoriously poor for small molecules, suggesting that the correlation contributions are not negligible.

Further analysis of the constrained μ^v computations provides insights into some deficiencies of the current semiempirical dipole treatment and into methods for its improvement. The largest discrepancy between μ^v matrix elements and those of the point charge model emerge for matrix elements containing either the bonding carbon $2p_\sigma$ or hydrogen 1s orbitals (or both). These two orbitals in full ab initio SCF computations have polarization contributions that somewhat affect computed energies but have dramatic influences on dipole matrix elements. The semiempirical point charge model dipole matrix elements completely neglect these important polarization contributions. On the other hand, polarization corrections enter into the constrained μ^v calculations through the perturbation corrections. Thus, it is not surprising that the greatest discrepancies between semiempirical and constrained μ^v matrix elements emerge for the most polarized valence shell orbitals.

The ab initio effective dipole operator calculations represent only an initial application of the effective operator approach, but they strongly suggest that an improved semiempirical dipole method must, at least, include corrections due to orbital polarization. Additional contributions from true "correlation" processes would also be desirable. The appropriate forms might be deduced by trial and error parametrizations using polarized SCF orbitals, but extensive ab initio μ^v computations should provide an alternative route to generate a data base from which to discern general trends.

5. Acknowledgment

This research is supported, in part, by NSF Grant CHE93-07489. I am grateful to the many talented graduate students and postdocs who have contributed to this research. Thanks go to Chuck Martin for his comments on the manuscript.

6. References

[1] Parr, R. G. (1963) The Quantum Theory of Molecular Electronic Structure, Benjamin, New York and references therein.

[2] Murrell, J. N. (1963) The Theory of the Electronic Spectra of Organic Molecules, Wiley, New York.

[3] Dewar, M. J. S. (1969) The Molecular Orbital Theory of Organic Chemistry, McGraw-Hill, New York.

[4] Pariser, R., and Parr, R. G. (1953) J. Chem. Phys. **21**, 446, 467; Pople, J. A. (1953) Trans. Faraday Soc. **49**, 1375.

[5] Murrell, J. N., and Harget, A. J. (1972) Semi-empirical Self-consistent Field Molecular Orbital Theory of Molecules, Wiley, NewYork.

[6] Schaefer, H. F. (1972) The Electronic Structure of Atoms and Molecules, Addison-Wesley, Reading, MA.

[7] Bauschlicher, C. W., Jr., Langhoff, S. R., and Taylor P. (1992) Adv. Chem. Phys. **87**, 103 and references therein.
[8] Dunning, T. H., Jr., (1989) J. Chem. Phys. **90**, 1007.
[9] Freed, K. F. (1972) in Segal, G. A. (ed) Modern Theoretical Chemistry, Plenum, New York, volume 7.
[10] Freed, K. F. (1983) Acc. Chem. Res. **16**, 137.
[11] Dewar, M. J. S. (1992) Int. J. Quantum Chem. **44**, 427.
[12] Zerner, M. C., (1991) Rev. Comp. Chem. **2**, 313.
[13] Dewar, M. J. S. (1975) Science **187**, 1037; Dewar, M. J. S., Zoebisch, E. G., Healy, E. F., and Stewart, J. J. P. (1985) J. Am. Chem. Soc. **107**, 3902.
[14] Thiel, W., (1988) Tetrahedron **44**, 7393; (1981) J. Am. Chem. Soc. **103**, 1413.
[15] Kolb, M., and Thiel, W. (1992) J. Comp. Chem. **14**, 775.
[16] Sadlej, J. (1985) Semiempirical Methods in Quantum Chemistry, Wiley, New York, p44.
[17] Stewart, J. J. P., (1989) J. Comp. Chem. **10**, 209, 221.
[18] Hirsh, J. E. (1990) Chem. Phys. Lett. **171**, 161.
[19] Iwata, S., and Freed, K. F. (1974) Chem. Phys. Lett. **38**, 425.
[20] McLachlan, A. D., (1959) Mol. Phys. **2**, 271; Koutecky, J. (1966) J. Chem. Phys. **44**, 3702.
[21] Csonka, G. I. (1993) J. Comp Chem. **14**, 895.
[22] Martin, R. L., (1993) in Paccioni, G. (ed) Cluster Models for Bulk and Surface Phenomena, Plenum, New York.
[23] See special issue of Int. J. Quantum. Chem. (1988) **43**, No. 1 and Springer Series in Solid state Sciences (1992) **102**; Prasad, P. N., and Williams, D. J. (1991) Introduction to Non-linear Optical Effects in Molecules and Polymers, Wiley, New York; Li, D., Marks, T. J., and Ratner, M. A. (1986) Chem. Phys. Lett. **131**, 370; de Melo, C. P., and Silbey, R. (1988) J. Chem. Phys. **88**, 2567.
[24] Chema, D. S., and Zyss, J. (1987) Nonlinear Optical properties of Organic Polymer Materials, Academic, New York; See also apecial issue on nonlinear optics in issue 1 of Int. J. Quantum. Chem. (1992) **43**.
[25] Feng, J., Li., J., Wang, Z., and Zerner, M. C. (1990) Int. J. Quantum. Chem. **37**, 599.
[26] Zerner, M. C. (1975) J. Chem. Phys. **62**, 2788.
[27] Nanda, D. N., and Jug, K. (1990) Theor. Chim. Acta **57**, 95.
[28] Kanzler, A. L., Freed, K. F., and Sun, H. (1992) J. Chem. Phys. **96**, 5245.
[29] Haser, M., and Amlof, J. (1991) J. Chem. Phys. **96**, 489; Amlof, J. (1991) Chem. Phys. Lett. **181**, 319; Neuheuser, T., von Arnim, M., and Peyerimhoff, S. D. (1992) Theoret. Chim. Acta **83**, 123.
[30] Freed, K. F. (1972) Chem. Phys. Lett. **13**, 181; **15**, 331; (1974) J. Chem. Phys. **60**, 1765.
[31] Freed, K. F., and Levy, M. (1982) J. Chem. Phys. **77**, 396.
[32] Herman, M. F., Freed, K. F., and Yeager, D. L. (1981) Adv. Chem. Phys. **48**, 1.
[33] Freed, K. F. (1989) Lect. Notes Chem. **52**, 1.
[34] Sheppard, M. G., and Freed, K. F. (1981) J. Chem. Phys. **75**, 4507.
[35] van Vleck, J. H. (1929) Phys. Rev. **33**, 467.
[36] Kato, T. (1949) Prog. Theor. Phys. **5**, 95, 207, 514.
[37] Bloch, C. (1958) Nucl. Phys. **6**, 329; Bloch, C., and Horowitz, J. (1958) *ibid.* **8**, 91.
[38] des Cloizeaux, J. (1960) Nucl. Phys. **20**, 321.
[39] Brandow, B. H. (1967) Rev. Mod. Phys. **39**, 771.
[40] Lindgren, I. (1974) J. Phys. B **7**, 2441.

[41] Kirtman, B. (1968) J. Chem. Phys. **49**, 3890, 3895; Kvasnicka, V. (1983) Int. J. Quantum. Chem. **24**, 335; Shavitt, I., and Redmon, L. T. (1980) J. Chem. Phys. **61**, 786.
[42] This important point has been missed in the discussion of third order contributions by Durand, P., and Malrieu, J. P. (1987) Adv. Chem. Phys. **67**, 321.
[43] Iwata, S., and Freed, K. F. (1976) J. Chem. Phys. **65**, 1071. The form of H_0 used in this paper is specified below in Eq. (3.2).
[44] Bauschlicher, C. W., Jr., Partridge, H., and Scuseria, G. E. (1992), J. Chem. Phys. **97**, 7471.
[45] Langhoff, S. R., and Davidson, E. R. (1974) Int. J. Quantum. Chem. **8**, 61.
[46] Stern, P. S., and Kaldor, U. (1975) J. Chem. Phys. **63**, 2199; (1976) *ibid.* **64**, 2002.
[47] Hose, G., and Kaldor, U. (1979) J. Phys. B **12**, 3827; (1982) J. Phys. Chem. **86**, 2133; (1987) J. Chem. Phys. **81**, 2406.
[48] Schucan, T. H., and Weidenmüller, H. A. (1972), Ann. Phys. (NY) **73**, 108; *ibid.* **76**, 483.
[49] Zarrabian, S., Laidig, W. D., and Bartlett, R. J. (1990) Phys. Rev. A **41**, 471; Zarrabian, S., and Paldus, J. (1990) Int. J. Quantum Chem. **S15**, 21.
[50] See, for instance, Daudey, J.-P., Heully, J. L., and Malrieu, J. P. (1993), J. Chem. Phys. **99**, 1240.
[51] Hoffmann, M. R. (1994) in Yarkony, D. R. (ed) Modern Electronic Structure Theory, World Scientific, Singapore, in press.
[52] Yeager, D. L., Sun, H., Freed, K. F., and Herman, M. F. (1978) Chem. Phys. Lett. **57**, 490; Sun, H., Freed, K. F., Herman, M. F., and Yeager, D. L. (1980) J. Chem. Phys. 72, 4158.
[53] Sheppard, M. G., and Freed, K. F. (1981) J. Chem. Phys. **75**, 4525; (1981) Int J. Quantum Chem. **S15**, 21.
[54] Lee, Y. S., and Freed, K. F. (1982) J. Chem. Phys. **77**, 1984; Lee, Y. S., Sun, H. Sheppard, M. G., and Freed, K. F. (1980) *ibid.* **73**, 1472.
[55] Freed, K. F., and Sun, H. (1980) Israel J. Chem. **19**, 99.
[56] Yeager, D. L., Sheppard, M. G., and Freed, K. F. (1980) J. Am. Chem. Soc. **102**, 1270.
[57] Oleksik, J. J., and Freed, K. F. (1983) J. Chem. Phys. **79**, 1396.
[58] Takada, T. Sheppard, M. G., and Freed, K. F. (1983) J. Chem. Phys. **79**, 325.
[59] Takada, and Freed, K. F. (1984) J. Chem. Phys. **80**, 3690; Oleksik, J. J., Takada, and Freed, K. F. (1985) Chem. Phys. Lett. **79**, 1396; Kanzler, A. W., and Freed, K. F. (1991) J. Chem. Phys. **94**, 3778.
[60] Wang, X. C., and Freed, K. F. (1987) J. Chem. Phys. **86**, 2899.
[61] Hurturbise, V., and Freed, K. F. (1993) Adv. Chem. Phys. **83**, 465.
[62] Wang, X. C., and Freed, K. F. (1989) J. Chem. Phys. **91**, 1142; (1991) *ibid.* **94**, 5253.
[63] Wang, X. C., and Freed, K. F. (1989) J. Chem. Phys. **91**, 1151.
[64] Finley, J. and Freed, K. F. (1994) manuscript in preparation.
[65] Graham, R. L., and Freed, K. F. (1992) J. Chem. Phys. **96**, 1304.
[66] Sun, H., Sheppard, M. G., and Freed, K. F. (1981) J. Chem. Phys. **74**, 6842
[67] Hoffmann, M. R., Wang, X. C., and Freed, K. F. (1987) Chem. Phys. Lett. **136**, 392.
[68] Serrano-Andres, L. Marchan, M., Nebot-Gil, I., Lindh, R., and Roos, B. J. (1993) J. Chem. Phys. **98**, 3151.
[69] Martin, C. H., Graham, R. L., and Freed, K. F. (1994) J. Phys. Chem. (in press).
[70] Martin, C. H., Graham, R. L., and Freed, K. F. (1993) J. Chem. Phys. **99**, 7833.

[71] Martin, C. H., and Freed, K. F. (1994) manuscript in preparation.
[72] Roos, B. O., Andersson, K., and Fülscher, M. P. (1992) Chem. Phys. Lett. **192**, 5.
[73] Kaldor, U. (1991) Theor. Chim. Acta **80**, 427; Hughes, S. R., and Kaldor, U. (1992) Chem. Phys. Lett. **194**, 99; Berkovic, S. and Kaldor, U. (1994) (in press).
[74] Takada, T., Sheppard, M. G., and Freed, K. F. (1983) J. Chem. Phys. **79**, 325.
[75] Wang, X. C., and Freed, K. F. (1989) J. Chem. Phys. **91**, 3002.
[76] Jensen, E., Keller, J. S., Waschewsky, G. C. G., Stevens, J. E., Graham, R. L., Freed, K. F., and Butler, L. J. (1993) J. Chem. Phys. **98**, 2882.
[77] Stevens, J. E., Graham, R. L., Freed, K. F., and Arendt, M. F. (1994) manuscript in preparation.
[78] Sun, H., and Freed, K. F. (1984) J. Chem. Phys. **80**, 779.
[79] Wang, X. C., and Freed, K. F. (1987) J. Chem. Phys. **86**, 2899.
[80] Kanzler, A. W., and Freed, K. F. (1991) J. Chem. Phys. **94**, 3778.
[81] Sun, H., Sheppard, M. G., Freed, K. F., and Herman, M. F. (1981) Chem. Phys. Let. 555.
[82] Kanzler, A. W., Freed, K. F., and Sun, H. (1992) J. Chem. Phys. **96**, 5245.
[83] Sun, H., and Freed, K. F. (1988) J. Chem. Phys. **88**, 2659; **89**, 5355.
[84] Sun, H., Lee, Y. S., and Freed, K. F. (1988) Chem. Phys. Lett. **150**, 529.
[85] Kanzler, A. W., Sun, H., and Freed, K. F. (1991) Int. J. Quantum Chem. **39**, 269.
[86] Sheppard, M. G., and Freed, K. F. (1981) Int. J. Quantum. Chem. **15**, 21.
[87] Bingham, R. C., Dewar, M. J. S., and Lo, D. H. (1975) J. Am. Chem. Soc. **97**, 1285.
[88] Moore, C. E. (1949) Atomic Energy Levels, Nat. Bur. Stand. (US) Circ. 467; Bashkin, S., and Stoner, J. O. (1978) Atomic Energy Levels and Grotrian Diagrams, North Holland, Amsterdam.
[89] Kanzler, A. W., Freed, K. F., and Sheppard, M. G. (1992) Int. J. Quantum Chem. **44**, 643.
[90] Kollmar, C. (19930 J. Chem. Phys. **98**, 7210; Pancif, J., and Zahradik, R. (1973) J. Phys. Chem. 77, 107, 114.
[91] Lee, Y. S., Freed, K. F., Sun, H., and Yeager, D. L. (1983) J. Chem. Phys. **79**, 3862.
[92] Wang, X. C. (1989) thesis, University of Chicago.
[93] Martin, C. H., and Freed, K. F. (1994) J. Chem. Phys. (in press).
[94] Martin, C. H., and Freed, K. F. (1994) J. Chem. Phys. (submitted).
[95] Hurtubise, V., and Freed, K. F. (1994) J. Chem. Phys. (submitted).
[96] Dewar, M. J. S., and Thiel, W. (1977) J. Am. Chem. Soc. **99**, 4899, 4915.

ELECTRON DELOCALIZATION IN THE THEORY OF INTERMOLECULAR AND INTERGROUP INTERACTIONS: CAUSE, EFFECT, PREVENTION

William H. Adams
Wright and Rieman Chemistry Laboratories
Rutgers University
New Brunswick, NJ 08903
U.S.A

1. Introduction

It is common to think about complex electronic systems as if they were made up of distinct groups of electrons such as core and valence shells, or atoms and molecules. If one wishes, however, to base a computational method on such physical concepts, one must address certain basic questions. The classic question is, what keeps the electrons of a group from delocalizing to its neighbors and collapsing into their cores? A second question is, what is the effect of transfer and collapse in terms of wave functions and energies? The energies will be wrong, but can one recognize that they are wrong? And, finally, can transfer or collapse be prevented? The Pauli Exclusion Principle "explains" why physically there is no transfer and collapse, but that does not prevent the Schrödinger equation from having solutions corresponding to transfer and collapse. To prevent transfer and collapse, strong-orthogonality has often been imposed in variational theories which distinguish between groups of electrons. In the Hartree-Fock theory screening or localizing potentials have been introduced without making approximations. In contrast, the problem has been largely ignored in the perturbation theory of intermolecular interactions, which we believe to be a serious mistake. In this paper these questions are considered anew within the framework defined by the exact eigenfunctions of the nonrelativistic Hamiltonian for *N*-electron systems and within that defined by exact solutions of the Hartree-Fock equation. We focus on the question of how electron delocalization can be prevented without resorting to approximations.

The second question above was answered in Claverie's analysis [1] of the most common form of intermolecular perturbation theory. He showed that the perturbation method itself made the electrons of one atom mathematically distinguishable from those of a second, and that the net effect of this was to make *unphysical* eigenfunctions of the Hamiltonian accessible, i.e., eigenfunctions that do not satisfy the Pauli Exclusion Principle. In addition, Claverie showed that He_2, and most other molecules, had unphysical solutions lying lower in energy than the physical ground state, and that the perturbation expansion must converge to the lowest unphysical state, if it converged at all. There are no unphysical solutions to the Schrödinger equation for H_2^+ and H_2, the customary test molecules of intermolecular perturbation theory. The molecules HeH and He_2 have, respectively, one and two unphysical states lower in energy than their physical ground states. Molecules containing just one atom of atomic number greater than 2, in contrast, have an infinite

J. L. Calais and E. S. Kryachko (eds.), Structure and Dynamics of Atoms and Molecules: Conceptual Trends, 69–95.

number of unphysical states, including a continuum, below the physical ground state energy [2]. In short, for all but the simplest molecules, there is the potential for great difficulties in applying perturbation theory to intermolecular interactions if the existence of unphysical solutions to the Schrödinger equation is ignored [3-5].

In variational theories which divide the electrons of an N-electron system into groups, unphysical states may also pose a problem. Whether or not they do, depends upon the computational methods and approximations used. In the Hartree-Fock approximation it might appear that unphysical states could not possibly play a role because the Hartree-Fock equation is derived by assuming a totally antisymmetric wave function. In fact, unphysical states are avoided because constraints are imposed on the occupied orbitals. The characteristic property of unphysical wave functions is that their antisymmetric projection is zero. A Slater determinant of spin orbitals will vanish if the orbitals are linearly dependent. Requiring the occupied orbitals to be mutually orthogonal, avoids linear dependence [6-10]. Requiring only that the occupied orbitals satisfy different self-consistent field equations, makes it possible for them to be linearly independent, but does not guarantee it [11-13]. In this paper, we use unphysical states to understand how localization may be maintained in N-electron theories and in one-electron theories.

Unphysical states play a major role in this paper. In the next Section we explain what they are and establish their most important properties. They can be understood in terms of orbital configurations in the same way that we understand physical eigenstates. It is noteworthy that in most polyatomic systems there are an infinite number of unphysical states which are lower in energy than the physical ground state. The smallest system for which this is true, is LiH. We begin Sec. 3 with a discussion of the concept of *separable N*-electron wave functions [7,8,10,14-18], i.e., wave functions that are antisymmetrized products of group functions. Wave functions can not be exactly separable due to the electron-electron Coulomb repulsion, but we explain how unphysical states may be combined with physical states to give N-electron wave functions which are approximately separable. One can calculate these almost separable wave functions without first calculating the eigenfunctions of the Schrödinger Hamiltonian, although not with completely conventional electronic structure programs. We present some numerical results for LiH, including a quantitative measure of the separability of its wave function as a function of nuclear distance. We turn in Sec. 4 to a more pragmatic method for dealing with unphysical states, the addition of a level shift operator to the N-electron Hamiltonian. Again we limit our discussion to what can be done without resort to approximation. Some results we have obtained for LiH suggest that level shift operators are worth further investigation. The Hartree-Fock approximation is taken up in Sec. 5. We review and compare the localized orbital theory [11-13] and the Huzinaga theory [10], and selected works based on these theories. It appears to us that the Huzinaga equation is particularly well suited to the calculation of valence shell orbitals; it effectively prevents collapse into the core. The localized orbital theory, on the other hand, seems better suited to studies of intermolecular interactions because it does not require orthogonality between orbitals belonging to different groups. We end the paper by considering what might be gained by combining some of the methods we have reviewed for preventing electronic delocalization and collapse.

It should be clear from the above summary that we are most interested in exact solutions of both the N-electron Schrödinger equation and of the Hartree-Fock equation. We realize that exact solution of either equation for polyatomic systems is not feasible, but we feel strongly that practical approximate theories for interacting electronic groups should be derived from exact theories. An exact theory provides not only a foundation upon which a practical/approximate theory may be built, but also a framework within which the approximate theory may be extended. We elaborate on this point in the concluding Section.

2. Unphysical States

Elaborate calculations are not needed to establish the nature of the unphysical electronic states that concern us, nor is extensive calculation needed to determine the relationship between unphysical state energies and the physical ground state energy. We derive in this section the essential features of the energy spectrum from orbital configurations and variational calculations with simple trial functions. We make no attempt to classify unphysical functions by irreducible representations of the symmetric group because we exploit only their distinguishing property: their antisymmetric projections are identically zero.

We limit the considerations of this paper to binary electronic systems, e.g., pairs of atoms or molecules. Assign electrons 1, 2, ..., N_a to group A; $N_a+1,\ldots,N=N_a+N_b$ to group B. All wave functions include spin and all are required to be antisymmetric under interchanges of pairs of electron coordinates belonging to the same group, but not under interchanges between groups. This requirement reduces the total number of unphysical functions that may play a role in our analysis without limiting its generality. Physical wave functions are totally antisymmetric, unphysical ones vanish when antisymmetrized. We begin by considering those unphysical states of atoms that play a role when one divides the electrons between core and valence shells. Understanding these states helps us to understand the unphysical states of interacting pairs of atoms and molecules.

One can learn nearly everything that one needs to know about the unphysical states of an atom in which the electrons are divided between core and valence shells, by considering the Be atom. Assign electrons 1 and 2 to the core, and 3 and 4 to the valence shell. The physical ground state of Be corresponds to the $1s^2 2s^2$ configuration, a singlet. The lowest energy unphysical singlet state corresponds to the configuration $1s^4$, the next to lowest, to $1s^3 2s$. The energies of these states in Hartrees are given below. For the ground state we quote the exact nonrelativistic energy [19]. The other energies were calculated variationally using the Slater type orbitals of Morse, *et al.* [20]. The $1s^4$ trial function was a product of four 1*s*-orbitals and the Be^+ function, a product of three 1*s*-orbitals. The $1s^3 2s$ function was determined by configuration interaction between the four functions that can be formed from a product of three 1*s*-orbitals with a 2*s*-orbital when all permutations of the coordinates are made. In each case, all 1*s*-orbitals had the same exponent. The lowest energy $1s^3 2s$ function does not interact with the $1s^4$ function due to its exchange symmetry.

Be $1s^4$	Be $1s^3 2s$	Be^+ $1s^3$	Be $1s^2 2s^2$
-18.7578	-17.3006	-17.0859	-14.6674

The Be^+ energy is included to show that the $1s^3 2s$ state ionizes well below the physical ground state energy. The exact Be^+ $1s^3$ energy is, of course, lower than our calculated value. Thus, if one could calculate the energies of unphysical states exactly, one would find an infinite number of discrete, bound states below the exact Be^+ $1s^3$ energy and a continuum of unbound states beginning at that energy.

The relationship between the physical and unphysical energy spectra of other atoms when their electrons have been split between the core and valence shells, is essentially the same as it is for Be: the physical ground state energy is higher than the energies of an infinite number of bound unphysical states and lies in a continuum of energies of unbound unphysical states. For example, one might wish to assign electrons 1 through 10 of Si to the core and the remainder, to the valence shell. The lowest energy unphysical state would correspond to $1s^4 2s^4 2p^6$, the next to lowest to $1s^4 2s^3 2p^7$, and so on. Note that no more than four electrons can occupy the same orbital when we distinguish between two groups of electrons. The energy of the Si^+ $1s^4 2s^3 2p^6$ state must be well below that of the physical ground state. It marks the beginning of the unphysical continuum. Buried in the unphysical continuum will be an infinite number of bound unphysical states, e.g., that corresponding to $1s^4 2s^2 2p^6 3s^2$. In short, the Si spectrum is much more complicated than the Be spectrum, but for both atoms the physical ground state is higher in energy than an infinite number of bound unphysical states and lies in a continuum of unbound, unphysical state energies.

The reasoning of the last two paragraphs is applicable to molecular and crystal wave functions when we divide the electrons between core and valence groups. The potential for error in calculations when such a division is made, is enormous since there is an infinity of unphysical state energies beginning far below the physical ground state energy. The problem of preventing the N-electron wave function from becoming unphysical when we distinguish between core and valence electrons, is exactly the problem of preventing the collapse of the valence shell wave function into the core.

When one would like to determine the interaction energy between two atoms or molecules, it is more attractive to assign specific electrons to each atom or molecule than to split them between core and valence shells. At very large separations, this is certainly a very good first approximation, but the question is, can the approximation be improved while maintaining the division of electrons between the interacting groups. This is, for example, what is attempted in intermolecular perturbation theory. Claverie [1] recognized that the existence of unphysical states is a threat to the validity of the perturbation method in this context. He examined the He-He problem in detail and concluded that the usual perturbation expansion for the physical ground state must converge to the lowest energy unphysical state of He_2 if it converged at all. In fact, our calculations show that below the He_2 ground state there are two unphysical states, corresponding to the $1\sigma_g^4$ and $1\sigma_g^3 1\sigma_u$ configurations. At large separations the existence of these states is numerically unimportant for energy calculations because they become degenerate with the He_2 physical ground state at infinite separation. In this regard, He_2 is exceptional [2].

A more typical molecule is LiH [3]. Consider the case of very large separation between the two atoms. Assign electron 4 to the H-atom, i.e., we are interested in those eigenfunctions of the molecular Hamiltonian that are antisymmetric only in the position-spin coordinates of electrons 1, 2 and 3. At infinite separation, we can solve this problem exactly in terms of the H-atom eigenfunctions, and the lowest energy unphysical and lowest energy physical eigenfunctions of the Li-atom. The lowest energy unphysical state of Li corresponds to the $1s^3$ configuration and is lower in energy than the physical ground state by more than 1.0 Hartree [3]. The lowest energy LiH state is unphysical and corresponds to the configuration Li $1s^3$ H $1s$. It may be interpreted as arising from the exchange of a pair of electrons between the two atoms, the H-electron then falling into the Li-core. Excitation of the Li-electron that was transferred to the H-atom gives additional bound unphysical states below the LiH physical ground state. Since the ionization potential of the H-atom is 0.5 Hartree, a continuum of unbound unphysical states starts below the physical ground state energy of LiH. Thus, the relationship between the unphysical states and the physical ground state energies in LiH is quite similar to what we found above when we divided the electrons of a system between core and valence levels.

Most molecular systems are no less complicated than LiH. Consider He-Be. Its lowest energy unphysical state at large separations corresponds to He $1s^2$ Be $1s^4$. It lies more than 4.0 Hartrees below the physical ground state according to the numbers given above for Be. The ionization potential of He is less than one Hartree. Thus, we expect to find an infinite number of bound unphysical states and a continuum of unbound unphysical states below the physical ground state energy of He-Be. We expect the same relationship between the unphysical states and the physical ground state for all molecules containing one or more atoms of atomic number greater than two. In each case, unphysical states can be interpreted as arising from the delocalization of the electrons of each group onto the neighboring groups, and the collapse of the transferred electrons into the neighbor's cores.

If one wants to base a computational method on the physical insight that many electron systems consist of weakly interacting groups of electrons, one must understand how the method distinguishes between the electrons of the groups. If it does so and, at the same time, insures that the Pauli Exclusion Principle is satisfied, unphysical states will not present a problem. If, however, the method yields unphysical solutions in addition to the physical ones, it is essential that one know how the unphysical energies are related to the physical ground state energy. If the method gives an infinite number of unphysical states below the lowest physical state, one needs a practical method for identifying the desired physical state. It is not practical to use the criterion that the physical function has a non-zero antisymmetric projection if there are an unlimited number of solutions to test. The usual Rayleigh-Schrödinger intermolecular perturbation theory fails in this regard.

3. Separability and Localizability of Electronic Wave Functions

The idea that N-electron wave functions should be separable in favorable cases is more easily stated than realized. Let Φ be a physical eigenfunction of the nonrelativistic, N-electron Hamiltonian $\hat{H}$ in the Born-Oppenheimer approximation. Imagine that the system consists of two groups of electrons, e.g., two atoms at a large nuclear separation R. Let ϕ_a be a function of the group A electron coordinates, 1 through N_a, and ϕ_b, a function of the group B coordinates N_a+1 through $N=N_a+N_b$. Let $\hat{\mathscr{A}}$ be the antisymmetric projection operator for the N electrons and let $\hat{\mathscr{O}}$ be a projector for spin and rotational symmetry. We call Φ separable if $\Phi \propto \hat{\mathscr{A}}\hat{\mathscr{O}}\phi_a\phi_b$. If Φ were separable, we could hope to calculate it more easily by calculating ϕ_a and ϕ_b individually, then forming Φ from them. Recognizing that a given Φ is separable, however, is not straightforward because $\hat{\mathscr{A}}$ effects an exchange coupling of ϕ_a and ϕ_b that makes it impossible to assign specific electrons to either group. There is no obvious way to recognize that a given Φ is separable, exactly or approximately. In fact, approximate separability is the best that one can hope to find because electrons repel each other. This Section is devoted to understanding how approximately separable, N-electron wave functions can be defined and calculated without resort to approximations.

If there were no electron-electron interactions, wave functions would be separable, of course, into products of spin orbitals, which in turn could be split into core and valence groups by their energies. To split Φ exactly between interacting atoms or molecules, even when there are no electron-electron interactions, requires either definition of what one means by the orbitals of an atom or molecule in a larger system., or consideration of the limit of infinite group separation. It should be no surprise that similar measures must be taken when looking for approximately separable, physical eigenfunctions of $\hat{H}$ when the electron-electron Coulomb repulsion is included.

To distinguish between electronic groups in a system without first introducing approximations, we introduce a function F that approximates the form $\hat{\mathscr{O}}\phi_a\phi_b$ rather than $\hat{\mathscr{A}}\hat{\mathscr{O}}\phi_a\phi_b$. In F electrons belong to specific electronic groups, they are distinguishable. We require that F be antisymmetric in the coordinates of A and, separately, antisymmetric in those of B. Most importantly, we require that $\Phi \propto \hat{\mathscr{A}}F$. These requirements do not define F, but they do limit the possibilities. Because F should not be antisymmetric under interchanges of pairs of electron coordinates belonging to different groups, it can not be a linear combination of physical state functions alone. The unphysical functions described in Sec. 2 must be included. This means that a doable computational method must include a means of distinguishing F from purely unphysical functions.

Functions which satisfy the constraint $\Phi \propto \hat{\mathscr{A}}F$ are called primitive functions in intermolecular perturbation theory. One can understand the implications of this constraint by expanding F in terms of the physical eigenfunctions Φ_k and unphysical eigenfunctions u_k of $\hat{H}$. Let E_k and E_k^u be, respectively, the physical and unphysical eigenvalues. Order the functions so that the lowest eigenvalues are E_1 and E_1^u, and that the $k+$ th is greater than or equal to the kth. Since F is not totally antisymmetric, it can not be represented by an expansion in the Φ_k alone. To satisfy $\Phi_1 \propto \hat{\mathscr{A}}F$, the most general form that F can have is

$$F = C_1\Phi_1 + \sum_k C_k^u u_k, \tag{3.1}$$

where the sum over the unphysical states includes integration over the unphysical continuum. We can specify the C_k^u in any way we please since $\hat{\mathscr{A}}u_k = 0$ for all k. The constraint $\Phi_1 \propto \hat{\mathscr{A}}F$ will be satisfied so long as $C_1 \neq 0$.

3.1 MINIMALLY DISTORTED LOCALIZED WAVE FUNCTIONS

Consider the question: if one actually knew Φ_1 and the u_k, how would one define F? If one is interested in the interaction between a pair of atoms or molecules, A and B, it would be advantageous if F were approximately equal to $\hat{\mathscr{O}}\phi_a\phi_b$, where ϕ_a° is the A ground state wave function in the absence of B, and ϕ_b°, the B ground state function in the absence of A. This function is customarily chosen as the unperturbed function in intermolecular perturbation theories. Note that ϕ_a° depends on electron coordinates 1 through N_a, and ϕ_b°, on N_a+1 through N. Let $\hat{h}_a$ and $\hat{h}_b$ be the Hamiltonians, respectively, of A in the absence of B, and B in the absence of A. Then $\hat{h}_a\phi_a^\circ = E_a^\circ\phi_a^\circ$ and $\hat{h}_b\phi_b^\circ = E_b^\circ\phi_b^\circ$, where E_a° and E_b° are the respective ground state energies. Since the lowest energy eigenfunction of $\hat{H}^\circ = \hat{h}_a + \hat{h}_b$ is $\phi_a^\circ\phi_b^\circ$, one would find $F = \hat{\mathscr{O}}\phi_a^\circ\phi_b^\circ$ by requiring that F minimize

$$\mathscr{E} = \langle F|\hat{H}^\circ|F\rangle / \langle F|F\rangle \tag{3.2}$$

under the lone constraint that F be square integrable. We proposed that one choose F to minimize with (3.1) as an additional constraint [21]. The resultant F is not equal to $\hat{\mathscr{O}}\phi_a\phi_b$, but it is minimally different from $\hat{\mathscr{O}}\phi_a^\circ\phi_b^\circ$ in the sense that it minimizes under the constraint (3.1).

The above definition of F would be of little value if one could calculate F only by first determining Φ_1 and all of the u_k. What is significant about the definition is that this F satisfies an eigenvalue equation which can be solved without prior knowledge of the eigenfunctions of $\hat{H}$. The equation is

$$\left(\hat{H}^\circ + \hat{V} - \hat{Q}\hat{V}\hat{Q}\right)F = \mathscr{E}F \quad \text{where } \hat{Q} = 1 - \hat{\mathscr{A}} + \hat{\mathscr{A}}|F\rangle\frac{1}{\langle F|\hat{\mathscr{A}}|F\rangle}\langle F|\hat{\mathscr{A}} \tag{3.3}$$

and $\hat{V} = \hat{H} - \hat{H}^\circ$. We proposed solving (3.3) iteratively to self-consistency for the lowest $\mathscr{E}$ [21]. This can be done in the configuration interaction approximation, but not with conventional programs using antisymmetric configuration functions. We solved (3.3) to high accuracy for H_2 [22], but before we understood how exceptional H_2 is. It does not have an infinite number of unphysical states below the physical ground state energy as do most molecules, it has none.

To appreciate the problems that one encounters in trying to solve (3.3) for a typical system, consider what would happen if we were to try to calculate the ground state function and energy by a full configuration interaction (FCI) method that gave unphysical as well as physical eigenfunctions of $\hat{H}^\circ + \hat{V}$. Choosing the eigenfunction of $\hat{H}^\circ + \hat{V}$ with the lowest eigenvalue would be inappropriate since there would be a large number of unphysical solutions lower in energy than the physical ground state. If the calculation were done exactly, the number would be an infinite. One might look instead for the lowest energy solution F that satisfies $\hat{\mathscr{A}}F \neq 0$, but that is not practical. In an exact calculation, the probability that any chosen F is antisymmetrizable, is zero; in an FCI calculation, the more configurations used, the smaller the probability that any solution picked satisfies $\hat{\mathscr{A}}F \neq 0$. One might hope to avoid this difficulty by a method that transforms the function $\hat{\mathscr{O}}\phi_a^\circ\phi_b^\circ$ into Φ_1, e.g., as the usual Rayleigh-Schrödinger perturbation theory for intermolecular interactions is supposed to do. Claverie [1] showed that if this method converged, it transformed $\hat{\mathscr{O}}\phi_a^\circ\phi_b^\circ$ into u_1, the lowest energy unphysical function, whenever $E_1^u < E_1$. Nonperturbative calculations on LiH support his analysis [4]. Finally, infinite order summation of the divergent perturbation expansion affords no solution; it must give u_1 if it does what it is supposed to do.

A clear understanding of the nature of the eigenvalue spectrum of (3.3) is essential for prescribing how to identify the desired solution of (3.3). Suppose that (3.3) has been solved. Since $\hat{\mathscr{A}}$ commutes with $\hat{H}^\circ + \hat{V}$, it follows from (3.3) and the definition of $\hat{Q}$, that

$$\left(\hat{H}^\circ + \hat{V}\right)\hat{\mathscr{A}}F = (\mathscr{E} + D)\hat{\mathscr{A}}F \tag{3.4}$$

where $D = \langle F|\hat{\mathscr{A}}\hat{V}|F\rangle / \langle F|\hat{\mathscr{A}}|F\rangle$ [21]. Thus, $\hat{\mathscr{A}}F$ is an eigenfunction of $\hat{H}^\circ + \hat{V}$ if $\hat{\mathscr{A}}F \neq 0$. To see that a solution of (3.3) can be selected such that $\hat{\mathscr{A}}F \neq 0$, first partition the effective Hamiltonian of (3.3) with the projection operators $\hat{Q}$ and $1 - \hat{Q}$.

$$\hat{Q}\left(\hat{H}^\circ + \hat{V} - \hat{Q}\hat{V}\hat{Q}\right)\hat{Q} = \hat{Q}\hat{H}^\circ\hat{Q} \tag{3.5a}$$

$$(1 - \hat{Q})\left(\hat{H}^\circ + \hat{V} - \hat{Q}\hat{V}\hat{Q}\right)\hat{Q} = (1 - \hat{Q})(\hat{H}^\circ + \hat{V})\hat{Q} \tag{3.5b}$$

$$(1 - \hat{Q})\left(\hat{H}^\circ + \hat{V} - \hat{Q}\hat{V}\hat{Q}\right)(1 - \hat{Q}) = (1 - \hat{Q})(\hat{H}^\circ + \hat{V})(1 - \hat{Q}) \tag{3.5c}$$

As a working hypothesis, assume that $\hat{\mathscr{A}}F \propto \Phi_1$. This implies that

$$\hat{Q} = |\Phi_1\rangle\langle\Phi_1| + \sum_k |u_k\rangle\langle u_k| \quad \text{and} \quad 1 - \hat{Q} = \sum_{k\neq 1} |\Phi_k\rangle\langle\Phi_k|.$$

It follows that the Φ_k for $k \neq 1$ diagonalize the operator on the right in (3.5c) and that its eigenvalues are the E_k, the physical state energies, **excluding** E_1. Furthermore, (3.5b) vanishes because $\hat{Q}$ commutes with $\hat{H}^\circ + \hat{V}$. Finally, the eigenvalues of (3.5a) are upper bounds to the corresponding eigenvalues of $\hat{H}^\circ$ by the separation theorem [23]. Thus, if $\mathscr{E}_1$ is the lowest eigenvalue of (3.5), $\mathscr{E}_1 \geq \mathscr{E}^\circ = E_a^\circ + E_b^\circ$, the lowest eigenvalue of $\hat{H}^\circ$. At

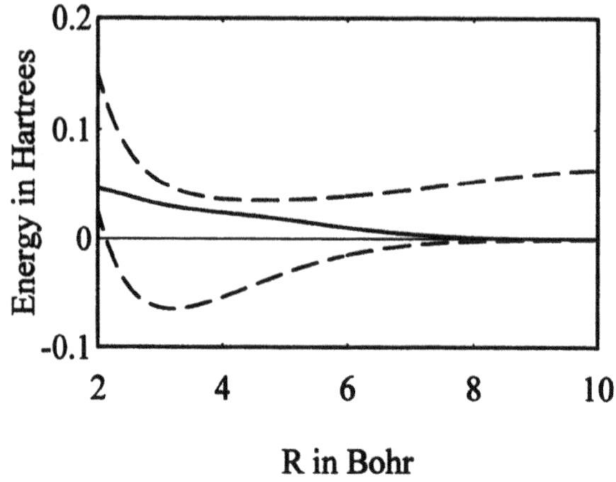

Fig. 3.1. The ground and first excited state energies, E_1 and E_2 (dashed lines), and the lowest eigenvalue $\mathcal{E}_1$ (solid line) of Eq. (3.3) as functions of the nuclear separation R for $^1\Sigma$ LiH. The energies were calculated by FCI using 1080 bideterminantal configurations. All energies are relative to $\mathcal{E}^\circ$, i.e, they are corrected for the basis set superpostion error.

infinite R, we have $\Phi_1 \propto \mathcal{A}\hat{\mathcal{O}}\phi_a^\circ\phi_b^\circ$ and $E_1 = \mathcal{E}^\circ = \mathcal{E}_1$. If the interaction between A and B is attractive, as R decreases from infinity, one must have $E_1(R) < \mathcal{E}^\circ < \mathcal{E}_1(R)$ until the minimum in $E_1(R)$ is passed. At infinite R, assuming a nondegenerate ground state, $E_2 > \mathcal{E}^\circ = \mathcal{E}_1$. As R decreases from infinity, the F that has the desired property $\Phi_1 \propto \hat{\mathcal{A}}F$ will be the one belonging to the lowest eigenvalue of (3.3) so long as $E_2(R) > \mathcal{E}_1(R)$. At sufficiently large R, the above working hypothesis will be true. What "sufficiently large" is, may be learned from experience.

In Fig. 3.1 we present some numerical results for LiH that suggest that the lowest energy solution of Eq. (3.3), is the desired solution for nuclear separations as small as the equilibrium bond length. We have written "suggest" because our calculations have used a relatively small orbital basis set and because LiH is the only molecule for which we have results at present. The calculations were done with an FCI program that is still under development. It uses unconventional configuration functions, *bideterminants*. Each bideterminant is the product of two Slater determinants, one for A group electrons and one for the B group. A bideterminantal basis can represent unphysical as well as well physical wave functions, a necessary condition for solving Eq. (3.3). More details on the calculations for Fig. 3.1 are in Appendix A.

To produce Fig. 3.1, we calculated F as well as $\mathcal{E}_1(R)$. This means that we can compare F to $\hat{\mathcal{O}}\phi_a^\circ\phi_b^\circ$ and to Φ_1. The integral $\langle\Phi_1|F\rangle$ measures the contribution of the physical ground state function to F. Its magnitude can equal a maximum value of one only if the two functions are identical. The integral is plotted against R in Fig. 3.2. It shows that the unphysical contribution to F is important.

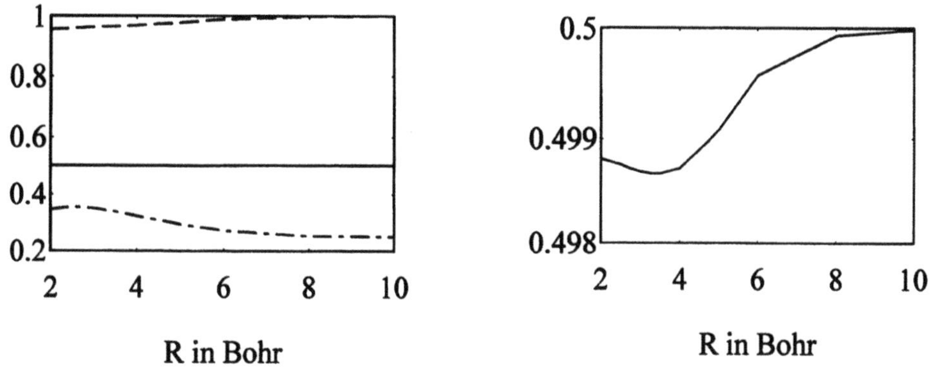

Fig. 3.2 The integral $\langle \Phi_1 | F \rangle$ is represented by the dot-dash curve in the left-hand graph. The dash curve represents the fourth root of $\langle \hat{\mathcal{O}} \phi_a^\circ \phi_b^\circ | F \rangle$, effectively a geometric average measure of orbital overlaps between the unperturbed LiH wave function and its localized wave function. The solid line in both graphs represents the occupation number of the first natural group functions.

The integral $\langle \hat{\mathcal{O}} \phi_a^\circ \phi_b^\circ | F \rangle$ measures the extent to which the uperturbed function has been changed by the interaction. It can not be greater than one in magnitude, the value it would have if F were identical to the unperturbed function. We have plotted in Fig. 3.2 the fourth root of this integral. We plotted the fourth root because if both functions could be represented by a product of orbitals, $\langle \hat{\mathcal{O}} \phi_a^\circ \phi_b^\circ | F \rangle$ would be a product of four orbital overlap integrals. Thus, the fourth root of $\langle \hat{\mathcal{O}} \phi_a^\circ \phi_b^\circ | F \rangle$ can be regarded as a geometric mean orbital overlap integral. It shows that F is not grossly different from the unperturbed function even in the neighborhood of the potential minimum.

We have also plotted in Fig. 3.2 a measure of the separability of F. We proposed some time ago to measure separability by analyzing F in terms of *natural* N_a- and N_b-electron functions [17]. These functions are analogs of Löwdin's natural orbitals [27]. One can readily calculate their occupation numbers when one calculates F by expansion in bideterminantal configuration functions. Each bideterminant is the product of two Slater determinants, one depending on electron coordinates 1 through N_a, the other on N_a+1 through N. The expansion coefficients are $C_{K,L}$, the subscripts referring respectively to the A and B group determinants. One can determines the A-group natural function occupation numbers by calculating the eigenvalues γ_K of the matrix of elements $\Gamma^A_{K,K'} = \sum_L C_{K,L} C^*_{K',L}$. The nonzero B-group occupation numbers, the eigenvalues of $\Gamma^B_{L,L'} = \sum_K C_{K,L} C^*_{K,L'}$, are identical to the nonzero γ_K [28]. The sum of each group's occupation numbers equals one when F is normalized to one. The significance of the γ_K for separability is found in the expansion of F in terms of the A- and B-group natural functions, A_K and B_K. Carlson and Keller [28] showed that $F = \sum_K c_K A_K B_K$ and that

$\gamma_K = |c_K|^2$. Thus, if one γ_K is equal to one, all of the others vanish and F is exactly separable. For $^1\Sigma$ LiH, the situation is more complicated because the natural functions must appear in pairs, e.g., for the H-group there must be up- and down-spin natural function pairs, identical except for their spin functions. Similarly, the occupation numbers will occur in identical pairs. Hence, in LiH an exactly separable F will have a pair of occupation numbers equal to 0.5. Figure (3.2) shows that the occupation number of the first natural up-spin function of the LiH function F differs little from 0.5 over a wide range of R-values. These first results suggest that F is highly separable for LiH even in the neighborhood of the potential minimum. We plan to carry out more extensive and accurate studies of the separability of localized wave functions for LiH and other small molecules.

We can not leave the subject of wave function separability without emphasizing that the measure of separability used here, has no intrinsic dependence on approximations. A feature common to many analyses of separability is the assumption that the A and B-group functions are strongly orthogonal [7,10,14, 18]. By strongly orthogonal, one means that

$$\int \phi_a^*(\ldots,i,\ldots)\phi_b(\ldots,i,\ldots)dV_i = 0$$

where the integral is over a single electronic coordinate i. This is a significant restriction on both group functions in that the one-electron basis used in the expansion of ϕ_a can not appear in the expansion of ϕ_b [33,34]. In short, by imposing the strong orthogonality condition, one introduces an approximation that already assumes a kind of separability of the group functions. In contrast, by calculating natural occupation numbers for localized wave functions, one can study the separability of wave functions without assuming separabilty in any form. The calculations, however, can not be done with completely conventional electronic structure programs.

3.2 GENUINE PRIMITIVE FUNCTIONS

From the standpoint of developing an intermolecular perturbation theory, Eq. (3.3) has a major disadvantage: it can not give in low order the exact coefficients of the leading terms in the asymptotic $1/R$ expansion of the energy. The perturbation theory is identical through second order in the energy to the theory of van der Avoird and Hirschfelder [24-26], which is known to have this defect. Kutzelnigg has explained what is required in order to have a perturbation theory which gives exactly the $1/R$-coefficients [29]. He recognized that there are unphysical solutions of the Schrödinger equation which have energy equal to $\mathscr{E}^\circ$ in the limit $R \to \infty$ and which have $1/R$-coefficients identical to those of Φ_1. He refers to such functions as *coasymptotic* with Φ_1. Let $u_{k,c1}$ be a member of the finite set of functions coasymptotic with Φ_1. Kutzelnigg classifies as *genuine* primitive functions, functions of the form

$$F = C_1\Phi_1 + \sum_k^{c1} C_{k,c1}^u u_{k,c1}, \tag{3.6}$$

where the sum is limited to those unphysical functions coasymptotic with Φ_1. The coefficients in (3.6) can be chosen any way one wishes so long as $C_1 \neq 0$. In view of the development based on (3.1) above, if one wants F to be related to $\hat{\mathscr{O}}\phi_a^\circ\phi_b^\circ$, one can require that F minimize $\mathscr{E}$ in (3.2) under the constraint (3.6). In fact, we proposed essentially this [30] before we proposed the theory outlined in Eqs. (3.1)-(3.5), which was before we began to understand the problem posed by unphysical functions. We showed that if F satisfied the constraint (3.6) and minimized $\mathscr{E}$ as defined by (3.2), then it would be a solution of

$$\left(\hat{H}^{\circ}+\hat{V}-\hat{Q}\hat{V}\hat{Q}\right)F=\mathscr{E}F, \tag{3.7}$$

where $\hat{Q}$ is the projector onto all of the linearly independent functions that can be constructed from F by interchanges of A with B coordinates [5]. Unfortunately, $\hat{Q}u_k=0$ for all u_k not in the coasymptotic set. It follows that all of these u_k are eigenfunctions of $\hat{H}^{\circ}+\hat{V}-\hat{Q}\hat{V}\hat{Q}$. Thus, the desired solution's eigenvalue, which is greater than $\mathscr{E}^{\circ}$, will be higher in energy than those of an infinite number of unphysical solutions and will lie in an unphysical continuum. Unmodified, the theory is unusable if (3.7) is to be solved directly. The perturbation theory derived from (3.7) is equivalent to the Hirschfelder-Silbey theory through second order in the energy [25,31,32]. We believe that our version and the original Hirschfelder-Silbey perturbation theory can not converge for most systems and that infinite order summation of the energy expansion must give the lowest unphysical energy.

In most of this Section we have focused on the division of an N-electron system into a pair of interacting atoms or molecules. In some applications a more useful division is between core and valence levels. While the insight expressed by Eq. (3.1) is applicable, the definition of F by minimizing (3.2) under the constraint (3.1) or (3.6), is inapplicable without a definition of $\hat{H}^{\circ}$ which distinguishes core from valence electrons. The essential feature of the $\hat{H}^{\circ}$ we have used for interacting atoms/molecules, is the restriction that the electrons of each group feel only the Coulomb potential due to the nuclei in their group. Valence and core electrons can not be distinguished this way; they share all of the nuclei and are not localized in distinct regions of position space. Core and valence electrons are distinguished by their ionization potentials, but it is not obvious how one can use this to define an $\hat{H}^{\circ}$ which gives wave functions that are manifestly separable. In this regard, it appears necessary to borrow from the one-electron theories which are surveyed in Sec. 5.

4. Level Shift Operators

If we were not interested in separating the wave functions of complex systems into more tractable pieces, we could exclude unphysical states from our calculations either by the conventional method, expanding N-electron wave functions in totally antisymmetric basis sets, or by adding to the Hamiltonian a level shift operator $\hat{\mathscr{L}}$ that increases the energies of the unphysical states. The latter device is useful, for example, when one expands wave functions in the bideterminantal basis functions defined in Sec. 3. The bideterminantal basis allows one to identify specific sets of coordinates with specific groups in a system, as we have done in studying intermolecular perturbation theory, but it does not make physical eigenfunctions manifestly separable because even in the bideterminantal basis, they are totally antisymmetric in the electronic coordinates. The physical eigenfunctions must still be totally antisymmetric. Addition of an $\hat{\mathscr{L}}$ operator to a Hamiltonian can have no effect on separability if it commutes with $\hat{\mathscr{A}}$, which is the case for all of the $\hat{\mathscr{L}}$s defined below. It is possible, however, to combine a level shift operator with other operators in theories which enhance separability, but have a limited range of applicability. We have discussed level shift operators in some detail elsewhere [4-5]; here we summarize and update that discussion.

For our purposes, the essential property of a level shift operator $\hat{\mathscr{L}}$ is that it makes the lowest unphysical state higher in energy than the physical ground state. If this were the only requirement, then we would recommend the simplest choice, $\hat{\mathscr{L}} = B(1-\hat{\mathscr{A}})$, where B is a positive constant. As long as B is large enough, it has the desired effect. An alternative choice, one not depending on an arbitrary constant, is $\hat{\mathscr{L}} = -(1-\hat{\mathscr{A}})\hat{\mathrm{v}}$, where $\hat{\mathrm{v}}$ is the total potential energy operator for the Coulombic interaction of nuclei with electrons. Since $\hat{\mathrm{v}}$ and $1-\hat{\mathscr{A}}$ commute, this $\hat{\mathscr{L}}$ is Hermitian. It makes all unphysical state energies positive by transforming them into unphysical states of an interacting, free-electron gas [5]. This suggests that it strongly perturbs all of the original unphysical states. A weaker $\hat{\mathscr{L}}$ might be better suited to numerical use, one that acts most strongly on those unphysical states that cause problems. Those are the states which lie below the physical ground state energy. We explained in Sec. 2 that they arise from the exchange of valence shell electrons between interacting atoms/molecules, followed by collapse of the transferred electrons into the cores. Let $\hat{\mathrm{v}}_{Ab}$ be the potential energy operator for the B group electrons in the Coulomb field of the A nuclei, and $\hat{\mathrm{v}}_{Ba}$, the operator for the A group electrons in the field of the B nuclei. We conjecture that $\hat{\mathscr{L}} = -(1-\hat{\mathscr{A}})(\hat{\mathrm{v}}_{Ab}+\hat{\mathrm{v}}_{Ba})(1-\hat{\mathscr{A}})$ will primarily raise the energies of those unphysical states which are a problem by killing for all unphysical states the potential energy operators responsible for electron transfer and collapse. We doubt that the three level shift operators defined above exhaust the possibilities. All three commute with $\hat{\mathscr{A}}$.

The above $\hat{\mathscr{L}}$ have a further shortcoming when viewed from the standpoint of intermolecular perturbation theory: $\hat{\mathscr{O}}\phi_a^\circ\phi_b^\circ$ is not an eigenfunction of $\hat{H}+\hat{\mathscr{L}}$ in the limit of infinite separation. This defect can be corrected with the aid of an operator $\hat{Q}^\circ$, which we define as the projector onto the space spanned by all of the linearly independent functions derivable from $\hat{\mathscr{O}}\phi_a^\circ\phi_b^\circ$ by interchanges of the A and B electron coordinates. We have shown that $\hat{\mathscr{O}}\phi_a^\circ\phi_b^\circ$ is an eigenfunction of $\hat{H}+(1-\hat{Q}^\circ)\hat{\mathscr{L}}(1-\hat{Q}^\circ)$ in the limit of infinite separation for any $\hat{\mathscr{L}}$ [5]. The effect of this modified level shift operator has been studied in a set of bideterminantal FCI calculations on $^1\Sigma$ LiH at R = 5 Bohr. We calculated the eigenvalues of the Hamiltonians (I) $\hat{H}^\circ+\lambda\hat{V}$, (II) $\hat{H}^\circ+\lambda\left[\hat{V}-B(1-\hat{\mathscr{A}})\right]$ and (III) $\hat{H}^\circ+\lambda\left[\hat{V}-B(1-\hat{Q}^\circ)(1-\hat{\mathscr{A}})(1-\hat{Q}^\circ)\right]$ by matrix diagonalization at a series of λ-values [4]. Perturbation theory would approximate the eigenvalues of these Hamiltonians by power series in λ. Our results for the lowest eigenvalues are presented in Fig. 4.1. The lowest eigenvalue of (I) behaves just as one would expect from Claverie's analysis. At $\lambda = 0$, it is equal to $\mathscr{E}^\circ$. As λ increases, the eigenvalue slowly decreases until $\lambda \approx 0.76$, where it curves abruptly downwards, becoming equal to the lowest unphysical energy E_1^u at $\lambda = 1$. The lowest eigenvalues of the other two operators, however, link $\mathscr{E}^\circ$ to the physical ground state energy E_1 as λ increases from zero to one. This is what perturbation methods are normally designed to do. The λ-dependence of the eigenvalue of Hamiltonian (II) is strong, that of (III), very weak until λ is almost equal to one, where it hooks abruptly downward. The differences between the two curves is due to the operator $1-\hat{Q}^\circ$. The lowest energy eigenfunction of Hamiltonian (III) is totally antisymmetric because $\hat{Q}^\circ$ commutes with $\hat{\mathscr{A}}$, and, consequently, (III) commutes with $\hat{\mathscr{A}}$. We interpret the abrupt

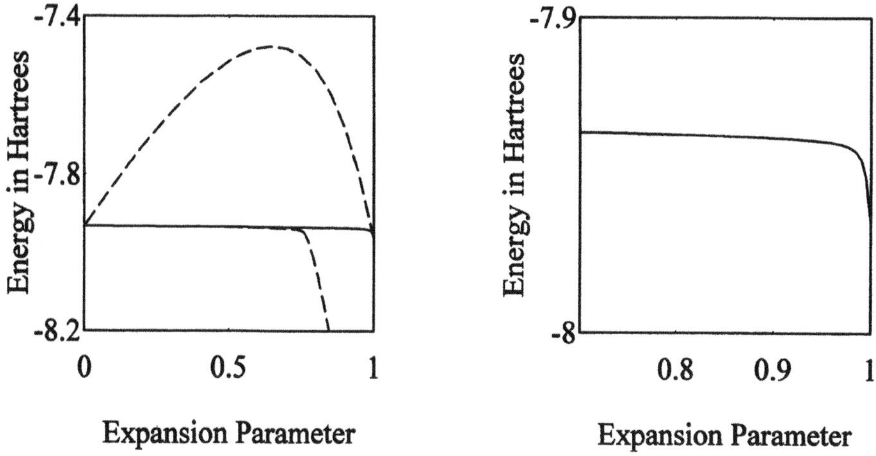

Fig. 4.1 On the left, the lowest eigenvalues of the Hamiltonians labeled I, II and III in the text are plotted as functions of the expansion parameter λ for $^1\Sigma$ LiH at R = 5.0 Bohr. The eigenvalue of Hamiltonian I is represented by the lower dashed line, that of II by the upper dashed line, and that of III by the solid line. The physical ground state energy of LiH is the point at $\lambda = 1$ where the II and the III curves intersect. On the right, a higher resolution plot of the eigenvalue of III shows how it hooks a downwards as $\lambda \to 1$.

downward bend of the eigenvalue curve for (III) in the vicinity of $\lambda = 1$ as an exchange effect.

We have argued that level shift operators can be used to extend the validity of the Hirschfelder-Silbey (HS) theory [5]. As explained in Sec. 3, the HS theory has the same defect that Claverie found in the usual intermolecular perturbation theory: if a system contains an atom with atomic number greater than two, there are an infinite number of unphysical eigenfunctions of the HS Hamiltonian of lower energy than the desired solution. Including an $\hat{\mathscr{L}}$ in the HS Hamiltonian makes the energy expansion more complicated, but it does so only beyond second order [5]. Numerical tests are planned in the near future.

Another reason for including level shift operators in this paper is that they can prevent the collapse of a valence shell into the atomic/molecular core. Let $\hat{v}_v$ be the Coulomb potential energy operator for the valence electrons in the field of the nuclei. If one adds $\hat{\mathscr{L}} = -(1-\hat{\mathscr{A}})\hat{v}_v(1-\hat{\mathscr{A}})$ to the Hamiltonian of the system, one kills the operation of $\hat{v}_v$ on all unphysical state functions and raises their energies. We have not yet carried out a bideterminantal study of this shifted Hamiltonian, but we have estimated its effect by calculating $\langle u_k | \hat{H} + \hat{\mathscr{L}} | u_k \rangle$ for Be with the same orbital approximations to u_1 and u_2 that we used in Sec. 2 to bound unphysical state energies. We have reoptimized the variational parameters of the orbitals for both matrix elements and found that they are nearly 12 Hartrees greater than the ground state energy. The implication is that one can carry out variational calculations on a system, with this $\hat{\mathscr{L}}$ added to the Hamiltonian, without requiring the basis to be totally antisymmetric, without imposing strong orthogonality between valence and core group functions, and without fear that the solution obtained will correspond to an unphysical state in which valence electrons have collapsed into the core. Note, however, that the exact solutions that interest us, will still be antisymmetric. The shift operator by itself again does not separate physical state wave functions into distinct group functions.

The level shift operators defined above exploit the symmetry of unphysical wave functions to shift the energies of unphysical states above those of the lowest physical states. We know that this is enough to prevent wave functions from collapsing as we calculate them, because we understand that thye collapse to unphysical states. Other kinds of level shift operators have been introduced which can prevent collapse. Huzinaga has discussed energy (level) shift operators defined in terms core orbitals [18], an idea used earlier by Kleiner and McWeeny [34]. Huzinaga founded his analysis on the strong orthogonality constraint, which, as we observed at the end of Sec. 3.1, imposes a kind of separability on the wave function. There are no constraints, no approximations, in the level shift operators defined above, but they do not make electrons distinguishable because they commute with $\hat{\mathscr{A}}$. We believe that they can be useful when combined with operators which do distinguish between electron groups, e.g., as in reference [5].

5. Hartree-Fock Approximation

In the preceding Sections we focused on exact solution of the nonrelativistic Schrödinger equation in terms of N-electron functions which are more nearly separable and/or localizable than are the eigenfunctions of the Hamiltonian. In this Section we consider the more practical question of how separability and localization can be realized in the Hartree-Fock (HF) approximation. We review, however, only theories in which non-canonical orbitals are used to solve the HF equations. The important studies based on the concept of equivalent orbitals [36] and the Ruedenberg-Edmiston energy minimum criterion, for example, are not covered [37]. Our objective is to understand how separability and localization are realized in the HF theory and to compare it to the N-electron theory.

The simplest form of HF theory approximates the total energy of an N-electron system by the expectation value of the Hamiltonian evaluated with a Slater determinant of spin-orbitals. Whether the spin-orbitals are assumed to be orthonormal, or just linearly independent, the resultant energy E is a functional only of $\hat{\rho}$, the Fock-Dirac density matrix, i.e., the projector onto the space spanned by the occupied spin-orbitals. That E depends only on $\hat{\rho}$ tells us that although it is sufficient for the occupied orbitals to be linearly independent, we must orthonormalize them in some way, before, or in, calculating E. If we do not, we are not doing a HF calculation, but introducing a *new* approximation that can be justified only by the results obtained. We emphasize this point because in calculating orbitals for parts of an electronic system, it is not always desirable, or possible, to maintain orthogonality between the orbitals of different groups. How one goes about orthonormalizing orbitals in the HF approximation is well understood [38-40].

The HF theory requires that $E[\hat{\rho}]$ be a minimum under the constraint that $\hat{\rho}$ is a projection operator, i.e., that $\hat{\rho}^2 = \hat{\rho}$, the trace of $\hat{\rho}$ equals N, and $\hat{\rho} = \hat{\rho}^\dagger$. We symbolize the Fock operator by $\hat{F}$. For our purposes, one need know only that $\hat{F}$ is Hermitian and a functional of $\hat{\rho}$, i.e., $\hat{F} = \hat{F}[\hat{\rho}]$. It has been shown that

$$(1-\hat{\rho})\hat{F}\hat{\rho}=0 \tag{5.1}$$

is the necessary condition for $E[\hat{\rho}]$ to be a minimum [41-43]. It follows from (5.1) that the commutator $[\hat{F},\hat{\rho}]=0$. Thus, the occupied, orthonormal, canonical HF orbitals ψ_k, the eigenfunctions of $\hat{F}$, diagonalize $\hat{\rho}$. To satisfy the constraints on $\hat{\rho}$, we must have

$$\hat{\rho}=\sum_{k=1}^{N}|\psi_k\rangle\langle\psi_k|. \tag{5.2}$$

The conventional procedure is to solve

$$\hat{F}\psi_k=\varepsilon_k\psi_k \tag{5.3}$$

iteratively until self-consistency is reached in the orbitals and the energies ε_k are fixed. A nontrivial aspect of the procedure is that one must select the occupied orbitals from all solutions of (5.3) at each step of the iterative procedure. While filling up the orbitals of lowest energy commonly makes E a minimum, it does not always work [43,44]. The problem of choosing the occupied orbitals to minimize E may be more complicated when one solves instead of (5.3), equations for the orbitals of distinct groups. Whatever those equations may be, their solution must assure that (5.1) is satisfied in order to be equivalent to solving the HF equations.

There are many variants of the HF theory. If we assume that $\hat{\rho}$ can be split into pure up-spin and down-spin components, then the description above encompasses both the restricted closed-shell HF theory and the unrestricted theory. In order to characterize the restricted open-shell HF theory we would have to introduce both an open and a closed-shell $\hat{\rho}$. They would be projectors onto the mutually orthogonal open and closed-shell, occupied orbital spaces. One commonly calculates the occupied orbitals of the restricted open-shell HF theory by solving for the eigenfunctions of both an open-shell and a closed-shell Fock operator [45], but they may be determined as the eigenfunctions of a single Fock operator [46]. The latter option may be advantageous when one wants to solve for noncanonical orbitals. We limit the present discussion to the restricted closed shell and the unrestricted HF approximations.

There are two principal approaches to introducing localized or group functions into the HF theory without introducing additional approximations. The first that we review [11-13] is probably best suited to calculating the properties of N-electron systems as collections of interacting atoms or molecules. The second, the Huzinaga approach [10], is well suited for implementing the division of systems into core and valence shells. The common assumption of both theories is that the occupied orbitals of the groups are linear combinations of the occupied canonical HF orbitals. The total number of occupied group orbitals is set equal to the total number of occupied canonical orbitals. The goal of both approaches is to determine group orbitals which are related to the occupied canonical orbitals by a nonsingular linear transformation. In this sense, both theories are equivalent to the HF theory.

5.1 LOCALIZED ORBITAL EQUATIONS

The questions that must be addressed in deriving equations within the HF approximation for atoms and molecules that are part of more complex systems, are essentially the same ones that were addressed in Sec. 3. (I) What do we mean by an atom or a molecule in a system made up of many atoms or molecules? (II) What keeps the electrons of one group from delocalizing to a neighboring group and falling into its core. There is nothing in the Fock operator to prevent delocalization. Only inner shell HF orbitals show significant localization. Question (II) is really asking, how must Eq. (5.3) be changed? The answers to these questions parallel those offered in Sec. 3.

One way to answer question (I) is to ask a related question: what is an isolated atom or molecule in the HF approximation? One answer: it is a set of occupied orbitals which satisfy the HF equation for the isolated group. Let $\hat{\eta}_a^\circ$ be the Fock-Dirac density matrix constructed from the occupied HF orbitals ϕ_{ak}° of the isolated group A. The HF group energy $E_a^\circ = E_a[\hat{\eta}_a^\circ]$ is a minimum and the occupied orbitals satisfy $\hat{F}_a[\hat{\eta}_a^\circ]\phi_{ak}^\circ = \varepsilon_{ak}^\circ \phi_{ak}^\circ$. These relationships suggest a definition of group orbitals in complex systems. If the A-group orbitals ϕ_{ak} of the complex system are to be linear combinations of the occupied, canonical HF orbitals of the system, then it seems reasonable to define the ϕ_{ak} by requiring that $E_a = E_a[\hat{\eta}_a]$ be a minimum under the constraint $\hat{\rho}\phi_{ak} = \phi_{ak}$. This implies that $\hat{\rho}\hat{F}_a[\hat{\eta}_a]\hat{\rho}\phi_{ak} = \varepsilon_{ak}\hat{\rho}\phi_{ak}$ [11], which is the HF equation in a form equivalent to the Roothaan matrix form [47], but with the occupied canonical orbitals ψ_k playing the role of the basis set. This definition makes the ϕ_{ak} minimally distorted from the ϕ_{ak}° in the sense that $E_a - E_a^\circ$ is a minimum. It has been verified in a few cases that the distortion is small, e.g., at the equilibrium bond length in NaCl, $E_a - E_a^\circ = 0.0124$ Hartrees for Cl^-, which is less than 7% of the binding energy [48].

The above definition is of special interest because it was shown that the ϕ_{ak} satisfy another set of equations, ones which can be solved without knowing the ψ_k [12]. Let the potential energy between the A-group and the rest of the system be $\hat{v}_a = \hat{F}[\rho] - \hat{F}_a[\hat{\eta}_a]$. The ϕ_{ak} satisfy

$$\left(\hat{F}_a[\hat{\eta}_a] + \hat{v}_a - \hat{\rho}\hat{v}_a\hat{\rho}\right)\phi_{ak} = \varepsilon_{ak}\phi_{ak}, \tag{5.4}$$

a particular form of the general AG equation. The operator $\hat{\rho}\hat{v}_a\hat{\rho}$ provides a nonlocal screening of the potential $\hat{v}_a$ seen by the A-group electrons and, thus, the ϕ_{ak} are localized on A. By self-consistently solving (5.4) for the occupied orbitals of each group in a system, one can determine the HF $\hat{\rho}$. From a practical, computational viewpoint, the screened potential $\hat{v}_a - \hat{\rho}\hat{v}_a\hat{\rho}$ is very complicated, but because it is exact, it makes sense to try to approximate it. Kunz has proposed and, with Gilbert, carefully analyzed one approximation [49,50]. It has been suggested, also, that if $\hat{v}_a$ itself is approximated, any inaccuracies introduced by the approximation will be damped by $\hat{\rho}\hat{v}_a\hat{\rho}$ [51]. We do not consider approximations further because in this paper we want to understand how localization/separation may be achieved in the HF approximation without additional approximations.

We wrote above that any equation proposed for the orbitals of groups in molecules, must pass the test that its solution implies Eq. (5.1) is satisfied. Assume that (5.4) is satisfied by the occupied orbitals of each group in a system. Multiply (5.4) from the left by $1-\hat{\rho}$ to get $(1-\hat{\rho})\hat{F}\phi_{ak} = 0$ for each group. This follows from the idempotency of $\hat{\rho}$ and the definition of $\hat{v}_a$. Note that in solving (5.4) we must construct $\hat{\rho}$ from the occupied ϕ_{ak} and, consequently, $(1-\hat{\rho})\phi_{ak} = 0$. So long as we can construct $\hat{\rho}$ in this way, we have $(1-\hat{\rho})\hat{F}\phi_{ak} = 0$ and (5.1) is satisfied.

Gilbert showed that Eq. (5.4) could be generalized [13]. One generalization is to replace $-\hat{\rho}\hat{v}_a\hat{\rho}$ by $\hat{\rho}\hat{U}_a\hat{\rho}$, where $\hat{U}_a$ is an arbitrary operator. One could define $\hat{U}_a$, for example, as an additional attractive potential seen by the A-group electrons in the region occupied by the A-group. This is quite different from the effect of $-\hat{\rho}\hat{v}_a\hat{\rho}$, which is to reduce the interaction of the A-group electrons with the surrounding groups. The most general one-electron equation proposed by Gilbert was

$$\left(\hat{F} - \hat{\rho}\hat{F}\hat{\rho} + \hat{\rho}\hat{G}_a\hat{\rho}\right)\phi_{ak} = \varepsilon_{ak}\phi_{ak} \tag{5.5}$$

where $\hat{G}_a$ is an arbitrary operator [13]. Equation (5.5) is sometimes referred to as the Adams-Gilbert (AG) equation. It is readily shown by the method applied to (5.4), that if Eq. (5.5) is satisfied by the occupied orbitals of each group in the system, the fundamental condition (5.1) will also be satisfied, provided that $\hat{\rho}$ can be constructed from the group orbitals. Note that $\left(\hat{F} - \hat{\rho}\hat{F}\hat{\rho}\right)\phi_{ak} = 0$ for all occupied ϕ_{ak} because (5.1) implies $\left[\hat{F}, \hat{\rho}\right] = 0$. Thus, the properties of the ϕ_{ak} and their eigenvalues are determined by $\hat{\rho}\hat{G}_a\hat{\rho}$. The definition of the $\hat{G}_a$s is critical, e.g., which solutions of (5.5) are to be identified as occupied. If ψ_v is a virtual (unoccupied) canonical HF orbital, then $\hat{\rho}\psi_v = 0$ and ψ_v is a solution to (5.5) with the eigenvalue ε_v. The ε_{ak} for the occupied ϕ_{ak} are a subset of the eigenvalues of the matrix of $\langle\psi_k|\hat{G}_a|\psi_l\rangle$, with k and l running over the indices of the occupied canonical orbitals. If the occupied ϕ_{ak} are to be the lowest energy solutions of (5.5), then $\hat{G}_a$ must be defined so that the highest energy occupied ϕ_{ak}, is lower in energy than the lowest energy virtual orbital.

Equation (5.5) encompasses many possibilities. Observe that setting $\hat{G}_a = \hat{F}_a[\hat{\eta}_a]$ in it gives (5.4). Since $\hat{\rho}\hat{\eta}_a = \hat{\eta}_a\hat{\rho} = \hat{\eta}_a$, one gets Matsuoka's modified AG equation by setting $\hat{G}_a = \hat{\eta}_a\hat{W}_a\hat{\eta}_a$ [52]. One could replace $\hat{\rho}$, where it appears explicitly in (5.4) and (5.5), by $\hat{\eta} = \sum_b \hat{\eta}_b$ because $\hat{\eta} = \hat{\rho}\hat{\eta} = \hat{\eta}\hat{\rho}$. One could use $\hat{\eta} - \hat{\eta}_a$ instead. In short, one has considerable flexibility in applying (5.4) and (5.5). This is because the only condition that must be satisfied in order to have a HF solution, is Eq. (5.1), which depends only on $\hat{\rho}$. It should be understood, however, that different forms of these equations may exhibit quite different numerical properties.

One can understand qualitatively how (5.4) maintains the localization of group orbitals by looking at the matrix elements of $\hat{v}_a - \hat{\rho}\hat{v}_a\hat{\rho}$ with different orbitals. Assume that A and B are closed shell molecules. One could use the molecular orbitals of isolated A and B as a basis set in calculating a first approximation to the eigenfunctions of $\hat{F}$. The well localized ϕ°_{ak} would delocalize into approximate canonical orbitals by combining with the ϕ°_{bk}, although this will not lower E. It can lower E because E is a functional of $\hat{\rho}$, which in this example can be constructed from the set of occupied ϕ°_{ak} and ϕ°_{bk} by orthonormalization. In contrast, $\langle\phi^{\circ}_{ak}|\hat{v}_a - \hat{\rho}\hat{v}_a\hat{\rho}|\phi^{\circ}_{bl}\rangle = 0$ and $\langle\phi^{\circ}_{bk}|\hat{v}_a - \hat{\rho}\hat{v}_a\hat{\rho}|\phi^{\circ}_{bl}\rangle = 0$ when $\hat{\rho}$ is constructed from the isolated molecule orbitals, i.e., the solutions of (5.4) will be the ϕ°_{ak}. Delocalization can only begin to occur in the solutions of (5.4) when add to the basis set functions which are orthogonal to the occupied ϕ°_{ak} and ϕ°_{bk}. If we solve (5.4) exactly to self-consistency in the A and B molecule orbitals, all matrix elements of $\hat{v}_a - \hat{\rho}\hat{v}_a\hat{\rho}$ evaluated with the ϕ_{ak} and the ϕ_{bl} will vanish, and these orbitals will be exactly the ones defined above in terms of the canonical HF orbitals. If one substitutes $\hat{\eta}_b$ for $\hat{\rho}$, as suggested in the preceding paragraph , then only the matrix elements $\langle\phi_{bk}|\hat{v}_a-\hat{\eta}_b\hat{v}_a\hat{\eta}_b|\phi_{bl}\rangle$ vanish. This means that the ϕ_{bl} do not see the Coulomb potential due to the B nuclei. Thus, although the ϕ_{ak} and the ϕ_{bl} are now coupled through off-diagonal elements of $\hat{v}_a-\hat{\rho}\hat{v}_a\hat{\rho}$, no great lowering of the ε_{ak} can be gained by delocalization of the A group orbitals onto the B group. Collapse into the B-group core is unlikely because the more tightly localized a ϕ_{bl} is in the core, the higher its kinetic energy will become, while there is no compensating decrease in potential energy due to concentration about the B-nuclei.

Equation (5.4) can be written down for the group orbitals of a complex system for any group defined by an energy minimum principle or by equations like $\hat{F}[\hat{\eta}_a^\circ]\phi_{ak}^\circ=\varepsilon_{ak}^\circ\phi_{ak}^\circ$. It should be obvious that care must be taken in doing this. Suppose that one wants to apply (5.4) to the calculation of valence shell orbitals. Matsuoka did this in calculations on HF, H_2O and CH_4. He replaced the actual charge on the heavy nucleus by an effective nuclear charge z in the Coulomb potential energy between the nucleus and the valence-electrons. He found that the lower valence orbitals resembled core orbitals and, consequently, studied two modified methods. To understand why his valence orbitals resembled core orbitals, consider HF. By Slater's rules, screening only by core electrons gives $z = 7.3$. Let $\hat{\eta}_V$ be the valence shell density matrix The lowest energy eigenfunction of $-(1/2)\nabla^2-7.3/r_F+\hat{J}[\hat{\eta}_V]-\hat{K}[\hat{\eta}_V]$ is a $1s$-orbital, the next to lowest are a $2s$ and a set of $2p$s. The $1s$ is not a valence shell orbital, its energy is much lower than those of the $2s$ and $2p$ orbitals. Reasonable approximations to the $2s$ and $2p$ orbitals should be Slater orbitals with Slater rule exponents appropriate for F. The $1s$ will have an exponent of about 7.0. We infer that Matsuoka's valence shell Fock operator had the form

$$\hat{F}_V=-(1/2)\nabla^2-(1/r_H)-(7.3/r_F)+\hat{J}[\hat{\eta}_V]-\hat{K}[\hat{\eta}_V].$$

From the definition of $\hat{v}_a$ above, it follows that $\hat{v}_V=-(1.7/r_F)+\hat{J}[\hat{\rho}-\hat{\eta}_V]-\hat{K}[\hat{\rho}-\hat{\eta}_V]$. The screening operator $\hat{\rho}\hat{v}_V\hat{\rho}$ screens $\hat{v}_V$, not the $-7.3/r_F$ potential. Thus, we must expect the lowest energy σ-eigenfunction of $\hat{F}_V+\hat{v}_V-\hat{\rho}\hat{v}_V\hat{\rho}$ will be predominantly a Slater 1s-orbital on F with an exponent of about 7.0. It will be about 7.0 because the valence shell electrons can not effectively screen the $-7.3/r_F$ potential. If one wants to use this $\hat{F}_V$ in calculations on HF, one should not include its lowest energy eigenfunction among the occupied, valence-shell orbitals.

In the broader context of what causes delocalization, and what can be done to prevent it, the analysis of this Section is reminiscent of the discussions in Secs. 2 and 3. Electrons delocalize to neighboring groups, then collapse into the core in the HF approximation, too. Assume that to obtain A- and B-group orbitals, respectively, it was sufficient to find eigenfunctions of $\hat{F}$ using first an A-center basis set, then a B-center set. As long as the basis sets are incomplete, one can hope to get linearly independent A- and B-group orbitals. If we let both basis sets become complete, however, all of the occupied orbitals of one group will be identical to occupied orbitals of the other group when we choose the occupied orbitals for both as those of lowest energy. Thus, the Slater determinant will vanish; only unphysical wave functions can be constructed from these occupied orbitals because they are linearly dependent. We saw that this happens, too, when we try to solve the Schrödinger equation without defining what we mean by electronic groups in a complex system. In that case, delocalization and collapse can be prevented by using unphysical states in the definition of the localized N-electron functions. In the HF approximation, the localized orbitals of the B-group are used analogously in defining the equations for the A-group orbitals. They prevent the A-orbitals from delocalizing and, ultimately, the Slater determinant wave function from vanishing.

5.2 HUZINAGA-EQUATION

Huzinaga examined the question of how to determine an optimum electronic wave function for the A group when a satisfactory wave function is known for the B group [10]. His starting point was McWeeny's theory of generalized product functions [7], which is based on the concept of strongly orthogonal functions. What is commonly called the Huzinaga equation, however, applies specifically to wave functions approximated by a single Slater determinant, or by a linear combination of Slater determinants in which the linear coefficients are fixed to give a specific spin and/or symmetry state. If the electronic energy is to be minimized with respect to all orbitals in such wave functions, one is working in the HF approximation in the sense that (5.1) must be satisfied. If one minimizes only with respect to the A group orbitals, (5.1) is replaced by another equation as the necessary condition for an energy minimum. We begin by deriving that condition, then we check that it is satisfied by the solutions of the Huzinaga equation. There are modifications of the Huzinaga equation for which it is not satisfied. We explain what can be done in these cases if one wants to obtain solutions which are equivalent to HF solutions.

Assume that one has a set of orthonormal orbitals for the B group that are accurate enough for the calculation one plans, but that one wants to determine an optimum set of A group orbitals. The B group might be the electronic core; the A group, the valence shell. We assume, as in the preceding subsection, that the N-electron wave function is a single Slater determinant. This means that $E = E[\hat{\rho}]$, with $\hat{\rho}$ the projector onto the space spanned by the A and B group occupied orbitals. Let $\hat{\rho}_b$ be the projector onto the B group orbitals. Since $\hat{\rho}\hat{\rho}_b = \hat{\rho}_b\hat{\rho} = \hat{\rho}_b$, it follows that $\hat{\rho}_a = \hat{\rho} - \hat{\rho}_b$ is also a projector and that $\hat{\rho}_a\hat{\rho}_b = 0$. To minimize E with respect to $\hat{\rho}_a$, we adopt McWeeny's method [42] for constructing $\delta\hat{\rho}$, the first variation of $\hat{\rho}$, and a formalism we have used before [43]. The most general form for $\delta\hat{\rho}$ that is consistent with the properties of the above projectors, when only the A group orbitals are varied, is $\hat{\rho} = \lambda(\hat{v} + \hat{v}^{\dagger})$, where $\hat{v} = (1 - \hat{\rho})\hat{\Delta}\hat{\rho}_a$ and $\hat{\Delta}$ is an arbitrary operator. The parameter λ defines the order of the variation. By the same reasoning that is used to obtain (5.1), we find that the necessary condition for an energy minimum is

$$(1 - \hat{\rho})\hat{F}\hat{\rho}_a = 0, \qquad (5.6)$$

which is equivalent to the result found by Huzinaga and Cantu [10]. We noted above that (5.1) implies that $[\hat{F}, \hat{\rho}] = 0$. Equation (5.6) implies only that

$$[\hat{F}, \hat{\rho}_a] = \hat{\rho}_b\hat{F}\hat{\rho}_a - \hat{\rho}_a\hat{F}\hat{\rho}_b, \qquad (5.7)$$

i.e., $\hat{\rho}_a$ commutes with $\hat{F}$ only if $\hat{\rho}_b$ does.

The original Huzinaga equation corresponds to

$$\left[\hat{F} - x(\hat{\rho}_b\hat{F} + \hat{F}\hat{\rho}_b)\right]\phi_{ak} = \varepsilon_{ak}\phi_{ak} \qquad (5.8)$$

with $x = 1$. The adjustable parameter x has been introduced by some to improve agreement between Model Potential approximations to the Huzinaga equation and the HF approximation. The questions we address are, does solution of (5.8) imply that (5.6) is satisfied, and should the occupied ϕ_{ak} be chosen as the lowest energy solutions of (5.8)? If (5.6) is not satisfied, then the solution (5.8) does not minimize the electronic energy.

We check first the original Huzinaga equation in which $x = 1$. Assume that we have determined an occupied set of ϕ_{ak} which satisfy (5.8). We can construct $\hat{\rho}$ from these ϕ_{ak} and the given set of ϕ_{bk} by symmetric orthogonalization [38]. We can not assume that the two sets of orbitals are orthogonal. Multiplication of (5.8) from the left by $1-\hat{\rho}$ gives $(1-\hat{\rho})\hat{F}(1-\hat{\rho}_b)\phi_{ak} = 0$. Note that $(1-\hat{\rho}_b)\phi_{ak}$ is orthogonal to all of the occupied ϕ_{bk}. Thus, if the $(1-\hat{\rho}_b)\phi_{ak}$ constructed from the occupied ϕ_{ak} form a linearly independent set, we can construct $\hat{\rho}_a$ from them. This means that (5.6) is satisfied. The possible ε_{ak} when (5.8) is satisfied are the eigenvalues of

$$\hat{\rho}_a\left[\hat{F} - \rho_b\hat{F} - \hat{F}\rho_b\right]\hat{\rho}_a = \hat{\rho}_a\hat{F}\hat{\rho}_a, \tag{5.9a}$$

$$(1-\hat{\rho})\left[\hat{F} - \rho_b\hat{F} - \hat{F}\rho_b\right](1-\hat{\rho}) = (1-\hat{\rho})\hat{F}(1-\hat{\rho}), \tag{5.9b}$$

$$\hat{\rho}_b\left[\hat{F} - \rho_b\hat{F} - \hat{F}\rho_b\right]\hat{\rho}_b = -\hat{\rho}_b\hat{F}\hat{\rho}_b. \tag{5.9c}$$

All other projections of $\hat{F} - \rho_b\hat{F} - \hat{F}\rho_b$ vanish. We assume that the ε_{ak} of occupied orbitals are negative. It is reasonable to assume that the occupied ϕ_{bk} have been chosen so that the nonzero eigenvalues of (5.9c) are positive. There is, therefore, no chance of picking a linear combination of the ϕ_{bk} as an occupied ϕ_{ak}. The eigenvalues of (5.9b) will be those of virtual orbitals. As long as the highest ε_{ak} for occupied orbitals is lower than the lowest virtual orbital energy, there is no chance of choosing a virtual orbital as one of the occupied ϕ_{ak}. Finally, if the equation for the ϕ_{bk}, corresponding to (5.8), is solved , we find $(1-\hat{\rho})\hat{F}\hat{\rho}_b = 0$. This, added to (5.6), gives (5.1). Thus, solving the original Huzinaga equation to self-consistency in both A and B group orbitals is equivalent to solving the canonical HF equations. This is exactly the conclusion of Huzinaga and Cantu [10].

When $x \neq 1$ in Eq. (5.8), we can not prove that (5.6) is satisfied. We assume that (5.8) has been solved, and we again assume that $\hat{\rho}$ can be constructed from the ϕ_{ak} and the given set of ϕ_{bk}. When we multiply (5.8) from the left by $1-\hat{\rho}$ we get

$$(1-\hat{\rho})\hat{F}(1-x\hat{\rho}_b)\phi_{ak} = (1-\hat{\rho})\hat{F}(1-\hat{\rho}_b)\phi_{ak} + (1-x)(1-\hat{\rho})\hat{F}\hat{\rho}_b\phi_{ak} = 0,$$

i.e., $(1-\hat{\rho})\hat{F}(1-\hat{\rho}_b)\phi_{ak} \neq 0$ when $x \neq 1$ unless $(1-\hat{\rho})\hat{F}\hat{\rho}_b\phi_{ak} = 0$. The latter may occur if $\hat{\rho}_b$ commutes with $\hat{F}$ or if $\hat{\rho}_b\phi_{ak} = 0$. We can not assume that $\hat{\rho}_b$ commutes with $\hat{F}$. If we multiply (5.8) from the left by $\hat{\rho}_b$ we find that

$$\left[\varepsilon_{ak}\hat{\rho}_b + (2x-1)\hat{\rho}_b\hat{F}\hat{\rho}_b\right]\phi_{ak} = (1-x)\hat{\rho}_b\hat{F}\hat{\rho}_a\phi_{ak}.$$

This tells us that if $\hat{\rho}_a\phi_{ak} \neq 0$, and if $\hat{\rho}_b$ does not commute with $\hat{F}$, then $\hat{\rho}_b\phi_{ak} \neq 0$ so long as the square bracketed operator on the left has an inverse in the space of the occupied ϕ_{bk}. Thus, solving (5.8) with $x \neq 1$ gives ϕ_{ak} which are not orthogonal to the ϕ_{bk}, and does not ensure that (5.6) is satisfied. On the other hand, $x < 1$ has been introduced when approximations are made in (5.8) by introducing model potentials [53-56]. In that context, we suspect that x partially compensates for errors in the model potentials.

If instead of looking for an equation for the ϕ_{ak} which implies that (5.6) is satisfied, one is willing to accept a modified form of (5.6) that implies (5.1) is satisfied, then one can use the localized orbital theory of Sec. 5.1. This gives

$$\left[\hat{F} - x(\hat{\rho}_b\hat{F}\hat{\rho} + \hat{\rho}\hat{F}\hat{\rho}_b)\right]\phi_{ak} = \varepsilon_{ak}\phi_{ak}, \tag{5.10}$$

which simplifies to (5.8) when self-consistency is achieved in both the A and B group orbitals [40]. Multiplication of (5.10) from the left by $1-\hat{\rho}$ gives $(1-\hat{\rho})\hat{F}\phi_{ak}=0$. If the corthese two results, one can show that (5.1) is satisfied.

The Huzinaga theory and the theory of Sec. 5.1 are not equivalent. The Huzinaga theory with $x=1$ assures a minimum energy with respect to the ϕ_{ak} with the ϕ_{bk} fixed. One responding equation for the ϕ_{bk} is solved, one finds that $(1-\hat{\rho})\hat{F}\phi_{bk}=0$. Combining is expected to have chosen the ϕ_{bk} realistically, e.g., such that $\langle\phi_{bk}|\hat{F}|\phi_{bk}\rangle<0$. Solving (5.8) with $x=1$ guarantees only that (5.6) is satisfied, not the HF condition (5.1). The theory of Sec. 5.1 is designed to be equivalent to the HF approximation when self-consistent solutions of (5.4) or (5.5) have been obtained for all groups. It does not assure an energy minimum with respect to A-group orbitals when the orbitals of other groups are frozen. One can make the Huzinaga theory equivalent to the HF theory by self-consistently solving for the orbitals of all groups. When this is done, the two theories become equivalent.

6. Correspondences and Extensions

We have discussed several theories in the preceding Sections as if invoking one excluded the others. In this concluding Section we wish to draw attention to the mathematical correspondences between the N-electron theory and the one-electron theory, and to explain how the one-electron theory might be useful in the N-electron theory. In addition, we argue that the theories of Secs. 3 and 5 provide nontrivial bridges between approximate theories and the exact N-electron and exact Hartree-Fock theories, respectively. These bridges may be used to refine approximate theories.

In the one-electron theories of Sec. 5, one sets out to calculate occupied orbitals which are, at best, linear combinations of the occupied set of canonical HF orbitals, eigenfunctions of the Fock operator $\hat{F}$. We have included the qualifier "at best" because the Huzinaga theory permits an intermediate level of optimization, one in which the energy is minimized with respect to the orbitals of just one group. In the N-electron theories of Sec. 3 one wants to calculate a single function which is a linear combination of the ground state wave function and an arbitrary number of unphysical eigenfunctions of $\hat{H}$. In both theories one can specify the linear combinations in any way one sees fit. The choice is limited only by the need for an equation which can be solved to give the desired linear combinations without first having to calculate the eigenfunctions of $\hat{F}$ or $\hat{H}$. The functions may be defined implicitly by the equation to be solved, but it seems to us advantageous to define the functions so that one can guess good first approximations to them. One needs good first approximations because the equations are nonlinear and must be iterated to a self-consistent solution. The reason one wants to calculate orbitals which are linear combinations of occupied HF canonical orbitals is that one can construct the Fock-Dirac density matrix $\hat{\rho}$ directly from the occupied, noncanonical orbitals by the relatively simple procedure of orthonormalization, then calculate the energy $E[\hat{\rho}]$. The situation is analogous in the N-electron theory. If the antisymmetric projection of the calculated function is guaranteed to be an eigenfunction of $\hat{H}$, one can simply antisymmetrize, then calculate the eigenvalue of $\hat{H}$ by integration. That Eqs. (3.3) and (5.4) are similar is a consequence of the correspondences we have noted and of the variational definitions we chose for localized orbitals and for localized, N-electron wave functions.

We ended Sec. 3 by explaining that it was not obvious how one could define an $\hat{H}^\circ$ which distinguished between valence and core electrons, and that, consequently, we could not apply the theory of Sec. 3 to obtain a wave function that was separated approximately into a product of core and valence group functions. This was one stimulus for the review in Sec. 5 of localization and separation in the HF approximation. The Huzinaga theory, and works based on that theory, seem most directly applicable to the definition of a valence shell Hamiltonian for N-electron systems. One can not, however, simply set $\hat{H}^\circ$ equal to the sum of the HF core and valence Hamiltonians because the electron correlation error is too large. We explained in Sec. 3 why the eigenvalue $\mathscr{E}$ should be lower than the energy of the first-excited physical state. This can not be the case if $\hat{H}^\circ$ is the sum of HF Hamiltonians. One may be able to circumvent this difficulty by adopting a hybrid approach like those outlined by Kleiner and McWeeny [35], and by Huzinaga [18]. One includes electron correlation within each shell, but neglects the inter-shell correlations. Collapse of the valence shell into the core is prevented by one-electron projection operators, as in the HF theory. We plan to investigate core-valence separability using the bideterminantal CI program to see if there is a definition of the valence shell Hamiltonian which enhances the separability of the wave function.

The N-electron localized wave function theory is undoubtedly too complicated to apply without approximation, but in this it is like the HF theories we have reviewed. They are, without approximation, no easier to apply to large systems than the canonical HF theory. If one is going to introduce approximations into the N-electron theory, we believe that it should be done by defining $\hat{H}^\circ$ so that $\hat{V}$ may be neglected. If the degree of separability we found for the lowest energy, localized wave function of LiH, is found in other systems, then choosing $\hat{H}^\circ$ as the sum of an A- and a B-group Hamiltonian should be a good approximation. What we have in mind is adapting the group functions to the system environment by including in the A-group Hamiltonian the Coulomb potential due to the B-group nuclei and electron density, but screened as $\hat{v}_a$ is screened in Eq. (5.4). This is quite similar to the theories of references [18] and [35]. We would define the B-group Hamiltonian similarly. One could determine the lowest energy eigenfunction of $\hat{H}^\circ$ by finding the eigenfunctions of the A- and B-group Hamiltonians separately, then use the spin-coupled, antisymmetrized product of A and B functions to calculate a correction to the sum of group energies. The simplification that is achieved is the resolution of the N-electron problem into N_a- and N_b-electron problems, which are presumably easier to solve than the original problem no matter what method we use. What is left out by an approximation like this is the dispersion interaction between groups. It may or may not be important at bonding distances. It is our intention to investigate approximations of this type with the bideterminantal CI program.

Approximations are the key to making exact theories practical, but they seldom provide final solutions. Eventually one wants, or needs, a more accurate solution. In this regard perturbation theory is attractive because it is a systematically refinable method of approximation. If one is not satisfied with the nth order approximation to the energy, for example, one can in principle calculate higher order corrections. If the perturbation theory is convergent, one can be confident that calculated higher order corrections improve the original approximation even when the disagreement between theory and experiment has widened. If one uses an approximate variational theory, however, there is no assurance that one can improve the results obtained, even in principle, by perturbation methods *if one has separated the electrons into distinguishable groups.* For example, suppose one has calculated the orbitals of the A- and B-groups using one-electron Hamiltonians $\hat{F}_a$ and $\hat{F}_b$, which may depend on approximate potentials. Suppose one then decides to improve the results obtained by treating $\hat{v}_a = \hat{F} - \hat{F}_a$ as a perturbation and applying perturbation theory. If $\hat{F}$ is the Fock-Dirac operator, and if the perturbation expansion converges, the orbitals determined by the expansion become, in infinite order, the lowest energy canonical HF orbitals, assuming that the true ground state $\hat{\rho}$ was used to construct $\hat{F}$. Under these conditions, the same solutions will be obtained for the B-group. Thus, the occupied A- and B-group orbitals will be linearly dependent if they are chosen as those of lowest energy. Consequently, $\hat{\rho}$ can not be formed from these orbitals, nor can the energy be calculated. These effects will not be manifest in low order in the perturbation expansion, but one will still be calculating corrections which ultimately can not contribute to the exact solution of the problem. One can circumvent this problem by basing the perturbation expansion on Eq. (5.4) with $\hat{v}_a - \hat{\rho}\hat{v}_a\hat{\rho}$ regarded as the perturbation. Equation (5.4) was derived by requiring that its eigensolutions be minimally distorted from the eigenfunctions of $\hat{F}_a$. In this sense, it provides a bridge between approximate orbital theories and the exact HF theory. A careful formal and numerical study of $\hat{v}_a - \hat{\rho}\hat{v}_a\hat{\rho}$ might also prove a useful guide to refining an approximate variational theory.

The situation regarding approximate N-electron theories and the theory of Sec. 3.1 is analogous. Suppose that $\hat{H}^\circ$ is an approximate N-electron Hamiltonian which does not commute with the antisymmetrizer $\mathscr{A}$, i.e., the Hamiltonian divides the electrons into distinguishable groups. One can not justify treating $\hat{V} = \hat{H} - \hat{H}^\circ$ as a perturbation even if the lowest eigenvalue of $\hat{H}^\circ$ is an accurate approximation to the ground state energy because the physical state energy is higher than an infinite number of bound unphysical state energies and lies in a continuum of unbound unphysical state energies. The situation is exactly that found in intermolecular perturbation theory [1-5]. One can, however, base a perturbation development on Eq. (3.3) with $\hat{V} - \hat{Q}\hat{V}\hat{Q}$ as the perturbing potential, provided that $\mathscr{E}_1$ is less than the energy of the next-to-lowest-energy physical state. In Sec. 3.1, we found this to be the case for the interaction between a lithium and a hydrogen atom. It can not be true for every choice of $\hat{H}^\circ$, e.g., if $\hat{H}^\circ$ is set equal to the sum of HF Hamiltonians. In those cases where the perturbation method is valid, one may hope to recognize ways in which to improve the approximate variational Hamiltonian $\hat{H}^\circ$ by studying the effect of $\hat{V} - \hat{Q}\hat{V}\hat{Q}$ on the zeroth order wave function. In short, the theory of Sec. 3.1 is a bridge that links approximate theories to the exact N-electron theory.

In this paper we have tried to understand why electrons delocalize, what effect it has in terms of wave functions and energies, and how it can be prevented without approximation, in the N-electron theory and in the HF one-electron theory. At both levels of electronic theory, it is the unphysical states which drive the delocalization. In the N-electron theory, the unphysical states appear explicitly as possible solutions when we try to distinguish between groups of electrons. In the one-electron theory, the unphysical states play an implicit role: the set of occupied orbitals becomes linearly dependent and can only be used to construct unphysical wave functions. Because we are interested only in physical solutions of the Schrödinger equation, we can combine the unphysical functions with a single physical function to produce a wave function in which groups of electrons are distinguishable and from which the exact energy can be calculated by antisymmetrization and integration.

Appendix A

The essential features of the bideterminantal FCI program that was used in the calculations for Secs. 3 and 4 have been sketched in previous papers [3-4]. We add here to those sketches a description of new features.

Instead of solving Eq. (3.3) for the eigenfunction F, it was more convenient to solve the alternative equation

$$\left[\hat{H}^{\circ} + \hat{\mathscr{A}}(\hat{V} - D)\right]F = \mathscr{E}F, \qquad \text{where } D = \frac{\langle F|\hat{\mathscr{A}}\hat{V}|F\rangle}{\langle F|\hat{\mathscr{A}}|F\rangle}, \tag{A.1}$$

because we had been investigating a perturbation theory based on an equation differing from that above only through the absence of D. The F that satisfies (A.1) is exactly the same function that one would get by solving (3.3). And the $\mathscr{E}$ above is also identical to that in (3.3). The properties of the two equations are otherwise rather different, but that is not important for the purposes of this article, namely, obtaining an approximation to the exact solution of (3.3) and numerical values for $\mathscr{E}_1$, E_1 and E_2 as functions of R.

We solved (A.1) in the bideterminantal FCI approximation by diagonalizing the real matrix corresponding to $\hat{H}^{\circ} + \hat{\mathscr{A}}(\hat{V} - D)$ using the International Mathematical Subroutine Library's program DE2LRG. Because that program requires the matrix to be in core, we modified our programs to contract the matrix in the bideterminantal representation to a matrix in a spin eigenfunction representation, in which each function was a linear combination of bideterminants. In this way the 1080 bideterminantal configurations were reduced to 420 orthonormal spin eigenfunctions.

Since D is defined in terms of F, (A.1) must be solved iteratively to self-consistency. It turned out that the solution of (A.1) is so sensitive to the value of D, that an extrapolation procedure had to be added to the program to obtain convergence. We based the extrapolation procedure on the idea that a function $f(D)$ existed such that when (A.1) was solved with $D = D_0$, the resultant F substituted into the integral above for D, gave $D_1 = f(D_0)$. Similarly, we calculated $D_2 = f(D_1)$ before extrapolating. The extrapolation was based on the conjecture that $f(D)$ was linear in D. We started the calculation by setting $F = \hat{\mathscr{O}}\phi_a^{\circ}\phi_b^{\circ}$. The calculations converged at the smallest R values in 10 iterations or less.

The LiH orbital basis set consisted of the accurate Li $1s$ and $2s$ Hartree-Fock orbitals, a Li SCF $2p\sigma$ orbital, and the hydrogen atom $1s$, $2s$ and $2p\sigma$ eigenfunctions. These are the same basis orbitals that were used in our earlier calculations. We also did a set of calculations in which the hydrogen $2s$ and $2p\sigma$ eigenfunctions were replaced with Slater type orbitals with exponents equal to one. This change had a noticeable effect on the quantities reported in Sec. 3, showing that we must carry out larger calculations to obtain definitive results. The differences between the results of the two calculations were not so large, however, to alter our general conclusion that F is separable to a high degree of approximation and that $\mathscr{E}_1 < E_2$, even at the potential minimum,

The results presented in Sec. 4 are a subset of the results reported in reference [4].

References

[1] Claverie, P. (1971) "Theory of Intermolecular Forces. I. On the Inadequacy of the Usual Rayleigh-Schrödinger Perturbation Method for the Treatment of Intermolecular Forces", Int. J. Quantum Chem. **5**, 273-296.

[2] Morgan, J. D. and Simon, B. (1980) "Behavior of Molecular Potential Energy Curves for Large Nuclear Separations", Int. J. Quantum Chem. **17**, 1143-1166.

[3] Adams, W. H. (1990) "Perturbation Theory of Intermolecular Interactions: What Is the Problem, Are There Solutions?", Int. J. Quantum Chem. **S24**, 531-547.

[4] Adams, W. H. (1991) "Perturbation Theory of Intermolecular Interactions: Are Second-Order Rayleigh-Schrödinger Energies Meaningful?", Int. J. Quantum Chem. **S25**, 165-181.

[5] Adams, W. H. (1992) "The Problem of Unphysical States in the Theory of Intermolecular Interactions", J. Math. Chem. **10,** 1-23.

[6] Fock, V., Wesselow, M. and Petrashen, M. (1940) "Partial separation of variables for bivalent atoms ", Zh. Eksp. Teor. Fiz. **10**, 723-739.

[7] McWeeny, R. (1959) "The Density Matrix in Many-electron Quantum Mechanics: I, Generalized Product Functions. Factorization and Physical Interpretation of the Density Matrices", Proc. Roy. Soc. (London) **A253**, 242-259.

[8] Szász, L. (1959) "Extension of the Hartree-Fock approximation by correlation functions", Z. Naturforsch. **14a**, 1014-1020.

[9] Weeks, J. D. and Rice, S. A. (1968) "Use of Pseudopotentials in Atomic-Structure Calculations", J. Chem. Phys. 49, 2741-2755.

[10] Huzinaga, S. and Cantu, A. A. (1971) "Theory of Separability of Many-Electron Systems", J. Chem. Phys. 55, 5543-5549.

[11] Adams, W. H. (1961) "On the Solution of the Hartree-Fock Equation in Terms of Localized Orbitals", J. Chem. Phys. **34**, 89-102.

[12] Adams, W. H. (1962) "Orbital Theories of Electronic Structure", J. Chem. Phys. **37**, 2009-2018.

[13] Gilbert, T. L. (1964) "Self-Consistent Equations for Localized Orbitals in Polyatomic Systems", in P.-O. Löwdin and B. Pullman (eds.), Molecular Orbitals in Chemistry, Physics and Biology, Academic Press, New York, pp. 405-420.

[14] Parr, R. G. (1965) "Quantum Theory of Molecular Electronic Structure", W. A. Benjamin, New York, pp. 41-45.

[15] Primas, H. (1965) "Separability in Many-Electron Systems", in O. Sinanoglu (ed.), Modern Quantum Chemistry, Vol. 2, Academic Press, New York, pp. 45-74.

[16] Huzinaga, S., McWilliams, D. and Cantu, A. A. (1973) "Projection Operators in Hartree-Fock Theory", Adv. Quantum Chem. 7, 187-220.

[17] Adams, W. H. (1975) "On the Separability of Electronic Wave Functions", Int. J. Quantum Chem. **S9**, 367-373.

[18] Huzinaga, S. (1991) "Effective Hamiltonian method for molecules", J. Mol. Struct.(Theochem.) **234**, 51-73.

[19] Bunge, C. F. (1976) "Accurate Determination of the Total Energy of the Be Ground State", Phys. Rev. A **14**, 1965-1978 .

[20] Morse, P. M., Young, L. A. and Haurwitz, E. S. (1935) "Tables for Determining Atomic Wave Functions and Energies", Phys. Rev. **48**, 948-. Corrected integrals appear in Slater, J. C. (1960) Quantum Theory of Atomic Structure, Vol. I, McGraw-Hill Book Company, New York, p. 350

[21] Adams, W. H. (1974) "Localized Wave Functions and the Interaction Potential between Electronic Groups", Phys. Rev. Lett. **32,** 1093-1095.

[22] Adams, W. H. and Clayton, M. M. (1985) "The Hydrogen Atom in the Hydrogen Molecule", Int. J. Qunatum Chem. **S19,** 333-348.

[23] Hylleraas, E. A. and Undheim, B. (1930) "Numerische Berechnung der 2 S-Terme von Ortho- und Par-Helium", Z. Physik **65**, 759-772. MacDonald, J. K. L. (1933) "Succesive approximations by the Rayleigh-Ritz variation method", Phys. Rev. **43**, 830-.

[24] van der Avoird, A. (1967) "Perturbation Theory of Intermolecular Interactions in the Wave Operator Formalism", J. Chem. Phys. **47**, 3649-3653; (1967) "Note on a Perturbation theory for intermolecular interactions in the wave operator formalism", Chem. Phys. Lett. **1**, 411-412.

[25] Hirschfelder, J. O. (1967) "Perturbation Theory for Exchange Forces II", Chem. Phys. Lett. **1**, 363-368.

[26] Polymeropoulos, E. E. and Adams, W. H. (1978) "Exchange Perturbation Theory. II. Eisenschitz-London Type", Phys. Rev. A **17**, 18-23.

[27] Löwdin, P.-O. (1955) "Quantum Theory of Many-Particle Systems. I. Physical Interpretations by Means of Density Matrices, Natural Spin-Orbitals and Convergence Problems in the Method of Configurational Interaction", Phys. Rev. **97**, 1474-1489.

[28] Carlson, B. C. and Keller, J. M. (1961) "Eigenvalues of Density Matrices", Phys. Rev. **121**, 659-661.

[29] Kutzelnigg, W. (1980) "The Primitive Wave Function in the Theory of Intermolecular Interactions", J. Chem. Phys. **73**, 343-359.

[30] Adams, W. H. (1971) "On the Solution of the Schrödinger Equation in Terms of Wave Functions Least Distorted from Products of Atomic Wave Functions", Chem. Phys. Lett. **11**, 441-444.

[31] Hirschfelder, J. O. and Silbey, R. (1966) "New Type of Molecular Perturbation Treatment", J. Chem. Phys. **45**, 2188-2192.

[32] Polymeropoulos, E. E. and Adams, W. H. (1978) "Exchange Perturbation Theory. III. Hirschfelder-Silbey Type", Phys. Rev. A **17**, 24-29.

[33] Arai, T. (1960) "Theorem on the Separability of Electron Pairs", J. Chem. Phys. **33**, 95-98.

[34] Löwdin, P.-O. (1961) "Note on the Separability Theorem for Electron Pairs", J. Chem. Phys. **35**, 78-81.

[35] Kleiner, M. and Mcweeny, R. (1973) "Valence-Electron-Only Calculations of Electronic Structure", Chem. Phys. Lett. **19**, 476-479.

[36] Lennard-Jones, J. E. (1949) "The molecular orbital theory of chemical valency. I. The determination of molecular orbitals", Proc. Roy. Soc. (London) **A198**, 1-13.

[37] Edmiston, C. and Ruedenberg, K. (1963) "Localized Atomic and Molecular Orbitals", Revs. Modern Phys. **35**, 457-473.

[38] Löwdin, P.-O. (1950) "On the Nonorthogonality Problem Connected with the Use of Atomic Wave Functions in the Theory of Molecules and Crystals", J. Chem. Phys. **18**, 365-375.

[39] Löwdin, P.-O. (1956) "Quantum Theory of Cohesive Properties of Solids", Advances in Phys. **5**, 1-171.

[40] Francisco, E., Pendás, A. M. and Adams, W. H. (1992) "Generalized Huzinaga building-block equations for nonorthogonal electronic groups: Relation to the Adams-Gilbert theory", J. Chem. Phys. **97**, 6504-6508.

[41] Dirac, P.A.M., (1930) "Exchange Phenomena in the Thomas Atom", Proc. Cambridge Phil. Soc. **26**, 376-385.

[42] McWeeny, R (1960) "Some Recent Advances in Density Matrix Theory", Revs. Modern Phys. **32**, 335-369.

[43] Adams, W. H. (1962) "Stability of Hartree-Fock States", Phys. Rev. **127**, 1650-1658.

[44] Thouless, D. J. (1961) The Quantum Mechanics of Many-Body Systems, Academic Press, New York. pp. 24-29.

[45] Roothaan, C. C. J. and Bagus, P. S. (1963) "Atomic Self-Consistent Field Calculations by the Expansion Method", in B. Alder, S. Fernback and M. Rotenberg (eds.), Methods in Computational Physics, Vol. 2, Academic Press, New York, pp.47-94.

[46] Adams, W. H. (1978) "Open shell self-consistent-field theory", J. Chem. Phys. **69**, 1924-1928.

[47] Roothaan, C. C. J. (1951) "New Developments in Molecular Orbital Theory", Revs. Modern Phys. **23**, 69-89.

[48] Adams, W. H. (1971) "Distortion of Interacting Atoms and Ions", Chem. Phys. Lett. **12**, 295-298.

[49] Kunz, A. B. (1973) "Approximation to the method of local orbitals", J. Phys. B **6**, L47-50.

[50] Gilbert, T. L. and Kunz, A. B. (1974) "Single-center orbital localization", Phys. Rev. B **10**, 3706-3710.

[51] Adams, W. H. (1971) "Least Distorted Localized Orbital Self-Consistent Field Equations", Chem. Phys. Lett.**11**, 71-74.

[52] Matsuoka, O. (1977) "Expansion methods for Adams-Gilbert equations. I. Modified Adams-Gilbert equation and common and fluctuating basis sets", J. Chem. Phys. **66**, 1245-1254.

[53] Gropen, O., Huzinaga, S. and McLean, A. D. (1980) "Model potential Scf calculations on Cl_2, Br_2 and I_2", J. Chem. Phys. **73**, 402-406.

[54] Gropen, O., Wahlgren, U. and Pettersson, L. (1982) "Effective Core Potential Calculations on the NiH_4^{2-} Ion as a Test Case for Studying Rotational Barriers", Chem. Phys. Lett. **66**, 453-458.

[55] Luaña, V. and Pueyo, L. (1987) "Core Projection Effects in Atomic Frozen-core Calculations: A Numerical Analysis", Int. J. Quantum Chem. **31**, 975-988.

[56] Luaña, V. and Pueyo, L. (1989) "Simulation of ionic transition-metal crystals: The cluster model and the cluster-lattice interaction in the light of the theory of electronic separability", Phys. Rev. B **39**, 11093-11112.

QUANTUM DYNAMICS OF DIATOMS IN EXTERNAL FIELDS

J. Broeckhove, B. Feyen [*] and P. Van Leuven
Universiteit Antwerpen (RUCA),
Department of Mathematics and Computer Sciences
Groenenborgerlaan 171, B2020 Antwerpen, Belgium

Abstract

In this contribution we present a quantum wavpacket propagation scheme suitable for diatomics interacting with a laser field. It is based on a hybrid representation for the nuclear wavefunction. It has an FFT-grid for the internuclear coordinate and an expansion in spherical harmonics for the angular coordinates. We demonstrate the usefulness of the scheme by considering two applications. One is the multiphoton excitation of Hydrogen Fluoride in an intense infrared laser. The other is the exciation dynamics of singlet states in Boron Hydride, making use of a two-color laser and the non-adiabatic coupling between the states. The latter application involves multiple electronic states and the introduction of a quasi-diabatic basis. In each instance we use ab initio or experimentally determined potentials and dipoles.

1 Introduction

A new trend in the study of molecular systems is the explicitely time-dependent description of the intramolecular dynamics. Quantum time-dependent theories in paricular have received a lot of attention, not in the least because they fit well with the

*Research associate IIKW, Belgium

J. L. Calais and E. S. Kryachko (eds.), Structure and Dynamics of Atoms and Molecules: Conceptual Trends, 97–131.

language and concepts used in discussing many experiments. For a an overview of the literature on time-dependent approaches to molecular theory, we refer to [DDLO94] and references therein.

The field of femtosecond spectroscopy provides a striking example of the need for a time-dependent theory. Recent developments in spectroscopical techniques have enabled time resolved investigation of intramolecular phenomena. These techniques rely on very intense laser pulses of subpicosend duration to excite the system and probe its response. Through the careful design of the pulses (frequency, duration, delay between pulses) one is able to achieve selectivity in the excitation or reaction path. The theoretical analysis of such experiments necessitates the solution of the time-dependent Schrödinger equation. Given the electronic potential energy surfaces and coupling matrix elements, one can then study the motion of the nuclei and their interacting with an external electromagnetic field. The choice of time propagation method is linked to the type of system one investigates. We use the pseudo-spectral method. It is limited to systems with up to three active degrees of freedom but can handle any kind of potential, including arbitrary, non-periodic time-dependence. The topic of this contribution, viz. quantum wavepacket propagation in diatomics interacting with laser fields, falls within this framework.

We consider a time propagation scheme based on a hybrid representation of the diatomic nuclear wavefunction. It has an FFT-grid representation for the internuclear coordinate and an expansion in spherical harmonic basis functions for the angular coordinates. Two applications that we consider in detail, demonstrate the usefulness of this scheme. The first is the multiphoton excitation of Hydrogen Fluoride in an intense infrared laser. The second is the dynamics of populating an excited $^1\Sigma^+$ state in Boron Hydride, making use of a two-color laser and the effects of non-adiabatic coupling between the $^1\Sigma^+$ states. In both instances, we have ab initio or experimentally determined potentials, coupligs and dipole functions. We also wish show that the time-propagation approach in no way limits one to use model potentials.

2 The Hamiltonian

In this section we introduce the notations and concepts related to the hamiltonian that will be used in the calculations discussed in this contribution. This includes the familiar Born-Oppenheimer approximation, and beyond that, the non-adiabatic coupling between potential energy surfaces. This is augmented with the interaction between the molecule and the external electromagnetic field, in the dipole approximation. Lastly, all of these expressions are specialized to the case of diatomic molecules.

2.1 Born-Oppenheimer and Beyond

Even with present day computational technology, it is impossible to consider the molecular problem explicitly as N interacting electrons and nuclei, except for the smallest of molecules. Already in 1927, Born and Oppenheimer suggested an approximation based on the assumption that the electrons can rearrange themselves around the nuclei instantaneously with respect to the timescales on which the nuclei move. The approximation implies that the motion of the nuclei does not influence the motion of the electrons, only the positions matter. One can then solve the problem of the electronic motion, for each configuration of the nuclei. Subsequently one can calculate the motion of the nuclei, using the electronice potentials.

The total Hamiltonian of a molecule is a sum of the kinetic energies of the nuclei and electrons and the Coulomb energies V of the various charges. Using a Greek index to sum over the nuclei and a Roman index to sum over the electrons, one has:

$$\hat{H} = \sum_\alpha \frac{\vec{P}_\alpha^2}{2m_\alpha} + \sum_i \frac{\vec{p}_i^2}{2m_i} + \hat{V}(\vec{r}, \vec{R}) \tag{1}$$

$$\hat{V}(\vec{r}, \vec{R}) = \sum_{i<j} \frac{q_i q_j}{|\vec{r}_i - \vec{r}_j|} + \sum_{\alpha<\beta} \frac{q_\alpha q_\beta}{|\vec{R}_\alpha - \vec{R}_\beta|} + \sum_{i,\alpha} \frac{q_i q_\alpha}{|\vec{r}_i - \vec{R}_\alpha|} \tag{2}$$

The m_α and q_α stand for the mass and charge of the αth nucleus, located at $\vec{R}_\alpha$, whereas $q_i = -e$ and $m_i = m_e$ denote the mass and charge of the ith electron with position $\vec{r}_i$. Planck's constant $\hbar$ has been set to unity throughout this text. In what follows we will also adhere to the notations $\vec{r} = (\vec{r}_i)_i$ and $\vec{R} = (\vec{R}_\alpha)_\alpha$ for the full set of electronic and nuclear coordinates respectively.

In general, the full wavefunction of the molecular system can be expanded with respect to a complete set of electronic wavefunctions η_n, that depend parametrically on the nuclear configuration $\vec{R}$ (referred to as the Born-Huang expansion)

$$\Psi(\vec{r}, \vec{R}) = \sum_n \chi_n(\vec{R}) \eta_n(\vec{r}; \vec{R}) \tag{3}$$

Using this basis set, the molecular Hamiltonian can be rewritten as a matrix-operator acting in the space of multi-component wavefunctions defined by equation 3

$$\hat{H} = \sum_n |\eta_n > \hat{H}_{mn} < \eta_n| \tag{4}$$

Each $\hat{H}_{mn}$ is to be considered an operator in the space of nuclear component wavefunctions $(\chi_n)_n$ and is given by

$$\hat{H}_{nn} = \left[\sum_\alpha \frac{1}{2m_\alpha} \vec{P}_\alpha^2 + < \eta_n|\hat{H}_{el}|\eta_n > + \sum_\alpha \frac{1}{2m_\alpha} < \eta_n|\vec{P}_\alpha^2|\eta_n >\right] \tag{5}$$

$$\hat{H}_{mn} = \left[\sum_\alpha \frac{1}{2m_\alpha} \left(2 < \eta_m|\vec{P}_\alpha|\eta_n > \vec{P}_\alpha + < \eta_m|\vec{P}_\alpha^2|\eta_n >\right) + < \eta_m|\hat{H}_{el}|\eta_n >\right] \tag{6}$$

The straightforward choice for the wavefunctions η_n are the eigenfunctions of the electronic Hamiltonian

$$\hat{H}_{el} = \sum_i \frac{\vec{p}_i^2}{2m_i} + \hat{V}(\vec{r}, \vec{R}) \tag{7}$$

They are the so-called adiabatic or Born-Oppenheimer wavefunctions and together they constitute the adiabatic electronic basis:

$$\hat{H}_{el}\phi_n(\vec{r}; \vec{R}) = W_n(\vec{R})\phi_n(\vec{r}; \vec{R}) \tag{8}$$

The eigenvalues $W_n(\vec{R})$ define the potential energy surfaces (PES) associated with each electronic state. For this choice of basis, the expressions for the hamiltonian (4)simplify because one has $< \eta_m|\hat{H}_{el}|\eta_n >= \delta_{mn}W_n(R)$. When dealing with electronic states which are well separated in energy, the adiabatic basis is a very good choice because then the corresponding nondiagonal matrix elements will be very small. In the Born-Oppenheimer approximation one simply neglects them. As a consequence, the nuclear and the electronic problems can be treated separately, the latter providing a potential energy surface for the former. This approximation assumes a simple product form wavefunction for the total wavefunction:

$$\Psi(\vec{r}, \vec{R}) = \chi_n(\vec{R})\phi_n(\vec{r}; \vec{R}) \tag{9}$$

The nuclear motion Hamiltonian then represents all of the nuclei moving on the Born-Oppenheimer potential energy surface W_n

$$\hat{H} = \sum_\alpha \frac{\vec{P}_\alpha^2}{2m_\alpha} + W_n(\vec{R}) \tag{10}$$

Notice that, compared to the full equations (4,5,6), one has not only negelected the coupling $\hat{H}_{mn}$ between electronic states but also dropped the third term in $\hat{H}_{nn}$ which is of the same order.

For electronic states with energy differences comparable with the nuclear vibrational energy, the Born-Oppenheimer approximation breaks down. The off-diagonal coupling matrix elements of the hamiltonian (4) are no longer negligible. A particular example are the so-called avoided crossings, electronic states of the same symmetry, which for some nuclear configurations are (nearly) degenerate. In the adiabatic representation each of these states is coupled to every other state through the nuclear momentum operator:

$$\begin{aligned} \hat{H} &= \sum_n |\phi_n> \left[\sum_\alpha \frac{\vec{P}_\alpha^2}{2m_\alpha} + W_n + g_{nn}\right] < \phi_n| \\ &+ \sum_{m,n}{}' |\phi_m> \left[\sum_\alpha \vec{f}_{mn}^{(\alpha)} \cdot \vec{P}_\alpha + g_{mn}\right] < \phi_n| \end{aligned}$$

$$\begin{aligned} \vec{f}^{(\alpha)}_{mn} &= \frac{1}{m_\alpha} < \phi_m | \vec{P}_\alpha | \phi_n > \\ g_{mn} &= \sum_\alpha \frac{1}{2m_\alpha} < \phi_m | \vec{P}^2_\alpha | \phi_n > \end{aligned} \tag{11}$$

The couplings g_{nm} and f_{nm} are therefore referred to as dynamical couplings. They are often strongly peaked as a function of $\vec{R}$ near the crossing. Moreover the composition of the adiabatic states in terms of molecular orbitals changes quite drastically in the neighbourhood of the crossing region. Both features make for a description that is not very amenable to numerical treatment. Therefore, one considers an alternative set of basis functions η_n that does not have these drawbacks. This leads to the introduction of the diabatic basis set [Smi69]. It is obtained through a unitary transformation of the adiabatic set, diagonalizing the momentum operator :

$$< \eta_m | \vec{P}_\alpha | \eta_n > = 0 \tag{12}$$

As a result the dynamical coupling terms disappear. The diabatic states are not eigenfunctions of the electronic Hamiltonian H_{el} and thus coupling terms are generated by the electronic Hamiltonian:

$$\hat{H} = \sum_n | \eta_n > \left[\sum_\alpha \frac{\vec{P}^2_\alpha}{2m_\alpha} + U_{nn} + g_{nn} \right] < \eta_n | + \sum_{m,n}' | \eta_m > U_{nm} < \eta_n | \tag{13}$$

We have used U_{mn} to denote the matrix elements of the electronic hamiltonian in the the diabatic basis

$$U_{mn} = < \eta_m | \hat{H}_{el} | \eta_n > \tag{14}$$

It was shown later [MT82] that the set of equations (12) has no solution in general, except the trivial one for which all the electronic wavefunctions are independent of R. As a result, a number of proposals have appeared in the litterature for the definition of quasi-diabatic bases. The main objective in each case is to construct a basis in which the coupling is smooth and its effect small.

2.2 The Dipole Interaction

The general (nonrelativistic) Hamiltonian for a molecule interacting with an electromagnetic field is obtained by replacing of the standard impuls operator $\vec{p}_j$ by $\vec{p}_j - q_j\vec{A}(\vec{r}, \vec{R})$, q_j being the charge and $\vec{A}(\vec{r}, \vec{R})$ the vector potential. The total Hamiltonian in Coulomb gauge is,

$$\hat{H} = \frac{1}{2m_\alpha} \sum_\alpha \left(\vec{P}_\alpha - q_\alpha \vec{A}(\vec{r}, \vec{R}) \right)^2 + \frac{1}{2m_i} \sum_i \left(\vec{p}_i - q_i \vec{A}(\vec{r}, \vec{R}) \right)^2 + \hat{V}(\vec{r}, \vec{R}) + H_{field} \tag{15}$$

and includes a field term. In principle, one should consider a quantized field in which the vector potential is expressed in terms of photon number operators. For laser fields, one enters the regime of high average photon number and strong coherence. This allows one [Mit93] to approximate (15) with a classical description for the vector potential and to eliminate the field term.

The Hamiltonian can be rewritten using the physical fields, the magnetic field $B(\vec{r})$ and the transversal part $E_T(\vec{r})$ of the electric field (the longitudinal part is contained in the Coulomb interaction between the charges). As a further approximation, one usually introduces the expansion of these fields around the centre of mass of the molecule. It allows one to split the Hamiltonian in multipole contributions which are of decreasing importance. In this work we will limit ourselves to the leading term in that expansion i.e. we will use the well known electric dipole approximation. This is equivalent to neglecting the spatial variation of the EM-field over the volume of the molecule, an approximation that is valid for the molecules, field-wavelengths and intensities involved in our study [LJD94]. The resulting approximate hamiltonian then becomes:

$$\hat{H} = \frac{1}{2m_\alpha}\sum_\alpha \left(\vec{P}_\alpha\right)^2 + \frac{1}{2m_i}\sum_i (\vec{p}_i)^2 + \hat{V}(\vec{r},\vec{R}) + \vec{E}\cdot\left(\sum_\alpha q_\alpha \vec{R}_\alpha + \sum_i q_i \vec{r}_i\right) \qquad (16)$$

where $\vec{E}$ represents the external (transversal) electric field at the center of the molecule. The first terms constitute the molecular hamiltonian in the absence of the field, the last term is the dipole interaction term. When this hamiltonian is represented in the diabatic basis for example one obtains

$$\hat{H} = \hat{H}_{mol} + \sum_{m,n} |\eta_m> \left[\vec{\mu}_{mn}(\vec{R}_\alpha)\cdot\vec{E}(t)\right] < \eta_n| \qquad (17)$$

Here $\vec{\mu}$ is the molecular dipole vector, whose matrix elements in the diabatic basis are

$$\vec{\mu}_{mn}(\vec{R}_\alpha) =< \eta_m| \left(\sum_\alpha q_\alpha \vec{R}_\alpha + \sum_i q_i\vec{r}_i\right) |\eta_n > \qquad (18)$$

In a similar fashion the hamiltonian may be written down in the adiabatic basis.

2.3 Diatoms in the Dipole Field

At this point we intend to focus on the specific case of diatomic molecules. After eliminating the center of mass motion, the nuclear coordinates reduce to the single inter-nuclear vector $\vec{R}$ and reduced mass M. The equations of motion associated with the dipole interaction Hamiltonian are, in the adiabatic representation:

$$i\hbar\frac{\partial\chi_n(\vec{R},t)}{\partial t} = \sum_m \left[(\frac{\vec{P}^2}{2M} + W_n)\delta_{nm} + \vec{f}_{nm}\cdot\vec{P} + g_{nm} + \vec{\mu}^{ad}_{mn}\cdot\vec{E}(t)\right]\chi_m(\vec{R},t)$$

$$\begin{aligned}
\vec{f}_{mn}(\vec{R}) &= \frac{1}{M} < \phi_m|\vec{P}|\phi_n > \\
g_{mn}(\vec{R}) &= \frac{1}{2M} < \phi_n|\vec{P}^2|\phi_n > \\
\vec{\mu}^{ad}_{mn}(\vec{R}) &= < \phi_m|\sum_i q_i\vec{r}_i|\phi_n > + \delta_{mn}\sum_\alpha q_\alpha \vec{R}_\alpha
\end{aligned} \tag{19}$$

We can also write down the equations of motion in the quasi-diabatic representation, were on has:

$$\begin{aligned}
i\hbar\frac{\partial\chi_n(\vec{R},t)}{\partial t} &= \sum_m \left[\frac{\vec{P}^2}{2M}\delta_{nm} + U_{nm} + \vec{\mu}^{di}_{mn}\cdot\vec{E}(t)\right]\chi_m(\vec{R},t) \\
\vec{\mu}^{di}_{mn}(\vec{R}) &= < \eta_m|\sum_i q_i\vec{r}_i|\eta_n > + \delta_{mn}\sum_\alpha q_\alpha \vec{R}_\alpha
\end{aligned} \tag{20}$$

with U_{nm} defined as in the previous section. Our approach to the rotational motion is guided by the following observations. In the absence of the dipole field, the Hamiltonian is rotationally invariant. As a consequence the diabatic or adiabatic electronic states depend only on the magnitude R of the inter-nuclear coordinate. The orientation of the diatom is completely defined by the two angular coordinates θ and ϕ of the internuclear axis. In the absence of the field, the nuclear wavefunction will have a spherical harmonic $Y_{lm}(\theta,\phi)$ for its angular dependence.

The dipole field introduces an angular dependence in the problem through the orientation of the diatom relative to the polarization axis of the field. This will be expressed in the $\cos\theta$ dependence in the scalar product of the dipole, oriented along the nuclear axis, and the field (assuming a linear polarization). Thus the wavefunction

$$\Psi(\vec{r},R,\theta,\phi,t) = \sum_n \chi_n(R,\theta,\phi,t)\eta_n(\vec{r};R) \tag{21}$$

must now be considered in all its coordinates and the equation of motion will be three-dimensional. We can expand the wavefunction using the spherical harmonic basis however,

$$\Psi(\vec{r},R,\theta,\phi,t) = \sum_{nlm} \frac{\chi_{nlm}(R,t)}{R} Y_{lm}(\theta,\phi)\eta_n(\vec{r};R) \tag{22}$$

thus defining a wavefunction made up of an abritrary number of electronic (index n) and angular momentum components (indices lm). The Hamiltonian becomes a matrix operator in the multi-component wavefunction space and couples all of the components:

$$i\hbar\frac{\partial\chi_{nlm}(R,t)}{\partial t} = \sum_{n'l'm'} \hat{H}_{nlm;n'l'm'}\chi_{n'l'm'}(R,t) \tag{23}$$

The structure of this operator in the diabatic basis is as follows:

$$\hat{H}_{nlm;n'l'm'} = \left[-\frac{\hbar^2}{2m}\frac{d^2}{dR^2} + \frac{l(l+1)}{R^2}\right]\delta_{nn'}\delta_{ll'} + U^{el}_{nn'}(R)\delta_{ll'} + E(t)\mu_{nn'}(R) < Y_{lm}|\cos\theta|Y_{l'm'} > \quad (24)$$

In this expression E and $\mu(R)$ represent the magnitudes of the corresponding vector quantities. The molecular hamiltonian couples electronic components for fixed lm, while the dipole term couples both electronic and nuclear angular momentum components. In the adiabatic representation, the dynamic coupling term $\vec{f}_{nm} \cdot \vec{P}$ would introduce angular momentum coupling in the molecular term also.

The dipole term has an interesting structure because it factorises in radial and angular contributions, and the angular contribution can be evaluated analytically. Using a standard expression from the theory of the rotational group [BL81]

$$\int_0^{2\pi} d\phi \int_0^{\pi} \sin\theta d\theta Y^*_{l_1 m_1} Y_{l_2 m_2} Y_{l_3 m_3} = (-1)^{m1}\sqrt{\frac{(2l_1+1)(2l_2+1)(2l_3+1)}{4\pi}} \times \begin{pmatrix} l_1 & l_2 & l_3 \\ 0 & 0 & 0 \end{pmatrix} \begin{pmatrix} l_1 & l_2 & l_3 \\ -m_1 & m_2 & m_3 \end{pmatrix}$$

and inserting $\cos\theta = \sqrt{\frac{4\pi}{3}} Y_{10}$ and explicit expressions for the $3j$ symbols, one finds for the matrix element in question

$$< Y_{lm}|\cos\theta|Y_{l'm'} > = \sqrt{\frac{(l-m)(l+m)}{(2l-1)(2l+1)}}\delta_{mm'}\delta_{l'l-1} + \sqrt{\frac{(l-m+1)(l+m+1)}{(2l+1)(2l+3)}}\delta_{mm'}\delta_{l'l+1} \quad (25)$$

These are straightforward to evaluate and confirm the well-known selection rule concerning the dipole which couples only neighbouring l-channels. Notice also that the axial symmetry, typical for a diatomic, is preserved due to the linear polarization. As confirmed by the explicit expressions, the hamiltonian only contains contributions with $m = m'$. Assuming that the initial condition has a definite value of m, one can limit the expansion of the wavefunction to a summation over l only. The strength of the coupling is, however, dependent upon m.

3 Quantum Time Evolution

In this contribution we will focus on the pseudo-spectral approach to the time-propagation of quantum wavefunctions. This approach provides an explicit numerical solution to the time dependent Schrödinger equation (TDSE):

$$\imath \frac{\partial \Psi}{\partial t} = \hat{H}\Psi = \left[\hat{T} + \hat{V}(t)\right]\Psi \tag{26}$$

With present day computing power, it may be used for systems with up to three active degrees of freedom or coordinates, but there are no restrictions on the potential. Specifically, the hamiltonian may be time-dependent and may contain arbitary pulsed fields. Because of these features, the pseudo-spectral approach is the method of choice for the study of diatomics in external (laser) fields.

The pseudo-spectral approach is well-established and we refer to the literature for an overview of its background [Kos88]. Several methods exist a.o. Second Order Differencing [KK83], Split Operator [FFS82], and Lanczos [QL90] that are applicable to problems with time-dependent hamiltonians. We will use the Split Operator method. The TDSE equation is formally solved by applying the time evolution operator to the initial condition

$$\Psi(t) = U(t, t_0)\Psi(t_0) \tag{27}$$

where U is to be constructed using the Dyson time-ordering operator

$$U(t, t_0) = \mathcal{T} \exp\left(-i \int_{t_0}^{t} ds \hat{H}(s)\right) \tag{28}$$

The time-ordering is required because of the presence of a time dependent potential $\hat{V}(t)$ in the hamiltonian. The key concept in the split-operator method is to subdivide the total time interval into short time slices

$$U(t, t_0) = \prod_{j=1}^{j=N} U(t_j, t_{j-1}) \qquad t - t_0 = N\epsilon, \quad t_j = t_0 + j\epsilon \tag{29}$$

and to approximate the evolution operator over each time slice

$$U(t+\epsilon, t) = \exp\left(-i\frac{\epsilon}{2}\hat{T}\right) \exp\left(-i \int_{t}^{t+\epsilon} ds \hat{V}(s)\right) \exp\left(-i\frac{\epsilon}{2}\hat{T}\right) \tag{30}$$

The power of this algorithm lies in the splitting of the potential $\hat{V}$ and kinetic $\hat{T}$ terms. The action on the wavefunction of the first is evaluated while the wavefunction is in a coordinate representation i.e. defined by its values on a grid in coordinate space. For the second, one transforms the wavefunction to the momentum representation, using the discrete Fourier transform associated with the grid. The kinetic energy operator is multiplicative in that representation, so its action is also easily evaluated. Afterwards one transforms the wavefunction back to the coordinate representation with an inverse Fourier transform. This is a viable approach because of the availability of fast Fourier transform (FFT) software that performs these transformations very efficiently.

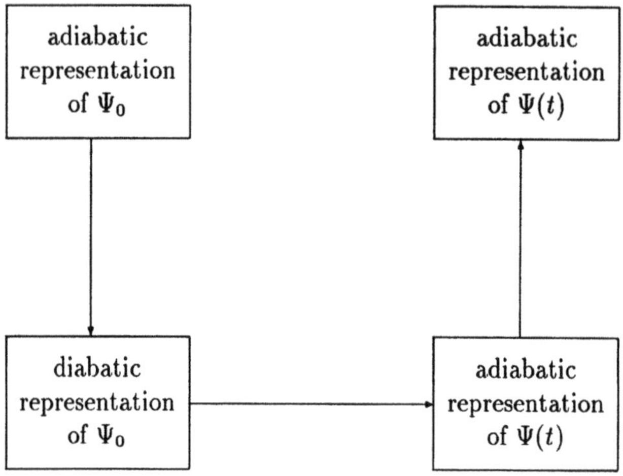

Figure 1: Time-evolution sceme in a multi-PES calculation: adiabatic-to-dibabatic transformation at initial time, propagation in the diabatic basis and diabatic-to-adiabatic transformation at final time.

The decomposition outlined above, including the elimination of the time-ordering operator, is correct up to second order in epsilon provided the potential has a well-behaved time-dependence [FMF76]. Higher order approximations to the propagator are also possible, see for example [TI93]. Their use allows one to extend the time slice ϵ based on the higher order precision. On the other hand, it leads more floating point operations per time propagation step due to the appearance of more than three factors in the expression (formula (30)) for the approximate propagator. This is a trade-off issue, involving cpu-time requirements, accuracy and programming effort. We have selected to use the second order scheme.

When using the time-propagation for problems with coupled electronic PES's, a choice also has to made concerning the representation of the wavefunction. By far the more common and familiar representation is that of the adiabatic electronic basis. It seems natural therefore to use this basis. However, as was indicated in the previous section, this basis is not well-suited for the propagation calculations because of the presence of the dynamical coupling terms in the hamiltonian and the strong variations in crossing regions of the PES. These drawbacks are eliminated in the quasi-diabatic basis. A straightforward approach to this situation is indicated in diagram (2.3). We specify initial conditions in the adiabatic basis and subsequently transform them to the quasi-diabatic basis. Time-propagation calculations are then performed in a quasi-diabatic representation in which we can neglect the dynamical couplings and all potentials have smooth variations. At final time (or possibly also at intermediate times of interest) the wavefunctions are transformed back into their adiabatic representation. The key equations, to be propagated numerically, are given by formulae (22) to (24).

An important aspect of the problem is the multi-component nature of the wavefunction and the corresponding matrix structure of the hamiltonian operator, and as a consequence the time evolution operator also has this structure:

$$\chi_{nlm}(t+\epsilon) = \sum_{nlm,n'l'm'} U_{nlmn'l'm'}(t+\epsilon,t)\chi_{n'l'm'}(t) \tag{31}$$

The evaluation of its elements requires some care because they now involve the exponentiation of matrix operators

$$U_{nlmn'l'm'}(t+\epsilon,t) = \left[\exp\left(-i\frac{\epsilon}{2}\hat{T}\right)\exp\left(-i\int_t^{t+\epsilon} ds\hat{V}(s)\right)\exp\left(-i\frac{\epsilon}{2}\hat{T}\right)\right]_{nlmn'l'm'} \tag{32}$$

The kinetic energy is represented by a diagonal matrix in the (nlm)-vector space (in the quasi-diabatic basis i.e. in the absence of the dynamic coupling terms) and the above simplifies to

$$U_{nlmn'l'm'}(t+\epsilon,t) = \exp\left(-i\frac{\epsilon}{2}\hat{T}\right)\left[\exp\left(-i\int_t^{t+\epsilon} ds\hat{V}(s)\right)\right]_{nlmn'l'm'}\exp\left(-i\frac{\epsilon}{2}\hat{T}\right) \tag{33}$$

Thus we can continue to use transformations to momentum space to evaluate the kinetic energy factors in the propagation formula. The only new feature is that for the potential term one now has to exponentiate a matrix. This can be done analytically in case of two channels or components in the (nlm)-vectorspace [BFLL90] or numerically [ML78] in case more channels are involved.

The scheme outlined above may be summarized by saying that it is a hybrid technique for the quantum time propagation of three-dimensional wavefunctions. For the radial coordinate we use the FFT-based grid-representation and for the angular coordinates we have a basis expansion with the sperical harmonics. This approach make sense (a) provided the potentials involved have a simple dependence on the angular variables and (b) the potential only couples channels over a small range of (lm) values. The first condition guarantees easy evaluation of the potential matrix in the hybrid representation, the second quarantees that the matrices involved do not get to be too large and thus the computational load remains resonable. The second condition also implies that the approach compares favourably with a grid-representation of the angular variables which is an alternative that has been used in the the literature [DM92, QL90]. The systems that we look at in this contribution, viz. diatomic molecules in a dipole field, certainly satisfy the aforementioned conditions.

4 Multiphoton Absorption in Hydrogen Fluoride

The phenomenon of infrared multiphoton excitation (MPE) and dissociation (MPD), is one of the many interesting spectroscopic discoveries of the previous decades. Irradiation of the molecule with an intense laser field generates excitation through discrete levels, and then on into the quasi-continuum and continuum. In the latter regime the transitions are resonant, but in the former much depends on the level spacing in the system. The spacing of the individual discrete levels determines the fraction of molecules to be excited up to the quasi-continuum, and it is therefore key to the efficiency of the overall process.

The above does not apply directly to diatomics [CB90], which possess a discrete spectrum over the entire range of bound states, and where one is taken directly from discrete levels to the continuum. And indeed, calculations have shown that even with extremely intense monochromatic lasers, dissociation probability is generally low in diatomics. This is due to the field frequency becoming completely off-resonance as one ladders through the anharmonically spaced higher levels. Several calculations have shown that the application of two fields [JA91, JB89, DS91] or the use of chirped fields [CBC90], alleviates the off-resonance condition and is capable of producing significant excitation and dissociation rates.

Many studies of infrared MPE in diatomics are based on a description of the molecule, containing only the internuclear vibrational coordinate. In a full description, which includes the angles specifying the orientation of the molecular axis with respect to the field, the transitions can be vibration-rotational with $\Delta j = 0, \pm 1$. Thus the anharmonic shift in the vibrational levels can be compensated by the addition or removal of rotational energy, in order to sustain near-resonance in a sequence of transitions. The reduced, purely vibrational, description eliminates the rotational splitting from the level scheme and in that way conceivably impacts significantly on the MPE process. One of our objectives is to find out whether this is indeed the case. To this end we investigate, in both descriptions, the MPE of Hydrogen Fluoride (HF), a molecule that is often considered as a reference case [WP77, DG82, GM88, LM91].

In our calculations we have used the HF potential V_0 given by Coxon and Hajigeorgiou [CH90]. It is defined by analytical expressions in the regions of small and large R and by tensioned cubic spline fits to experimental data for the inetrmediate region. The dipole function μ was taken from [ZSL+91]. It is linear at small distances, zero for $R > 20au$ and in between it is defined by a tensioned cubic spline fit to ab initio data. As a second objective, we will look into the effect of using model potentials and dipoles i.o. the the ab initio functions. The electric field is taken to be a purely harmonic continuous wave with strength E and frequency Ω. These are the two parameters in our calculation. We consider a value of E corresponding to a field intensity of $43.7TW/cm^2$ as in [WP77]. The values of Ω are taken in the range relevant for multiphoton resonances in the vibrational spectrum i.e. from $0.015au$ to $0.021au$. Throughout the calculations we have used the same initial condition Ψ_0 viz. the HF ground state calculated with the renormalized Numerov method [Joh77]. In the full description, i.e. angular variables included, the rotational quatum numbers for Ψ_0 are $l = 0, m = 0$. As explained in the previous section, only the latter is good quantum number.

4.1 Vibrational Motion

One-photon processes can only induce transtions between levels which are separated approximately by the energy of one photon. Consequently the field frequency must be near resonant with the relevant transition frequencies. In multiphoton processes transitions occur through virtual transitions to intermediate levels. Conservation of energy still requires that the energy difference between initial and final states equal the net absorbed energy from the exciting field i.e. the energy of n-photons. Therefore the relevant field frequencies for transitions from the ground state are the multiphoton frequencies $\Omega_n = (E_n - E_0)/n$, defined as the frequencies at which the n-photon process is in exact resonance with the transition from $v = 0, \ldots, n$. In the HF vibrational spectrum, the vibrational spacing narrows from $0.0183au$ for $E_1 - E - 0$ to $0.0129au$ for $E_8 - E_0$ due to the anharmonicity of yhe potential. As a consequence the multiphoton frequencies decrease as the multiphoton number n increases: from 0.0183 for the one-photon transition frequency to $0.0155au$ for the eight-photon frequency.

We have calculated the time evolution of the occupation probabilities

$$\mathcal{P}_n(t) = | < \Psi(t) | \Phi_n > |^2 \tag{34}$$

of the stationary vibrational levels Φ_n of the molecule, with a field at the 4-photon frequency $\Omega_4 = 0.01695a.u.$. The probabilities display the details of the dynamics. They are shown in figure 2 We have limited ourselves states $n = 0$ through $n = 5$. The multiphoton resonance time evolution is clearly quite complicated. It reveals the basic field frequency and a modulation by beat signals with periods of the order of multiphoton periods. Figure 2 shows that in the initial phase of populating the vibrational levels, there is an increasing delay for the successive excitation of the higher levels. This is consistent with the picture of a sequential process of exciation through intermediate levels. However, after this short initial phase, the wave packet is distributed in a complex way over the various levels.

It is interesting to investigate the time averaged value of the excitation energy. We define

$$\bar{\mathcal{E}} = \frac{1}{T} \int_0^T \left[< \Psi(t) \mid \hat{H}_{mol} \mid \Psi(t) > - E_0 \right] dt. \tag{35}$$

which, in view of the fact that the initial condition has $E = E_0$, represents the energy absorbed by the molecule. As a function of the interval T, the average tends

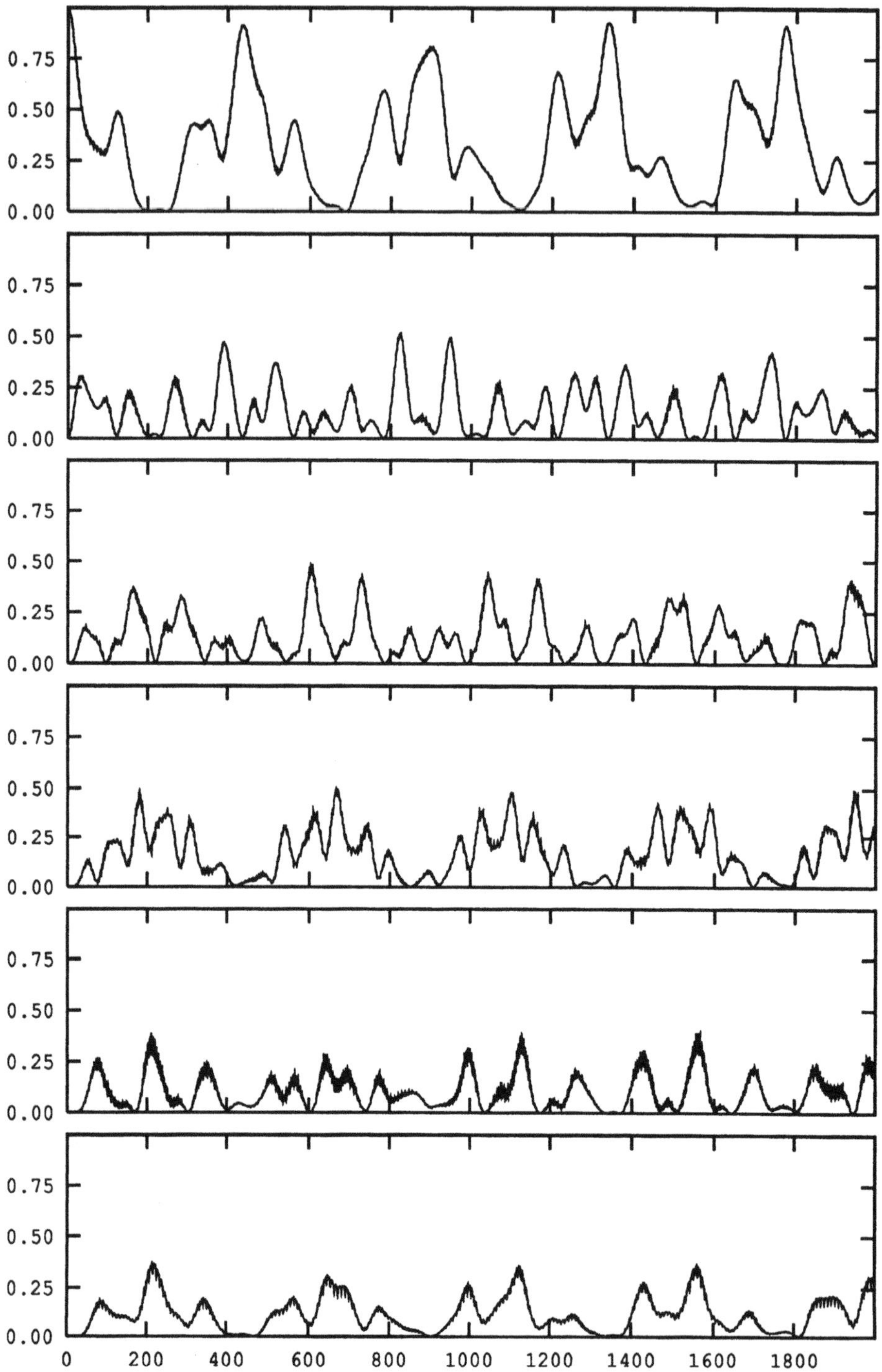

Figure 2: Time evolution of the occupation probabilities at the 4-photon frequency 0.01695 a.u. The vibrational quantumnumber n goes from 0 to 5 from top to bottom. Time in fs.

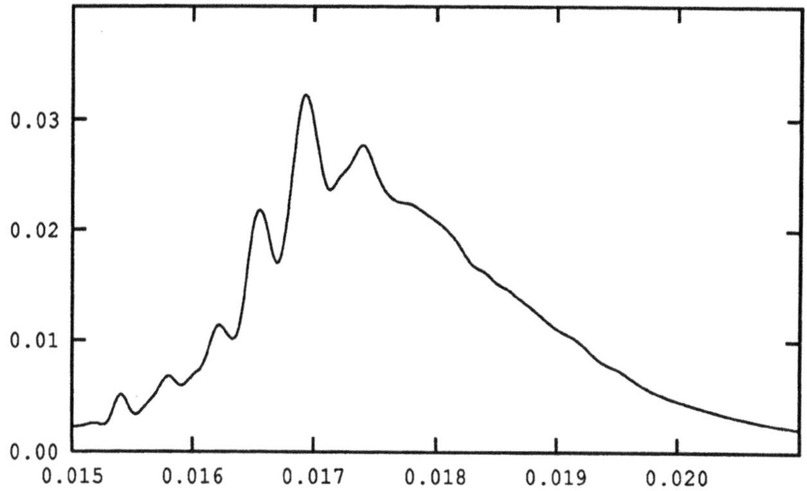

Figure 3: Time averaged energy versus frequency of the driving field at intensity of $43.7TW/cm^2$. Energy and frequency in a.u.

to vary strongly in the first few hundred femto-second due to the transient effects of effectively switching on the the field at time zero. After that, it stabilizes at an asymptotic value which we will refer to as the average. It is in this sense that define the average absorbed energy. In figure 3 we present the absorption spectrum, i.e. the aborbed energy as a function of the frequency Ω of the driving field. It shows a broad absorption region with superimposed peaks at $\Omega_3 = 0.01735$, $\Omega_4 = 0.01695$, $\Omega_5 = 0.01657$, $\Omega_6 = 0.01620$, $\Omega_7 = 0.01583$ and $\Omega_8 = 0.01546$, the multiphoton resonance frequencies. In figure 4 one finds the absorption spectrum, obtained with a field of intensity $10.9TW/cm^2$ intensity, compared to $43.7TW/cm^2$ previously. At the lower intensity, the 3-photon resonance is dominant and whereas at the higher intensity the 4-photon resonance dominates. This demonstrates the well-known intensity dependence: the maximum of absorption shifts to lower frequencies with increasing intensity. It indicates that transitions with higher multiphoton number, i.e. to levels higher in the spectrum, become more important.

In figure 5 we compare the energy absorption profile (as a function of field frequency) with the same quantity obtained using the Morse potential taken from refernces ([WP77, GM88]) i.o. the one fitted to experimatal data. The figure shows that the profiles differ mainly by a global shift. This shift is caused by a corresponding change in the vibrational spacing associated with the respective potentials. The other changes are small and can be attributed to differences in anharmonicities and transition frequencies of the two potentials. Similar conclusions can be drawn from

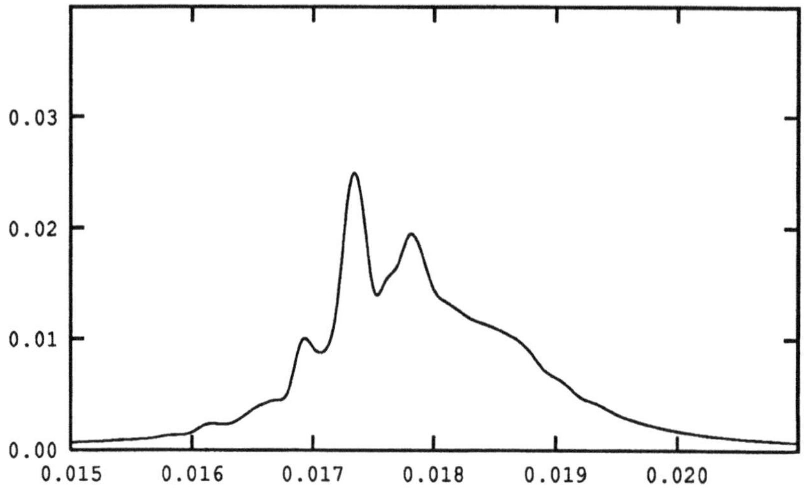

Figure 4: Time averaged energy versus frequency of the driving field at intensity of $10.9TW/cm^2$. Energy and frequency in a.u.

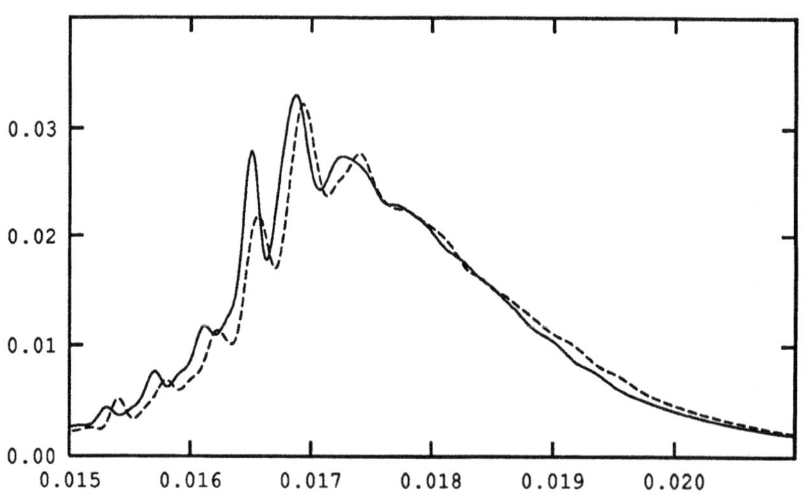

Figure 5: Time averaged energy versus driving field frequency. The electronic potential is approximated by a Morse well (full line) and a spectroscopically determined curve (dashed line). Intensity is $43.7TW/cm^2$. Energy and frequency in a.u.

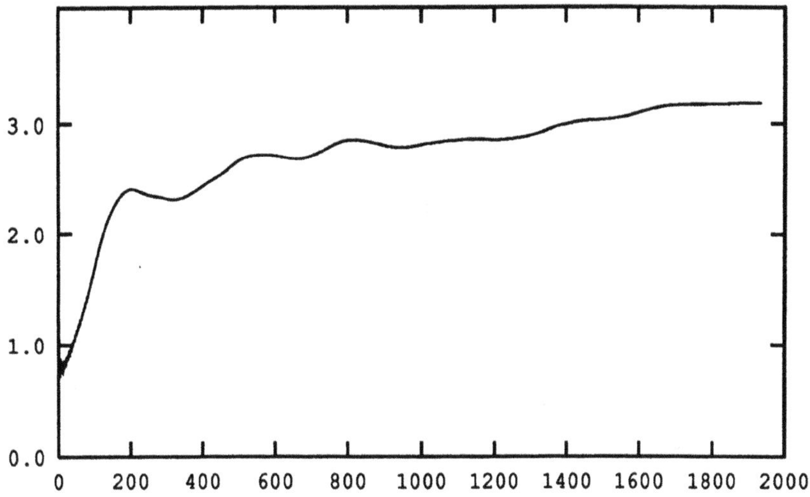

Figure 6: Time averaged $< l^2 >$ versus averaging interval. $\Omega = 0.01695 a.u.$ Intensity $= 43.7 TW/cm^2$. Time in fs.

comparing with a calculation using an expontial dipole function and a linear dipole function. These approximations reproduces the qualitative aspects of the absorption profile, but differs significantly in the quantitative features. Details of this analysis may be found in [BFL92].

4.2 Rovibrational Motion

Here we will focus on the angular momentum features of the rovibrational calculation. In view of the strength of the external field, a significant occupation probability for a number of l-channels can be expected a priori. As a first step we need to determine which l-channels are to be included in the calculation. To this end we have performed a calculation with channels up to $l = 7$ and found that starting with $l = 4$ the maximum probability per channel is less than 5% at any point in time and lower than 2% on the average. At $l = 7$, those numbers have dropped by an order of magnitude. Thus we have used $l = 3$ as a cut-off in all subsequent calculations. The predominance of the lower angular momentum components is illustrated in figure 6. It displays the time-averaged expectation value of the l^2 operator. This quantity tends towards a numerical value of approximately 3.0 indicating a limited presence at $l = 2$ (where $l(l+1) = 6$) and higher.

Again the relevant field frequencies for transitions from the ground state are the multiphoton frequencies $\Omega_n = (E_{v=n,l=0} - E_{v=0,l=0})/n$, i.e. the frequencies at which the n-photon process is in resonance with the transition from $v = 0$ to $v = n$ vibrational states. Figure 7 shows the vibrational occupation probabilities

$$P_v = \sum_l P_{vl} = \sum_l |< \Psi(t) \mid \Phi_{vl} >|^2 \quad (36)$$

at the four-photon frequency $\Omega_4 = 0.01695au$. The rovibrational stationary states are determined using the renormalized Numerov method. One clearly notices a significant amount of probability at $v = 4$ during the time intervals that $v = 0$ is depleted. The presence of non-negligible probability in other vibrational channels is consistent with the picture of a higher-order transition process that proceeds in a sequential manner [NM92]. The time-averaged absorbed energy as a function of the frequency Ω of the field is shown in figure 8 and should be compared to figure 3 which represents the same quantity calculated in the purely vibrational approach. In the latter one can distinguish the multiphoton transitions from $v = 3$ to $v = 8$. In the comparison of the figures two things stand out. There is an overall decrease in the absorbed energy and a more pronounced structure of the dominant peaks. We attribute the first effect to a stronger effective field strength in the purely vibrational calculation. Indeed, in the rovibrational calculation, the molecule will experience different orientations with respect to the field, at which there is lower effective interaction strength than in case it is alligned with the field. Thus a lowering of the absorbed energy is to be expected. It is not clear how to define field strengths that are comparable in this situation. Secondly, he center of gravity of the groups of transition frequencies defined by $\Omega_{vl} = (E_{vl} - E_{v=0,l=0})/n$ correspond to the four- and five-photon peaks in figure 3. Also the the peaks in the profile are sharper than in the nonrotating case. The figure clearly suggests that an in-depth analysis of the energy absorebd in each of the angular momentum channels is required.

5 Laser Assisted Dynamics in Boron Hydride

Boron Hydride has been studied extensively, both spectroscopically and theoretically. From each of the lowest three $^1\Sigma^+$ states X, B, C, at least four vibrational bands have been observed spectroscopically [Bau] and one for the fourth $^1\Sigma^+$ state $E^1\Sigma^+$. Several ab initio calculations [Bro] show that the $^1\Sigma^+$ state B has a double minimum in its potential energy curve. This is due to a Rydberg-ionic crossing, which also affects the shapes of the other curves. Experimentally however, the vibrational states which should be accomodated in the outer minimum have not yet been observed. Our objective is to look for a method to demonstrate the existence of the outer minimum. The idea is to populate significantly its vibrational levels, so that transitions from these levels could be observed spectroscopically. The associated transition frequencies differ enough from those of the other bands for them to be distinguishable.

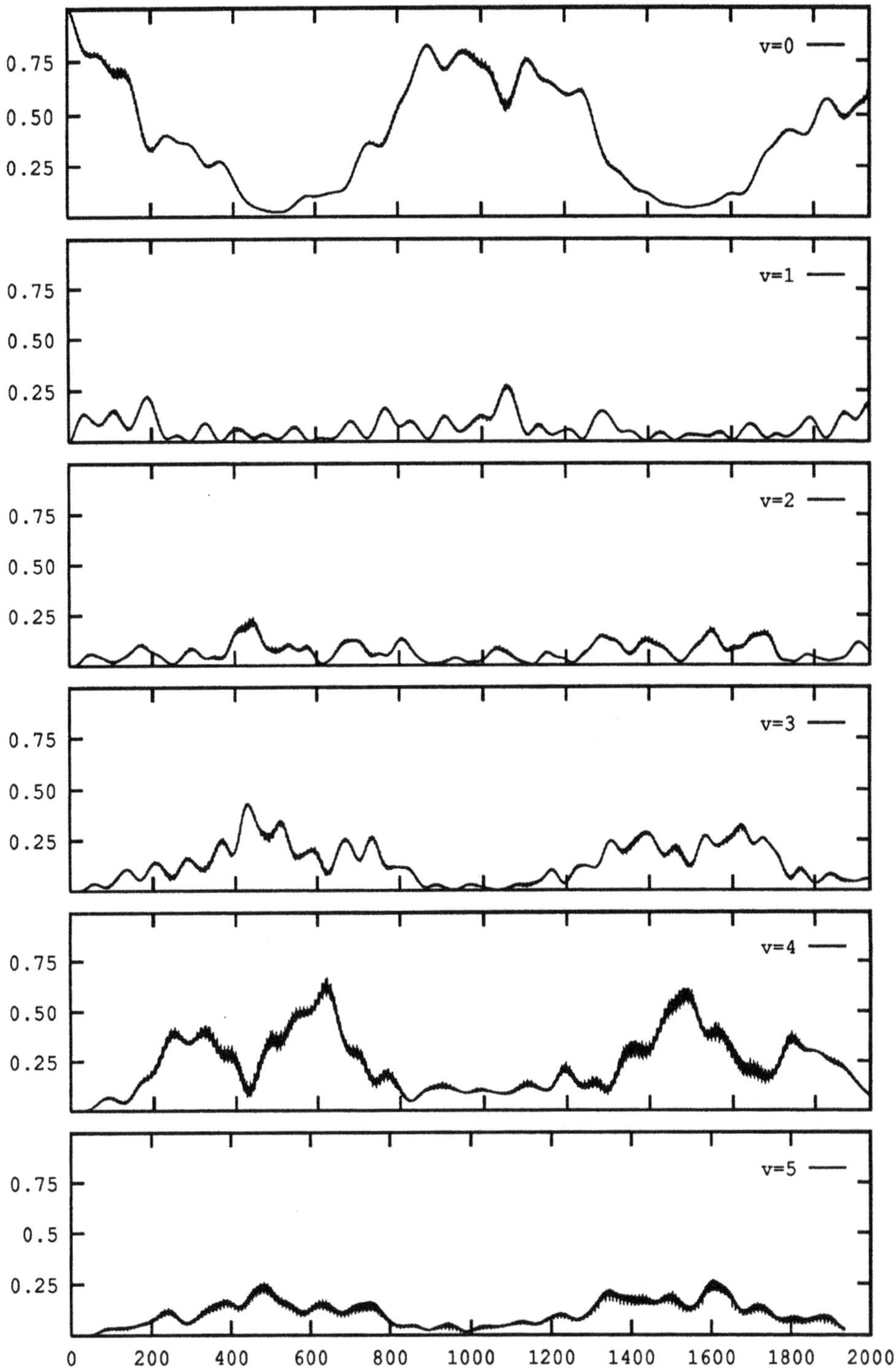

Figure 7: Vibrational occupation probabilities. $\Omega = 0.01695 a.u.$ Intensity $= 43.7 TW/cm^2$. Time in fs.

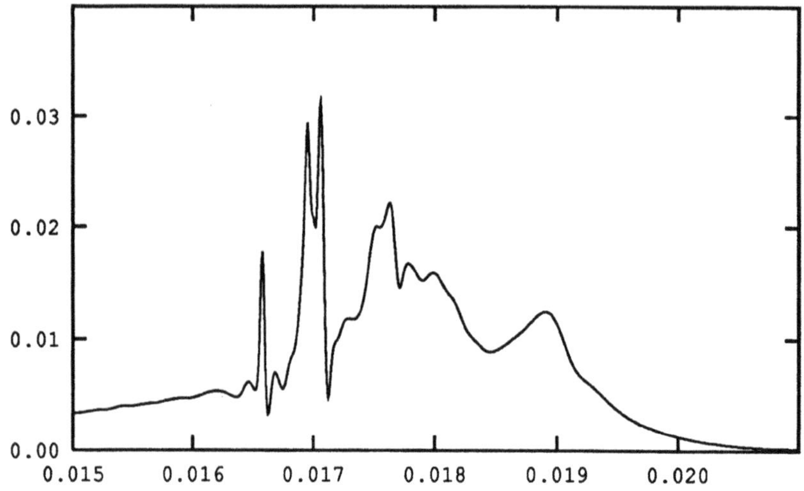

Figure 8: Time averaged energy versus frequency of the driving field at intensity of $43.7TW/cm^2$ for the full description. Energy and frequency in a.u.

The ab initio potentials, coupling functions and dipole moments used in this section, are described in [Cim92]. The ground state of BH has a predominantly $[B(^2S, 3s)H(^2S, 1s$ character at all distances with some mixing with $[B(^2D, 3s)H(^2S, 1s)]$ at short distances. As can be seen in figure 9, the first excited state appears to have a double minimum. This is due to Rydberg-ionic crossing with the B^+H^- state. It dissociates into $B(^2S, 3s)$ and $H(^2S)$ but acquires a strong ionic character in the outermost well. Around the top of the barier between the two wells, it regains its $B(^2S, 3s)$ character. The second excited state exhibits a strictly avoided crossing around 17.7 a.u. and a very shallow double minimum. It dissociates into $B(^2D, 2s2p^2)$ and $H(^2S)$. In the outer minimum its character is a mixture of B^+H^- and $B(^2P, 3p)$ while in the interior region it acquires the same character as in the dissociation region. The third excited state undergoes two avoided crossings, a broad one around 19 a.u. with the ionic state and and a narrow one around 17.7 a.u. with the $B(^2D, 2s2p^2)$ state. The $2p^2$ is dominant down to $R \approx 4.5a.u.$ where it changes again to $B(^2P, 3p)$.

As explained before, our time propagation is computed in the quasi-diabatic basis. The first step in the construction of this basis is the calculation of a zeroth order adiabatic basis. This adiabatic basis was obtained by a SCF-CI calculations at some twenty internuclear distances, starting from 1.2 a.u. A gaussian atomic basis set was used, similar to that of Jaszunski et al. [JRW81]. Some diffuse s and p functions were added to correctly represent higher Rydberg terms of the B atom and the H^- anion. Electron correlation was treated by the CIPSI multireference perturbation

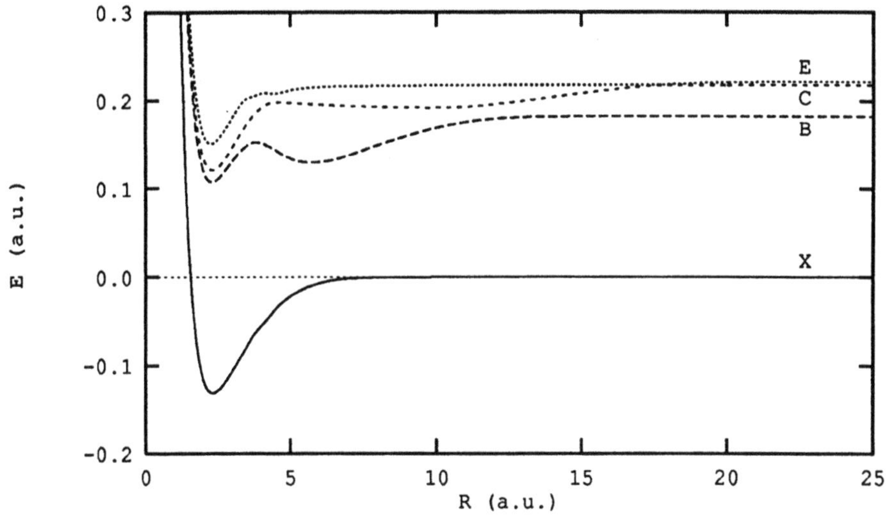

Figure 9: Adiabatic potentials of $X^1\Sigma^+$, the electronic ground state and $B^1\Sigma^+, C^1\Sigma^+$ and $E^1\Sigma^+$, the three lowest excited singlet states, from ab initio calculations [CMPS85, Cim92]. The $B^1\Sigma^+$ state has a double minimum due to a Rydberg-ionic crossing which also affects the potentials of the other states.

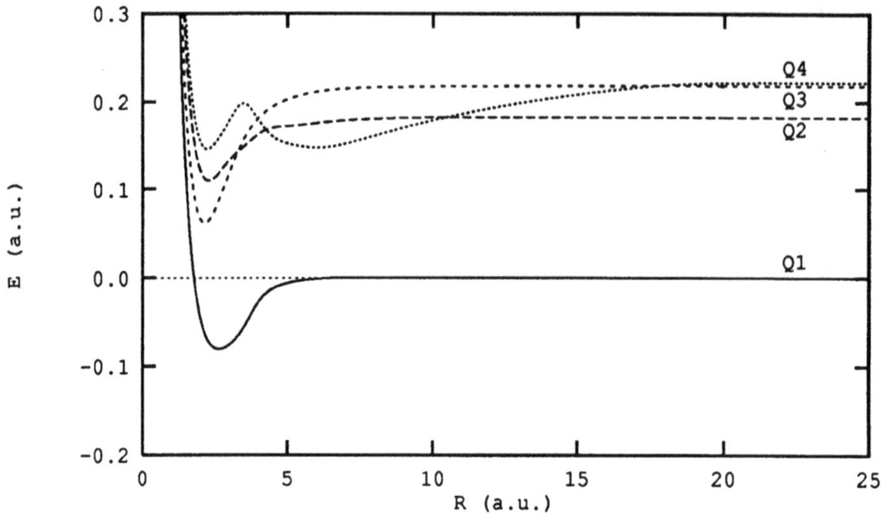

Figure 10: Quasi-diabatic potentials for the first four $^1\Sigma^+$ states of BH, obtained in [CMPS85, Cim92] using a maximum overlap approach.

algorithm [HMR73, CP87]. Spline interpolation was used to obtain the calculated functions at intermediate values of R. The zeroth order quasi-diabatic basis is obtained by unitary transformation. The transformation is such that it maximizes the overlap of the zeroth order quasi-diabatic functions with a set of independent reference functions. Quasi Degenerate Perturbation Theory is then applied to obtain the final quasi-diabatic basis set. For a more detailed description we refer to [CMPS85, Cim92]. The quasi-diabtic potentials and couplings are shown on figures (10) and (11) respectively. In what follows, we will refer to quasi-diabatic quantities using a subscripts $1, 2, 3, 4$ while we denote adiabatic quantities with a subscripts X, B, C, E. In comparing with the adiabatic potentials of figure (9), on notices rather pronounced differences. The quasi-diabatic ground-state minimum is less deep and less tight than the corresponding adiabatic one. The order of the second and third state is interchanged in the two representations. The large coupling elements (figure 11), in paticular $U - 13$, are consistent with this. The double minimum pertains to reference state 4 which is the combination of the ionic state B^+H^- and the $B(^2P, 3p)$ state. It has a 3p character in the inner minimum as well as in the dissociation region while in the second minimum it neatly fits the ionic state. The dipole moments μ_{ii} and the dipole transition matrix elements μ_{ij} that we use are also ab initio results [CMPS85, Cim92]. The influence of the ionic contribution in the electronic wavefunction is also present in these quantities. In the regions where the electronic states

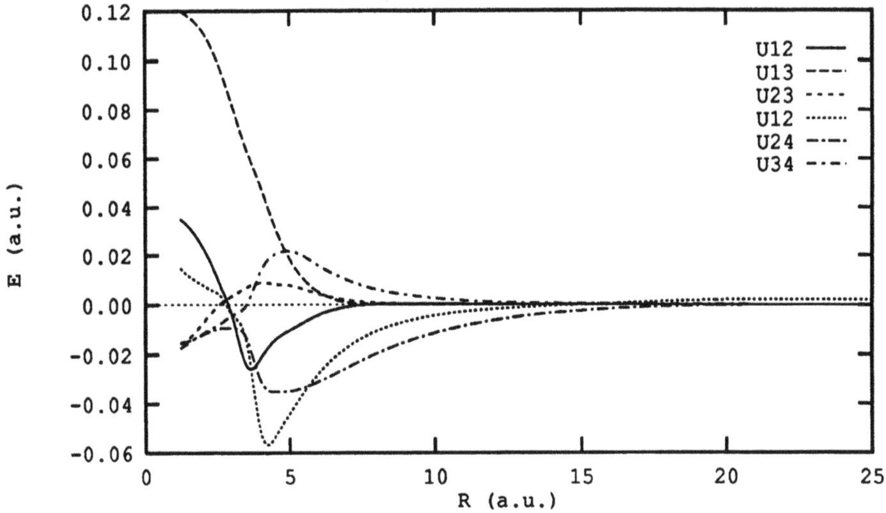

Figure 11: Off-diagonal matrix elements U_{ij}, in the quasi-diabatic basis of [CMPS85, Cim92].

are characterised as ionic, the dipole moment is large and varies almost linearly with distance.

5.1 The Excitation Pathway

A priori on can consider several possibilities for populating the second minimum of B by a combination of electromagnetic and nonadiabatic interactions. It is fairly obvious that one cannot directly excite a vibrational state of the outer minimum. The Franck-Condon factors are too small: there is no appreciable overlap between the respective wavefunctions due to the difference in position of the minima ($R^1_{inner} \approx 2.3$ a.u., $R^2_{outer} \approx 5.8$ a.u.). Only the vibrational states of $B^1\Sigma^+$ which are significantly localised in the inner well, can be excited directly from the ground state. To create population in both wells, we will have to consider highly excited vibrational states of B. The oscillator strengths [Per] of the corresponding transitions show, however, that these states are also not accessible by direct electromagnetic excitation from the vibronic ground state ($f \approx 10^{-10}$). In our time-dependent calculations this manifests itself in the absence of any appreciable population in the respective vibrational state even though we chose the correct transition frequency and a rather high intensity.

An alternative pathway of occupying B is through excitation to other electronic states and subsequent nonadiabatic population transfer. From our calculations we see that excitations to vibrational states of C are effective in populating B appreciably, starting from $v = 4$. Transitions to these states are, however, again prohibited due to the smallness of oscillator strengths [Per]. These strengths vary in magnitude from 10^{-9} to 10^{-5}.

The only remaining possibility is an electronic excitation to the E state. Oscillator strenghts leave us no choice but to make a transition to its vibrational ground state ($f \approx 10^{-2}$ for this transition). The process we propose consists of three steps: the first one is an electronic excitation by an UV pulse from the X ground state to the E electronically excited state. This is followed by a nonadiabatic interelectronic population transfer to the B state. The final step is a deexcitation of the B vibrational states by an IR pulse.

For the EM-field we choose gaussian harmonic pulses. The field parameters such as frequency and pulse duration are chosen on the basis of simple physical arguments concerning frequency matching. We do not use techniques to optimise the field characteristics as e.g. developed by Tannor and Rice [TKR86, KRG+89] and also by Rabitz and coworkers [GNR92, SR92]. As before, we will first perform the calculations in a purely vibronic picture, omitting rotational degrees of freedom. Next we look at the changes when going to the rovibronic picture i.e. the complete description of the system.

5.2 The Vibronic Picture

Let us first consider the effect of the UV pulse. The initial condition we is ψ_{X0} i.e. the electronic and vibrational ground state the wave packet (computed with the Renormalized Numerov method referred to before). The harmonic frequency of the field pulse $\omega = 0.28a.u.$. The Gaussian envelope has a Full Width Half Maximum (FWHM) of $50fs$, a peak intensity of $0.4GW/cm^2$ and is centered at $t = 100fs$. This guarantees a smooth switching on of the field. Figure 12 shows the occupation probabilities of the adiabatic electronic states as a function of time. Once the field reaches an appreciable strength, its starts exciting towards the E state. $P_E(t)$ is initially zero and grows to its maximum value during the time the field is active. During that time the nonadiabatic coupling transfers probablity to the B state. The order of magnitude of $P_B(\approx 10^{-4})$ is sufficient for spectroscopic purposes. However, at some point in time the same coupling works in the other direction to deplete the population of the B in favour of the E. We need to find a way to trap the probality in the B state, and more specifically in the vibrational levels in the outer minimum of that state.

If we look at figure 13, we see that the probability for the B electronic state to be to the right of the barrier is proportional to the total probability to be in that electronic

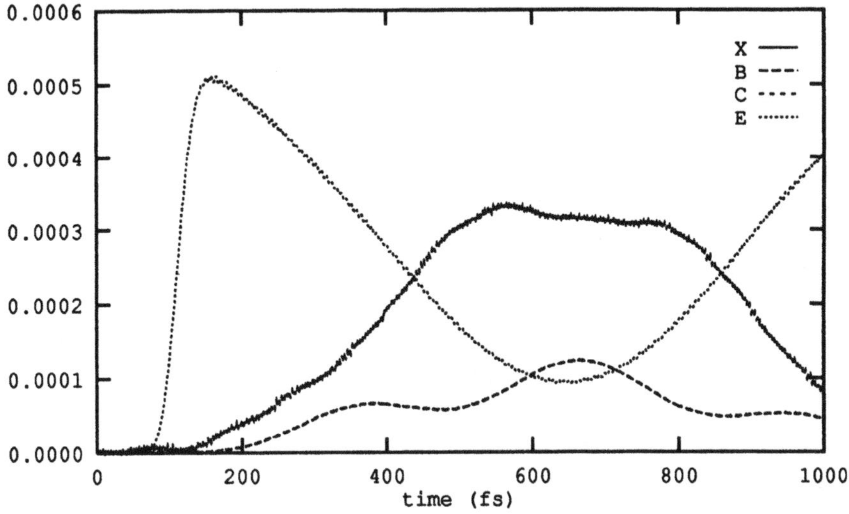

Figure 12: Occupation probabilities of electronic states as a function of time. The system is excited from ψ_{X0} (the vibrational ground state of the electronic ground state) by a Gaussian UV pulse. The FWHM of the pulse is $50fs$ and the maximum intensity is $0.4GW/cm^2$.

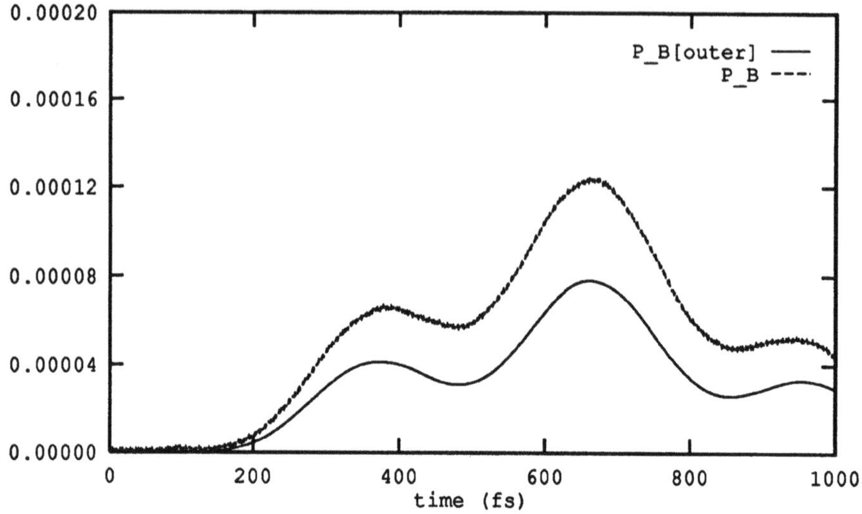

Figure 13: Probability to be in the outer minimum for the $B^1\Sigma^+$ state and total probability to be in that electronic state, as a function of time. The molecule is excited from ψ_{X0} using a Gaussian UV pulse.

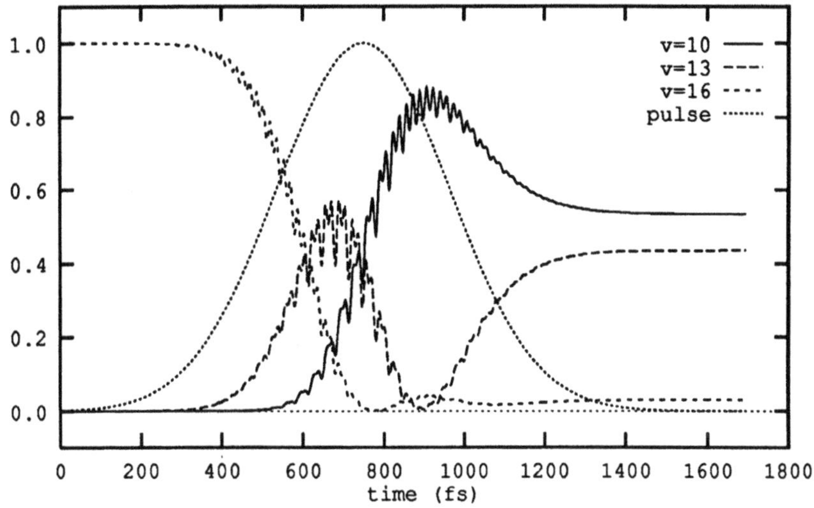

Figure 14: Time evolution of the occupation probabilities of vibrational states of $B^1\Sigma^+$ using a gaussian IR pulse with a FWHM of $500fs$. Initial condition is ψ_{B16}.

state. This can only be explained if ψ_B is not a wave packet moving back and forth between the inner and outer wells, or even in the outer well, but is almost a pure vibrational eigenstate. By projecting ψ_B on the adiabatic vibrational states of B, we indeed learn that $\psi_B \approx \psi_{B16}$. The populations of the other vibrational states are at least 2 orders of magnitude smaller. The ψ_{B16} state, however, which we thus succeed to populate through an UV pulse, is not localised in one of the two wells exclusively. It is our goal to trap some probability in states which are located in the outer well exclusively. ψ_{B10} is the highest of such states. This suggests the vibrational deexcitation from the state ψ_{B16} to the ψ_{B10} through an applied IR field.

To investigate the deexcitation, we sidestep for a moment and perform a single-electronic state calculation, with an initial condition of ψ_{B16}. Through selection of the IR field parameters, we want to maximize the deexcitation probability to the vibrational states located in the outer well. The key parameter is of course the frequency of the field. Scanning the frequency domain, we found that the optimal frequency for our purpose equals $\approx 0.0046au$. This corresponds to a 2-photon transition from ψ_{B16} to ψ_{B10}. The efficiency of this transition can be understood by the observation that the intermediate level ψ_{B13} lies halfway between the two and thus serves as an efficient doorway state. Figure 14 shows the evolution of the occupation probabilities of ψ_{B16}, ψ_{B13} and ψ_{B10} under the application of a gaussian laser pulse with a FWHM of $500fs$, a peak intensity of $40GW/cm^2$, centered at $t = 750fs$. The

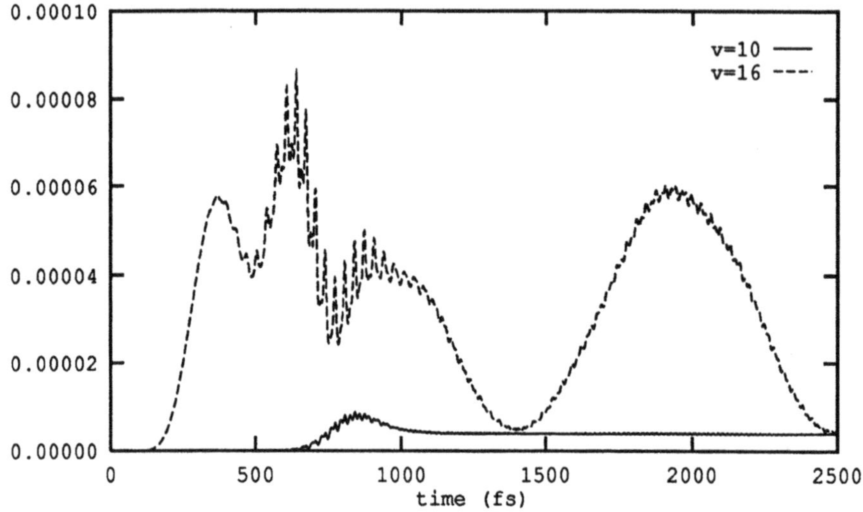

Figure 15: Time evolution of the occupation probabilities of $B^1\Sigma^+$ vibrational states using the UV and IR pulses. Initial condition is $\psi = \psi_{X0}$.

harmonic frequency equals $\omega = (E_{16} - E_{10})/2 = 0.004526a.u.$. The other vibrational levels only contribute negligibly to this process of vibrational deexcitation. Once the pulse reaches a sufficient strength to introduce transitions, the ψ_{B16} is depopulated in favour of the intermediate level ψ_{B13}. Then, in turn, this state is depleted in favour of ψ_{B10}. It is interesting to note that this all happens at some $150fs$ after the IR pulse reaches its maximum intensity. After that, the occupation of ψ_{B10} decreases somewhat again, mostly in favour of the intermediate level. The final probability to be in anyone of the three levels nearly equals one, which confirms that tghe other levels play a minor part in this process.

We now turn to the full calculation in which both the UV pulse, exciting the packet to the $(E^1\Sigma^+, v = 0)$ state and the IR pulse, deexciting from the $B^1\Sigma^+$ ψ_{B16} state, are included. The initial condition is $\psi = \psi_{X0}$. The UV pulse has the characteristics mentioned previously. The IR deexcitation pulse ($\omega = 0.0045a.u.$) has a FWHM of $400fs$ and is centered at $t = 750fs$. The delay between the pulses is chosen such that the second pulse reaches its full strength the moment the occupation of ψ_{B16} is maximum. From figure 15 we see that we succeed in populating ψ_{B10} to $\approx 10^{-5}$. Moreover, we can see that after the IR pulse has vanished, the molecule remains trapped in ψ_{B10}, while the probability of ψ_{B16} starts to oscillate again due to the nonadiabatic interactions with the other electronic states. the probality associated with ψ_{B10} is 4% of the total probabilty of the outer minimum on the average. In figure

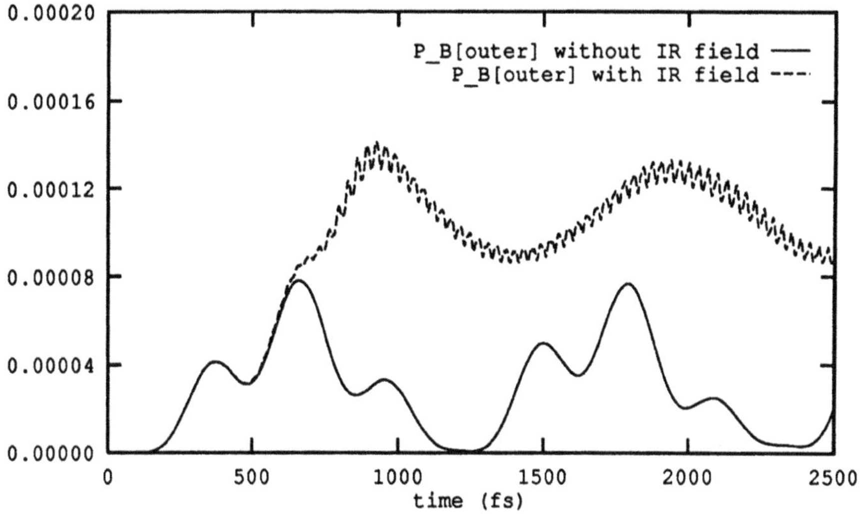

Figure 16: Time evolution of the occupation probabilities of the outer minimum using a gaussian UV (excitation) pulse with and without a gaussian IR pulse (deexcitation).

(16) we compare in detail the population of the outer well with and without the IR pulse included in the excitation scheme. We see that IR pulse leads to an increase by about a factor of two, reaching a time averaged value of 1.2×10^{-4}. From these two facts we conclude that it should be possible to observe the vibrational states of the second minimum experimentally.

5.3 The Rovibronic Picture

We will now consider the effect of including of the rotational degrees of freedom upon the excitation pathway described above. Since the rotational invariance of the hamiltonian is destroyed only during the presence of the electromagnetic field, only the UV excitation and the IR deexcitation step of the excitation scheme will be affected by the inclusion of the rotational degrees of freedom. The nonadiabatic coupling is not affected by the rotation. It couples only channels with equal l-values and thus merely propagates any l-components introduced by the fields. To determine whether the outer minimum of B will still be populated in the full rovibronic picture, it will therefore suffice to study the rotational effects associated with the UV excitation and IR deexcitation. In these calculations we have of course used the hybrid representation, outlined in a previous section, and included angular momentum components up

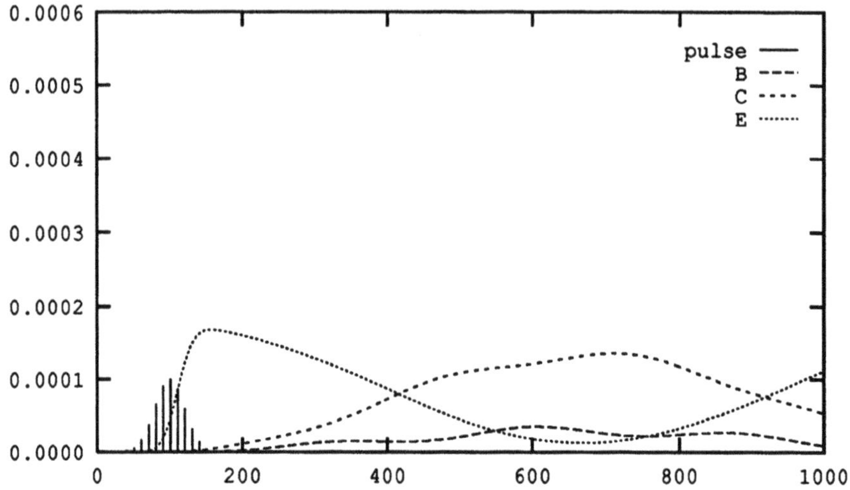

Figure 17: Occupation probabilities of electronic states in the $l = 1$ channel as a function of time. The system is excited from ψ_{X00} by a Gaussian UV pulse.

to $l = 7$. Inspection of the individual l-components of the wavefunctions indicated this to be quite sufficient as cut-off value.

Starting from the rovibronic ground state, i.e. the $X^1\Sigma^+(n = 0, l = 0)$ state ψ_{X00}, we propagate the system with the same UV pulsed field used previously. Figure (17) shows the associated electronic occupation probabilities for $l = 1$. This result should be compared to figure 12. It is clear that, apart from a scaling factor, the evolving probalities are almost identical with the vibronic results. We attribute the downscaling to the lower effective interaction strength of the dipole field, as compared to the one-dimensional vibronic model. All the other rotational channels have a negligeable occupation probability for the excited electronic states. This excitation step is dominantly one-photon transition and changes the $l = 0$ initial state to an $l = 1$ state. The figure also confirms, in the time frame after the pulse has disappeared, that the nonadiabatic electronic coupling propagates the $l = 1$ component without altering the rotational population.

For the deexcitation step, we propagate the $B^1\Sigma^+(n = 16, l = 1)$ rovibronic state, in the presence of the IR field (same parameters as before). Figure 18 displays the time evolution of the vibrational level occupation probabilities i.e. summed over all angular moemntum components. These should be compared with figure (14). The vibrational deexcitation to the $n = 10$ state appears to be even more effective in

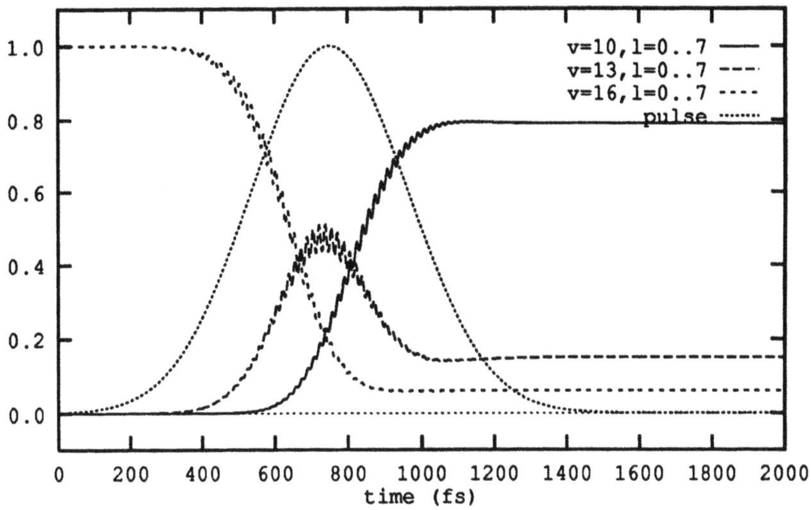

Figure 18: Time evolution of the occupation probabilities of vibrational states (summed over all l) of $B^1\Sigma^+$ using an IR field. Initial condition is $\psi_{B16,1}$ state.

the rovibronic picture. The higher density of states, due to the unfolding of the rotational bands, create more possibilties for near-resonant transitions in the IR field and enhances the deexcitation. Thus the full, rovibronic calculation also leads to the conclusion that the excitation pathway that we have put forward, provides a sensible scheme for achieving our stated objective.

6 Conclusions

In this contribution we have considered the quantum dynamics of diatoms in (dipole) laser fields. The propagation scheme we have developed is based on a hybrid representation of the wavefunction. It has an FFT-grid representation for the internuclear coordinate and an expansion in spherical harmonics for the angular coordinates. The usefulness of the representation derives from the fact that the dipole field is a rank one rotation tensor. It has explicitly known matrix elements in the spherical harmonics basis and couples only neighboring l values. A small number of l-components suffice to approximate the wavefunction. In a calculation involving multiple electronic potentials, the scheme is augmented by a transformation to a quasi-diabatic electronic basis. This simplifies the electronic couplings and smoothes the peaked potentials, thus reducing the required density for the internuclear grid.

We have demonstrated the viability of our approach with two applications. The infrared multiphoton excitation of Hydrogen Fluoride is calculated using a potential fitted to experimental data. We have computed the energy absorption profile, with and without the rotational degrees of freedom. Comparison reveals significant quantitative differences. In Boron Hydride we have investigated the dynamics of excitation to the outer minimum of the $B^1\Sigma^+$ from the $X^1\Sigma^+$ ground state, using ab initio potentials and dipoles. The exciation path involves a pulsed two-colour laser and the electronic non-adiabatic coupling. An UV laser pulse excites from X to E, non-adiabatic coupling transfers from E to B and an IR pulse vibrationally deexcites in B, thus trapping wavepacket probability in the outer minimum of the B state.

References

[Bau] Bauer, S., Herzberg, G. and Johns W. C. (1964) J. Mol. Spectrosc. **13**, 256; Johns, W. C., Grimm, G. A. and Porter, R. F., (1967) J. Mol. Spectrosc. **22**, 435 ; Pianalto, F.S., O'Brien, L. C., Keller P. C. and Bernath, P. F., (1988) J. Mol. Spectrosc. **129**, 248 .

[BFL92] Broeckhove, J., Feyen, B. and Van Leuven., P. (1992) J. Mol. Spectrosc. **261**, 265–276.

[BFLL90] Broeckhove, J., Feyen, B., Lathouwers, L. and Van Leuven., P. (1990) Chem. Phys. Lett. **174**, 504.

[BL81] Biedenharn, L. C., and Louck, J. D. (1982), Angular Momentum in Quantum Mechanics. Addison-Wesley Publishing Company.

[Bro] Browne, J. C. and Greenawalt, E. M. (1970) Chem. Phys. Lett. **7**, 363; Pearson, P. K., Bernder C. F. and Shaefer, H. F. (1971) J. Chem. Phys. **55**, 5235; Houlden, S. A. and Csizmadia, I. G. (1974) Theor. Chim. Acta **35**, 173; Meyer, W. and Rosmus, P. (1975) J. Chem. Phys. **63**, 2356; Jaszunski, M., Roos B. O. and Widmark, P.-O. (1981) J. Chem. Phys. **75**, 306; Botschwina, P. (1986) Chem. Phys. Lett. **129**, 279.

[CB90] Chelkowski, S. and Bandrauk, A. D. (1990) Phys. Rev., **A41**, 6480.

[CBC90] Chelkowski, S., Bandrauk, A. D. and Corkum, P. B. (1990) Phys. Rev. Lett., **65**, 2355.

[CH90] Coxon, J. A. and Hajigeorgiou, P. G. (1990) J. Mol. Spectrosc., **142**, 254.

[Cim92] Cimiraglia, R., in J. Broeckhove and L. Lathouwers, (eds), Time-Dependent Quantum Molecular Dynamics, volume **299** of Nato ASI series B: Physics, pp 11–26. Plenum Publishing, 1992.

[CMPS85] Cimiraglia, R., Malrieu, J.-P., Persico, M. and Spiegelmann, F (1985) J. Phys. B: At. Mol. Phys., **18**, 3073.

[CP87] Cimiraglia, R. and Persico, M. (1987) J. Comput. Chem., **39**, 39.

[DDLO94] Deumens, E., Diz, A., Longo, R. and Ohrn, Y. (1994) Rev. Mod. Phys., in press..

[DG82] Dardi, P. S. and Gray, S. K. (1982) J. Chem. Phys., **77**, 1345.

[DM92] Dateo, C. E. and Metiu, H. (1992) J. Chem. Phys., **95**, 7392–7400.

[DS91] Dibble, B. G. and Shirts, R. B. (1991) J. Chem. Phys., **94**, 3451.

[FFS82] Feit, M. D., Fleck and J. A., Steiger, A. (1982) J. Comput. Phys., **47**, 412.

[FMF76] Fleck, J. A., Morris, J. R. and Feit, M. D. (1976) Appl. Phys., **10**, 129.

[GM88] Goggin, M. E. and Milonni P. W. (1988) Phys. Rev., **A37**, 796.

[GNR92] Gross, P., Neuhauser D. and Rabitz, H. (1992) J. Chem. Phys., **96**, 2834.

[HMR73] Huron, B., Malrieu, J.-P. and Rancurel, P. (1973) J. Chem. Phys., **58**, 5745.

[JA91] Jolicard G. and Austin, E. (1991) J. Chem. Phys., **95**, 5056.

[JB89] Jolicard, G. and Billing, G. D. (1989) J. Chem. Phys., **90**, 346.

[Joh77] Johnson, B. R. (1977) J. Chem. Phys., **67**, 4086.

[JRW81] Jaszunski, M., Roos, B. O., Widmark P.-O. (1981) J. Chem. Phys., **75**, 306.

[KK83] Kosloff, D. and Kosloff, R. (1983) J. Comput. Phys., **52**, 35.

[Kos88] Kosloff, R. (1988) J. Phys. Chem., **92**,2087.

[KRG+89] Kosloff, R., Rice, S. A., Gaspard, P., Tersigni, S. and Tannor, D. J. (1989) Chem. Phys., **139**, 201.

[LJD94] Latinne, O., Joachain, C. J. and Dorr, M. (1994) Europhysics Letters, **26**, 333.

[LM91] Lin, F. J. and Muckerman, J. T. (1991) Comp. Phys. Comm., **63**, 538.

[Mit93] Mittleman, M. H. Theory of Laser-Atom Interactions. Plenum Press, New York and London, 1993.

[ML78] Moler, C. and Van Loan C. (1978) SIAM Review, **20**, 801.

[MT82] Mead, C. A. and Truhlar, D. G. (1982) J. Chem. Phys., **77**, 6090.

[NM92] Nakai, S. and Meath, W. J. (1992) J. Chem. Phys., **96**, 4991–5008.

[Per] Persico, M. private communication.

[QL90] Le Quéré, F. and Leforestier, C. (1990) J. Chem. Phys., **92**, 247.

[Smi69] Smith, F. T. (1969) Phys. Rev., **179**, 111.

[SR92] Shi, S. and Rabitz, H. (1992) J. Chem. Phys., **97**, 276.

[TI93] Takahashi, K. and Ikeda, K. (1993) J. Chem. Phys., **99**, 8680.

[TKR86] Tannor, D. J., Kosloff, R. and Rice, S. A. (1986) J. Chem. Phys., **85**, 5805.

[WP77] Walker, R. B. and Preston R. K. (1977) J. Chem. Phys., **67**, 2017.

[ZSL+91] Zemke, W. T., Stwalley, W. C., Langhoff, S. R., Valdemara, G. L. and Berry M. J. (1991) J. Chem. Phys., **95**, 7846.

DIMENSIONAL SCALING IN QUANTUM THEORY

John Avery
H.C. Ørsted Institute
University of Copenhagen

1.Introduction

During the last few years, application of the dimensional scaling technique to the quantum theory of atomic and molecular structure has yielded important new insights [1-41]. In this new technique, pioneered by Professor Dudley Herschbach and his co-workers at Harvard University, one solves the many-particle Schrödinger equation in a D-dimensional space. Exact solutions can be found in the limiting cases, $D \to \infty$ and $D = 1$. In the dimensional scaling procedure, coordinates and energies are scaled in such a way as to remove violent dependence on D. Only very gentle and smooth dependence remains, so that extrapolation and interpolation of properties as a function of D becomes possible. Wave functions and energies at the physical dimension, $D = 3$, are found using a perturbation series based on the parameter $1/D$, and by interpolation between exactly-known solutions at $D = 1$ and $D = \infty$.

In treating the large-D limit, it is convenient to transform the many-particle Schrödinger equation by means of a Jacobian weighting factor. Letting $\psi = J^{-\frac{1}{2}}\phi$, one solves, not for the wave function, ψ, but for ϕ, the square root of the distribution function. This transformation gives rise to an effective centrifugal potential, U. To see how the effective potential U arises, let us consider the transformation $\psi = \chi\phi$, where χ is a function of the coordinates. Instead of solving the Schrödinger equation for ψ, we now try to solve it for ϕ. Thus the original Schrödinger equation,

$$(T + V - E)\psi = 0 \tag{1}$$

becomes

$$\chi^{-1}(T + V - E)\chi\phi = 0 \tag{2}$$

J. L. Calais and E. S. Kryachko (eds.), Structure and Dynamics of Atoms and Molecules: Conceptual Trends, 133–154.

where T and V are respectively operators representing kinetic and potential energy, and where we have multiplied (1) on the left by χ^{-1}. Noting that χ commutes with V and E, we obtain:

$$(T' + U + V - E)\phi = 0 \tag{3}$$

where

$$\chi^{-1} T \chi \equiv T' + U \tag{4}$$

Here T' contains differential operators, while U is an effective potential. The transformation function, χ, can be chosen in any way which we think will simplify the task of solving the Schrödinger equation. In dimensional scaling, we choose χ in such a way that, in the large-D limit, the transformed kinetic energy T' becomes progressively less important, and the function ϕ becomes sharply localized at the minimum of the effective total potential, W, where

$$W \equiv U + V \tag{5}$$

If χ is chosen to be $J^{-1/2}$, where J is the Jacobian of the transformation from Cartesian coordinates to a set of internal and external coordinates, then ϕ corresponds to the distribution function of the system; and in the large-D limit, ϕ is sharply localized at the minimum of W. For example, the D-dimensional analogue of the hydrogen atom has a ground-state wave function of the form:

$$\psi = N\ e^{-2R/(D-1)} \tag{6}$$

where N is a normalization constant, and where R is the hyperradius:

$$R^2 \equiv \sum_{j=1}^{D} x_j^2 \tag{7}$$

As D becomes large, the radial distribution function

$$|\ \phi\ |^2 = J\ |\ \psi\ |^2 = N^2 R^{D-1} e^{-4R/(D-1)} \tag{8}$$

becomes more and more sharply localized at the point

$$R = \left(\frac{D-1}{2}\right)^2 \tag{9}$$

If we let

$$\chi = J^{-1/2} = R^{(D-1)/2} \tag{10}$$

then the point at which $|\ \phi\ |^2$ is localized will be a minimum of the total effective potential, W. Because of the sharp localization of the distribution function which occurs in the large-D limit, it has been called the "pseudo-classical limit". At slightly smaller values of D, expansion of the potential in a Taylor about the minimum of W yields a vibrational normal mode problem for large values of D; and at still smaller values of D, it yields a systematic procedure for constructing solutions, using harmonic oscillator wave functions as a basis. The procedure provides a convenient and intuitively meaningful way of treating both correlation and corrections to the Born-Oppenheimer approximation.

2. The Jacobian Weighting Factor

When dimensional scaling is applied to S-states of quantum mechanical systems, the coordinates can be divided into a set of internal coordinates, $X_1, X_2, ..., X_\tau$, (which do not include the Eulerian angles), and a set of external coordinates, $X_{\tau+1}, X_{\tau+2}..., X_d$, where $d = DN$. As $D \to \infty$, the external coordinates become infinitely numerous, while τ, the number of internal coordinates, remains fixed. For an N-particle system, there are N position vectors:

$$\begin{aligned} \mathbf{x}_1 &= (x_{11}, x_{12}, ..., x_{1D}) \\ \mathbf{x}_2 &= (x_{21}, x_{22}, ..., x_{2D}) \\ \vdots \;&\;\vdots\;\;\;\vdots \\ \mathbf{x}_N &= (x_{N1}, x_{N2}, ..., x_{ND}) \end{aligned} \tag{11}$$

The D-dimensional Schrödinger equation of the system is:

$$\left(-\sum_{i=1}^{N} \frac{1}{2m_i} \Delta_i + V \right) \psi = E\psi \tag{12}$$

where

$$\Delta_i \equiv \sum_{t=1}^{D} \frac{\partial^2}{\partial x_{it}^2} \tag{13}$$

For S-states, the wave function, ψ, depends only on the internal coordinates, $X_1, X_2, ..., X_\tau$. If we express Δ in terms of the internal coordinates, we can write

$$\Delta_i \psi = \sum_{\mu=1}^{\tau} \left(A_i^\mu \frac{\partial}{\partial X_\mu} + \sum_{\nu=1}^{\tau} G_i^{\mu\nu} \frac{\partial^2}{\partial X_\mu \partial X_\nu} \right) \psi \tag{14}$$

where

$$A_i^\mu \equiv \sum_{t=1}^{D} \frac{\partial^2 X_\mu}{\partial x_{it}^2} \tag{15}$$

and

$$G_i^{\mu\nu} \equiv \sum_{t=1}^{D} \frac{\partial X_\mu}{\partial x_{it}} \frac{\partial X_\nu}{\partial x_{it}} \tag{16}$$

Alternatively, we can write

$$\Delta_i \psi = J^{-1} \sum_{\mu,\nu=1}^{\tau} \frac{\partial}{\partial X_\nu} J G_i^{\mu\nu} \frac{\partial}{\partial X_\mu} \psi \tag{17}$$

where J is the Jacobian associated with the transformation to the internal coordinates. Since (17) can be rewritten in the form:

$$\Delta_i \psi = \sum_{\mu,\nu=1}^{\tau} \left[J^{-1} \left\{ \frac{\partial}{\partial X_\nu} (J G_i^{\mu\nu}) \right\} \frac{\partial \psi}{\partial X_\mu} + G_i^{\mu\nu} \frac{\partial^2 \psi}{\partial X_\mu \partial X_\nu} \right] \tag{18}$$

we can make the identification:

$$A_i^\mu J = \sum_{\nu=1}^{\tau} \frac{\partial}{\partial X_\nu}(J G_i^{\mu\nu}) \tag{19}$$

The many-particle Schrödinger equation, (12), can be transformed by letting

$$\psi = \chi\phi \tag{20}$$

where χ is a function of the internal coordinates. From (14), we have:

$$\begin{aligned}\Delta_i\psi &= \sum_{\mu=1}^{\tau} A_i^\mu \left(\chi\frac{\partial\phi}{\partial X_\mu} + \phi\frac{\partial\chi}{\partial X_\mu}\right) \\ &+ \sum_{\mu,\nu=1}^{\tau} G_i^{\mu\nu}\left(\chi\frac{\partial^2\phi}{\partial X_\mu \partial X_\nu} + \phi\frac{\partial^2\chi}{\partial X_\mu \partial X_\nu} + 2\frac{\partial\chi}{\partial X_\mu}\frac{\partial\phi}{\partial X_\nu}\right)\end{aligned} \tag{21}$$

We can see that if we choose χ to be a function which satisfies:

$$A_i^\mu \chi + 2\sum_{\nu=1}^{\tau} G_i^{\mu\nu}\frac{\partial\chi}{\partial X_\nu} = 0 \tag{22}$$

then the terms proportional to $\partial\phi/\partial X_\mu$ in (21) will vanish. The transformed Schrödinger equation then becomes:

$$\left(-\sum_{i=1}^{N}\frac{1}{2m_i}G_i^{\mu\nu}\frac{\partial^2}{\partial X_\mu \partial X_\nu} + U + V\right)\phi = E\phi \tag{23}$$

(where we have also multiplied on the left by χ^{-1}). The effective potential, U, which appears in the transformed Schrödinger equation is given by:

$$U = -\sum_{i=1}^{N}\frac{1}{2m_i}\chi^{-1}\Delta_i\chi \tag{24}$$

Combining equations (14), (17) and (22), we can rewrite U in the form:

$$U = \sum_{i=1}^{N}\frac{1}{4m_i J\chi}\sum_{\mu=1}^{\tau}\frac{\partial}{\partial X_\mu}(A_i^\mu J\chi) \tag{25}$$

In order to illustrate equations (11)-(25), let us consider the simplest possible example: the case where $N = 2$.Then there is only one internal coordinate,

$$X_1 = \left[\sum_{t=1}^{D}(x_{1t} - x_{2t})^2\right]^{\frac{1}{2}} \equiv R \tag{26}$$

Since there is only one internal coordinate, the sums in equation (14) are not needed, and the equation reduces to:

$$\Delta_i\psi = \left(A_i^1\frac{\partial}{\partial R} + G_i^{11}\frac{\partial^2}{\partial R^2}\right)\psi \tag{27}$$

Since

$$\frac{\partial R}{\partial x_{1t}} = \frac{x_{1t} - x_{2t}}{R}$$
$$\frac{\partial R}{\partial x_{2t}} = \frac{x_{2t} - x_{1t}}{R} \tag{28}$$

we have:

$$A_i^1 = \sum_{t=1}^{D} \left[\frac{1}{R} - \frac{(x_{1t} - x_{2t})^2}{R^3} \right] = \frac{D-1}{R} \qquad i = 1,2 \tag{29}$$

and

$$G_i^{11} = \frac{1}{R^2} \sum_{t=1}^{D} (x_{1t} - x_{2t})^2 = 1 \qquad i = 1,2 \tag{30}$$

Thus we obtain:

$$\Delta_i = \frac{D-1}{R} \frac{\partial}{\partial R} + \frac{\partial^2}{\partial R^2} \qquad i = 1,2 \tag{31}$$

Equation (19) requires that the Jacobian, J, should satisfy:

$$\left(\frac{D-1}{R} \right) J = \frac{\partial J}{\partial R} \tag{32}$$

from which we find that

$$J \sim R^{(D-1)} \tag{33}$$

Equation (22) becomes:

$$\left(\frac{D-1}{R} \right) \chi + 2 \frac{\partial \chi}{\partial R} = 0 \tag{34}$$

which will be satisfied if

$$\chi = R^{-\frac{1}{2}(D-1)} \tag{35}$$

The transformed Schrödinger equation, (23), becomes

$$\left(-\frac{1}{2\mu} \frac{\partial^2}{\partial R^2} + U + V \right) \phi = E\phi \tag{36}$$

where μ is the reduced mass,

$$\mu \equiv \frac{m_1 m_2}{m_1 + m_2} \tag{37}$$

and

$$U = -\frac{1}{2\mu} \chi^{-1} \left(\frac{D-1}{R} \frac{\partial}{\partial R} + \frac{\partial^2}{\partial R^2} \right) \chi \tag{38}$$

Substituting (25) into (28), we obtain:

$$U = \frac{f}{2\mu R^2} \tag{39}$$

where

$$f = \frac{1}{4}(D-1)(D-3) \tag{40}$$

For $D > 3$, the effective potential, U, is repulsive, and because of its form it is appropriate to call it the *quasicentrifugal potential.*

3.The large-D limit

For a two-particle system interacting through a Coulomb potential, the transformed Schrödinger equation, (34), becomes

$$\left(-\frac{1}{2\mu}\frac{\partial^2}{\partial R^2} + \frac{f}{2\mu R^2} + \frac{Z_1 Z_2}{R}\right)\phi = E\phi \tag{41}$$

where Z_1 and Z_2 are the charges on particles 1 and 2. In discussing the large-D limit for the two-particle system, it is convenient to introduce the scaled interparticle distance,

$$l \equiv f^{-1} R \tag{42}$$

and the scaled energy,

$$\mathcal{E} \equiv fE \tag{43}$$

where f is defined by equation (40). Then the transformed and scaled Schrödinger equation for the two-particle system becomes:

$$(T + W)\phi = \mathcal{E}\phi \tag{44}$$

where

$$T = -\frac{1}{2f\mu}\frac{\partial^2}{\partial l^2} \tag{45}$$

and

$$W = \frac{1}{2\mu l^2} + \frac{Z_1 Z_2}{l} \tag{46}$$

For the case where $Z_1 = -1$ and $Z_2 = Z$, this becomes:

$$W = \frac{1}{2\mu l^2} - \frac{Z}{l} \tag{47}$$

As D becomes large, the kinetic energy terms become progressively less and less important because they are multiplied by a factor $1/f$; and thus as $D \to \infty$, the scaled ground-state energy of the system becomes:

$$\mathcal{E}_\infty = W(\bar{l}) = -\frac{1}{2}\mu Z^2 \tag{48}$$

where

$$\bar{l} = \frac{1}{\mu Z} \tag{49}$$

denotes the value of l for which W is a minimum. The structure of the system at this minimum has been called the *Lewis structure*. For slightly smaller values of D, we can expand W in a Taylor series about this minimum:

$$W = \mu Z^2 \sum_{j=0}^{\infty} (-1)^j \left(\frac{j-1}{2}\right) q^j \tag{50}$$

where

$$q \equiv \frac{l - \bar{l}}{\bar{l}} \tag{51}$$

The leading term after the constant, $\mathcal{E}_\infty$, is the quadratic term, and we are thus led to an harmonic oscillator problem:

$$\left(-\frac{1}{2f}\frac{\partial^2}{\partial q^2} + \frac{1}{2}q^2\right)\phi = \frac{\mathcal{E} - \mathcal{E}_\infty}{\mu Z^2}\phi \tag{52}$$

where we have truncated the Taylor series for W after the quadratic term. The small vibrations of the system about the minimum of W for large but finite D have been called the *Langmuir vibrations*. Since $\chi = J^{-\frac{1}{2}}$, ϕ can be interpreted as the square root of the distribution function:

$$|\phi|^2 = J|\psi|^2 \tag{53}$$

Thus, in the large-D limit, as the kinetic energy terms become less and less important, the distribution function becomes sharply localized in the neighbourhood of the Lewis structure. For large but finite values of D, we can express ϕ as a linear combination of the the harmonic oscillator functions found by solving equation (52), and include higher terms in the Taylor series expansion of W. Of course, the procedure just described is unnecessary for the case where $N = 2$ because exact solutions can be found. However, the fact that exact solutions can be found for $N = 2$ can be used to check the validity and accuracy of the dimensional scaling procedure, since we can compare the exact eigenvalues with those obtained using dimensional scaling.

4.Three-particle systems

For a three-particle system, going directly to scaled coordinates, we can choose

$$\begin{aligned}
X_1 &= f^{-1}\left[\sum_{t=1}^{D}(x_{2t} - x_{3t})^2\right]^{\frac{1}{2}} \equiv l_1 \\
X_2 &= f^{-1}\left[\sum_{t=1}^{D}(x_{3t} - x_{1t})^2\right]^{\frac{1}{2}} \equiv l_2 \\
X_3 &= f^{-1}\left[\sum_{t=1}^{D}(x_{1t} - x_{2t})^2\right]^{\frac{1}{2}} \equiv l_3
\end{aligned} \tag{54}$$

as the internal coordinates, where f is defined by

$$f = \frac{1}{4}(D-1)(D-5) \tag{55}$$

rather than by equation (40). The transformation, $\psi = \chi\phi$, with the weighting factor,

$$\chi = (-l_1^4 - l_2^4 - l_3^4 + 2l_1^2 l_2^2 + 2l_1^2 l_3^2 + 2l_2^2 l_3^2)^{-\frac{1}{4}(D-1)} \tag{56}$$

can then be shown to remove the terms in $\partial\psi/\partial X_\mu$ from the transformed Schrödinger equation. To see this, we write the kinetic energy operator in terms of the internal coordinates l_1, l_2, and l_3. If we let

$$\Delta_1 \equiv \sum_{t=1}^{D} \frac{\partial^2}{\partial x_{1t}^2} = \sum_{\mu=2}^{3} A_1^\mu \frac{\partial}{\partial l_\mu} + \sum_{\mu,\nu=2}^{3} G_1^{\mu\nu} \frac{\partial^2}{\partial l_\mu \partial l_\nu} \tag{57}$$

where

$$\begin{aligned} A_1^\mu &\equiv \sum_{t=1}^{D} \frac{\partial^2 l_\mu}{\partial x_{1t}^2} \\ G_1^{\mu\nu} &\equiv \sum_{t=1}^{D} \frac{\partial l_\mu}{\partial x_{1t}} \frac{\partial l_\nu}{\partial x_{1t}} \end{aligned} \tag{58}$$

then, with f chosen as indicated in equation (55), we obtain:

$$f^2 \Delta_1 = \frac{D-1}{l_2}\frac{\partial}{\partial l_2} + \frac{D-1}{l_3}\frac{\partial}{\partial l_3} + \frac{\partial^2}{\partial l_2^2} + \frac{\partial^2}{\partial l_3^2} + 2cos\theta_1 \frac{\partial^2}{\partial l_2 \partial l_3} \tag{59}$$

where

$$cos\theta_1 \equiv \frac{l_2^2 + l_3^2 - l_1^2}{2l_2 l_3} \tag{60}$$

Δ_2 and Δ_3 are given by similar expressions, with the appropriate cyclic permutations of the particle indices. We now let P be the polynomial

$$P = -l_1^4 - l_2^4 - l_3^4 + 2l_1^2 l_2^2 + 2l_1^2 l_3^2 + 2l_2^2 l_3^2 \tag{61}$$

It is easy to verify that P satisfies the relationship

$$\frac{\partial P}{\partial l_2} + cos\theta_1 \frac{\partial P}{\partial l_3} = \frac{2P}{l_2} \tag{62}$$

from which it follows that χ fulfills equation (22). Similar considerations, with cyclic permutations of the particle indices, hold for Δ_2 and Δ_3. Notice that

$$\chi = P^{-\frac{1}{4}(D-1)} \sim A^{-\frac{1}{2}(D-1)} \tag{63}$$

where A is the area of the triangle formed by the three particles. It is easy to verify that, besides satisfying equation (62), P also satisfies the relation,

$$\sum_{\mu,\nu=2}^{3} G_1^{\mu\nu} \frac{\partial^2 P}{\partial l_\mu \partial l_\nu} = \frac{\partial^2 P}{\partial l_2^2} + \frac{\partial^2 P}{\partial l_3^2} + 2cos\theta_1 \frac{\partial^2 P}{\partial l_2 \partial l_3} = 0 \tag{64}$$

Using equations (57)-(62), we can calculate the contribution of particle i to the quasicentrifugal potential:

$$-\frac{1}{2m_i}\chi^{-1}\Delta_i\chi = \frac{1}{2fm_ih_i^2} \tag{65}$$

where f is defined by equation (55) and

$$h_i^2 \equiv \frac{P}{4l_i^2} = \left(\frac{2A}{l_i}\right)^2 \tag{66}$$

Thus the scaled effective potential, W, is given by

$$W = \sum_{i=1}^{3} \frac{1}{2m_ih_i^2} + \frac{Z_1Z_2}{l_3} + \frac{Z_2Z_3}{l_1} + \frac{Z_3Z_1}{l_2} \tag{67}$$

In the transformed and scaled Schrödinger equation,

$$(T + W)\phi = \mathcal{E}\phi \tag{68}$$

the kinetic energy operator, T, is a sum of three contributions:

$$T = T_1 + T_2 + T_3 \tag{69}$$

where

$$T_1 = -\frac{1}{2fm_1}\left(\frac{\partial^2}{\partial l_2^2} + \frac{\partial^2}{\partial l_3^2} + 2cos\theta_1 \frac{\partial^2}{\partial l_2 \partial l_3}\right) \tag{70}$$

while T_2 and T_3 are given by similar expressions, with appropriate permutations of the particle indices.

To find the Langmuir vibrations of the system, we need to represent $W \equiv U + V$ by a Taylor series expansion about its global minimum. To calculate the coefficients in the expansion, we notice that

$$P\sum_{i=1}^{N} \frac{1}{2m_ih_i^2} \equiv PU = \sum_{i=1}^{3} \frac{2l_i^2}{m_i} \tag{71}$$

from which it follows that

$$U\frac{\partial P}{\partial l_\mu} + P\frac{\partial U}{\partial l_\mu} = \frac{4l_\mu}{m_\mu} \tag{72}$$

and

$$\frac{\partial U}{\partial l_\mu} = \frac{1}{P}\left(\frac{4l_\mu}{m_\mu} - U\frac{\partial P}{\partial l_\mu}\right) \tag{73}$$

Similarly, differentiating (72) a second time, we obtain:

$$\frac{\partial^2 U}{\partial l_\mu \partial l_\nu} = \frac{1}{P}\left(\frac{4\delta_{\mu\nu}}{m_\mu} - \frac{\partial U}{\partial l_\mu}\frac{\partial P}{\partial l_\nu} - \frac{\partial U}{\partial l_\nu}\frac{\partial P}{\partial l_\mu} - U\frac{\partial^2 P}{\partial l_\mu \partial l_\nu}\right) \tag{74}$$

Since

$$V = \frac{Z_1 Z_2}{l_3} + \frac{Z_1 Z_3}{l_2} + \frac{Z_2 Z_3}{l_1} \tag{75}$$

we have

$$\frac{\partial^2 V}{\partial l_1^2} = \frac{2 Z_2 Z_3}{l_1^3} \tag{76}$$

and so on, the mixed partial derivatives being zero. Having calculated the second derivatives of the effective potential at its minimum, so that we can write

$$\begin{aligned} W &\approx \frac{1}{2}\sum_{\mu\nu} W_{\mu\nu}(l_\mu - \bar{l}_\mu)(l_\nu - \bar{l}_\nu) \\ W_{\mu\nu} &\equiv \frac{\partial^2 W}{\partial l_\mu \partial l_\nu} \end{aligned} \tag{77}$$

we transform $W_{\mu\nu}$ to a set of coordinates in which $T_{\mu\nu}$ is a unit matrix, $T_{\mu\nu}$ being defined by

$$T \equiv \frac{1}{2}\sum_{\mu\nu} T_{\mu\nu}\frac{\partial^2}{\partial l_\mu \partial l_\nu} \tag{78}$$

Finally, we diagonalize the transformed (and mass-weighted) matrix of force constants to find the normal modes of the system. The Langmuir frequencies are then given by the square roots of the eigenvalues.

5.An atom-to-molecule transformation

As Berry [43-45] and his coworkers have pointed out, correlation can give the electrons in an atom a molecule-like structure. Conversely (as the Chicago group have demonstrated experimentally), large amounts of vibrational energy can cause a molecule to "melt", so that the nuclear wave functions become delocalized, as they would be in a tiny drop of liquid; and in such a state, the concepts of molecular structure and the molecular point group fail completely. The similarity between atoms and molecules has also been discussed by Burden [46], and by Feagin and Briggs [47-49]. Both the Hartree-Fock approximation, in the case of the electrons of an atom, and the Born-Oppenheimer approximation, in the case of molecules, are only approximations. Because of correlation in atoms and "melting" in vibrationally-excited molecules, our very different approaches for treating the two types of systems should perhaps be replaced by single, uniform theory. Dimensional scaling gives us such a uniform approach, since it avoids both the Hartree-Fock approximation and the Born-Oppenheimer approximation; and it treats all particles on equal footing regardless of their masses.

In order to explore the way in which the characteristics of an atom go over smoothly into those of a molecule, we can think of a helium atom in which the mass of the nucleus (m_3) gradually becomes lighter, until it is equal to an electron's mass (Table 1). We can then examine what happens when the masses of the two electrons of the original helium atom (m_1 and m_2) gradually become heavier, until they are equal to the mass of a proton (Table 2). Finally, we can ask what happens as the charge Z_3 of the particle which was initially heavy, and which now is light, is gradually lowered from 2 to 1 (Table 3). At the end of these three steps, the system, which originally was a helium atom, will have become a charge-reversed H_2^+ ion. We treat it here by means of dimensional scaling partly to increase our understanding of the molecule-like properties of atoms and the atom-like properties of molecules, and partly to illustrate the discussion of the three-particle problem given in the previous section.
Looking at Table 1, we can see that in the pseudoclassical limit, the structure of helium is a roughly equilateral triangle, where the two electrons are slightly farther from each other than they are distant from the nucleus. The angle at the nuclear vertex of the triangle is

$$\theta_3 = 2\sin^{-1}\left(\frac{l_3}{2l_1}\right) = 95.29717^\circ \tag{79}$$

i.e., slightly larger than 90° (a result which was first derived Herschbach [8]). We might at first be surprised that the dimensional scaling technique gives us a pseudoclassical structure which is bent rather than linear: After all, it is known that, because of correlation, the square of the wave function of helium is largest when the two electrons are on opposite sides of the nucleus; and this would correspond to $\theta_3 = 180^\circ$. The reason for this seeming contradiction is that, in the pseudoclassical limit, the particles are localized at the maximum of the distribution function, $|\phi|^2 = J|\psi|^2$, rather than at the maximum of the squared wave function, $|\psi|^2$. In the case of our three-particle system, (equation (63)),

$$J \sim \chi^{-2} \sim A^{D-1} = \left(\frac{1}{2}l_1 l_2 \sin\theta_3\right)^{D-1} \tag{80}$$

where A is the area of the triangle. If a Hartree-Fock wave function is used in calculating the maximum of $J|\psi|^2$ for helium, the maximum occurs at $\theta_3 = 90^\circ$, since the Hartree-Fock wave function is independent of θ_3, while J contains the factor $\sin^{D-1}\theta_3$. As D increases, this factor has an increasingly sharp maximum at a right angle. Because of angular correlation, the value of θ_3 in the Lewis structure of helium is shifted to a slightly larger angle, but it is still closer to 90° than to 180°. The difference seen here between the maximum of $|\psi|^2$ and that of $J|\psi|^2$ is similar to that which we noted earlier in the ground-state radial distribution function of a D-dimensional hydrogenlike atom, equation (8), which has its maximum at the radius $R = (D-1)^2/4$, although the square of the wave function is largest at $R = 0$.

Looking once more at Tables (1-3), we can see that if we start with helium and decrease the mass of the nucleus, m_3, while holding the charges constant, as well as the masses of the two electrons, at first there is very little change; but as m_3 approaches the electron mass, $m_1 = m_2 = 1$, the Lewis structure increases in size, and its binding energy, $|\mathcal{E}_\infty|$, decreases. If we then increase $m_1 = m_2$ to the proton mass while holding m_3 constant at the value $m = 3 = 1$, still with constant charges, the Lewis structure shrinks in size, and its binding energy becomes large (Table 2). If we finally decrease Z_3 gradually to 1 (Table 3), so that the system approaches charge-reversed H_2^+, the Lewis structure grows in size, and its binding energy decreases. We can obtain a qualitative understanding of these features of the atom-to-molecule transformation by making the simplifying approximation:

$$l_1 = l_2 = l \qquad l_3 = \sqrt{2}\, l \qquad \theta_3 = 90° \tag{81}$$

so that

$$h_1 = h_2 = l \qquad h_3 = \frac{l}{\sqrt{2}} \tag{82}$$

With this simplifying assumption, the effective potential, (67), can be written in a form similar to equation (47):

$$W = \frac{1}{2\mu l^2} - \frac{\zeta}{l} \tag{83}$$

where

$$\frac{1}{\mu} \equiv \frac{1}{m_1} + \frac{1}{m_2} + \frac{2}{m_3} \tag{84}$$

and

$$-\zeta \equiv \frac{1}{\sqrt{2}} Z_1 Z_2 + Z_2 Z_3 + Z_3 Z_1 \tag{85}$$

The simplified effective potential, (84), has its minimum when

$$l = \bar{l} = \frac{1}{\zeta\mu} \tag{86}$$

If we substitute this value into (84), we obtain:

$$\mathcal{E}_\infty \approx -\frac{\zeta^2 \mu}{2} \tag{87}$$

From equations (87) and (88), we can see that when the charges are kept constant, then in the rough approximation where we assume the pseudo-classical structure to be an isosceles right triangle, the size of the Lewis structure is inversely proportional to the three-particle effective mass, μ, while the system's binding energy is directly proportional to μ. These approximate predictions are reflected in the features of the atom-to-molecule

Table 1.
Lewis structures and Langmuir frequencies for 3-particle systems
$Z_1 = Z_2 = -1,\ Z_3 = 2,\ m_1 = m_2 = 1,\ (\mathcal{E}_\infty = 4\epsilon_\infty)$

m_3	l_3	$l_1 = l_2$	$\mathcal{E}_\infty$	ω_A	ω_θ	ω_S
∞	0.897149	0.606964	−2.737769	0.44363	1.47033	0.67830
7344	0.897254	0.607051	−2.737361	0.44353	1.47013	0.67818
100	0.904802	0.613337	−2.708242	0.43664	1.45581	0.67024
50	0.912411	0.619683	−2.679455	0.43164	1.43928	0.66269
20	0.934992	0.638569	−2.597237	0.41119	1.40119	0.63968
10	0.971867	0.669576	−2.472493	0.38408	1.33970	0.60489
4	1.077803	0.759694	−2.168731	0.32460	1.18926	0.51871
2	1.242977	0.902893	−1.812842	0.26596	1.01086	0.41642
1	1.547208	1.173373	−1.381326	0.20758	0.78955	0.29376

ϵ_∞ in the papers by Prof. Hershbach and co-workers is related to our $\mathcal{E}_\infty$ by $\mathcal{E}_\infty = 4\epsilon_\infty$

Table 2.
Lewis structures and Langmuir frequencies
for 3-particle systems
$Z_1 = Z_2 = -1,\ Z_3 = 2,\ m_3 = 1,\ (\mathcal{E}_\infty = 4\epsilon_\infty)$

$m_1 = m_2$	l_3	$l_1 = l_2$	$\mathcal{E}_\infty$	ω_A	ω_θ	ω_s
1	1.547207	1.173372	−1.381325	0.20758	0.78955	0.29376
2	1.050927	0.840969	−1.902439	0.31184	1.12136	0.35607
4	0.779953	0.661242	−2.383549	0.44022	1.45046	0.37105
10	0.595493	0.540701	−2.859263	0.60657	1.80349	0.32213
20	0.525068	0.495425	−3.084678	0.70188	1.98349	0.25984
50	0.478443	0.465808	−3.248558	0.77905	2.12116	0.17954
100	0.461808	0.455327	−3.309744	0.80975	2.17437	0.13100
1836	0.445411	0.445049	−3.371331	0.84180	2.22905	0.03152
∞	0.444444	0.444444	−3.375000	0.84375	2.23235	0.00000

Table 3.
Lewis structures and Langmuir frequencies
for 3-particle systems
$Z_1 = Z_2 = -1,\ m_3 = 1,\ m_1 = m_2 = 1836,\ (\mathcal{E}_\infty = 4\epsilon_\infty)$

Z_3	l_3	$l_1 = l_2$	$\mathcal{E}_\infty$	ω_A	ω_θ	ω_S
2.0	0.445411	0.445049	−3.371331	0.84180	2.22905	0.03152
1.8	0.538264	0.519338	−2.537037	0.56007	1.70343	0.02307
1.6	0.669109	0.620809	−1.830018	0.32907	1.25093	0.01610
1.5	0.756077	0.686612	−1.523332	0.22994	1.05190	0.01350
1.4	0.863798	0.766648	−1.247293	0.13725	0.87090	0.01054
1.3	0.999881	0.865823	−1.001405	0.00884	0.70780	0.00826
1.2	1.175893	0.991480	−0.785102	–	0.56248	0.00630
1.1	1.410193	1.155104	−0.597733	–	0.43476	0.00465
1.0	1.733451	1.375553	−0.438538	–	0.32445	0.00329

transformation shown in Tables 1-3. From the definition of μ, equation (85), we can see that its value is dominated by the lightest particles in the system. This explains why the size and binding energy of the pseudoclassical structure are relatively insensitive to the value of m_3 as long as $m_3 >> m_1 = m_2$. When the masses of the particles are kept constant and their charges are changed, equations (87) and (88) predict that the size of the pseudoclassical structure should be inversely proportional to ζ, while the pseudoclassical binding energy should be directly proportional to ζ^2, and this approximately describes the features of Table 3.

Tables 1-3 also show the Langmuir frequencies as functions of the charges and masses during the atom-to-molecule transformation. The symmetry with respect to interchange of particles 1 and 2 is preserved throughout the transformation. The normal mode denoted by A is independent of Δl_3 and depends only on $\Delta l_1 - \Delta l_2$. In the atomic limit, this mode corresponds to an antisymmetric stretch, with one electron moving inwards towards the nucleus while the other moves outwards. In the molecular limit, this mode corresponds to motion of particle 3 (now the light particle) parallel to the line joining the two heavy particles. In Table 3 we can see that in the pseudoclassical limit, this mode becomes progressively softer as the charge Z_3 is decreased. Finally, when $Z_3 \approx 1.3$, the frequency ω_A becomes imaginary. This is because a symmetry-breaking occurs in the pseudoclassical limit for $Z_3 < 1.3$. For lower values of Z_3, $l_1 \neq l_2$ at the true minimum of W. The minima shown in Table 3 for $Z_3 = 1.2$, $Z_3 = 1.1$, and $Z_3 = 1.0$ are subject to the constraint $l_1 = l_2$.

The mode denoted by θ is a bending mode in the atomic limit, with the two electrons moving towards each other, while the nucleus moves very slightly away from them. In the molecular limit, this mode corresponds to a motion of particle 3 (now the light particle) perpendicularly outwards from a point halfway between the two heavy particles, 1 and 2.

Finally, the mode denoted by S corresponds in the atomic limit to a symmetric stretch, with both electrons moving simultaneously outward from the nucleus. In the molecular limit, this mode corresponds to a vibration of the two heavy particles, 1 and 2. It depends almost entirely on Δl_3, and contains only an extremely small admixture of $\Delta l_1 + \Delta l_2$. Throughout the atom-to-molecule transformation, this mode retains the character of a breathing mode, with all three particles moving simultaneously outward from the center of mass.

6. N-particle systems

In a treatment of N-electron atoms in the high-D limit, Loeser [preprint] chose the internal coordinates to be the scaled radii defined by

$$r_i^2 \equiv f^{-2} \sum_{t=1}^{D} x_{it}^2 \tag{88}$$

and the cosines of the angles between the position vectors

$$\gamma_{ij} \equiv cos\theta_{ij} \equiv \frac{\mathbf{r_i} \cdot \mathbf{r_j}}{r_i r_j} \tag{89}$$

He proposed that a scaling function of the form

$$\chi = (r_1 r_2 ... r_N)^{-\alpha} \Gamma^{-\alpha/2} \tag{90}$$

be used to transform the Schrödinger equation, where

$$\Gamma \equiv \mid \gamma_{ij} \mid \tag{91}$$

is the Gramian determinant [50], (i.e., the determinant of the matrix of cosines) and where

$$\alpha = \frac{D-1}{2} \tag{92}$$

In a previous paper generalizing Loeser's approach (Avery, Goodson and Herschbach, [41]) it was shown that with the transformation function of equation (91) the transformed and scaled Schrödinger equation becomes

$$(T + W)\phi = \mathcal{E}\phi \tag{93}$$

where $T = T_R + T_\gamma$ with

$$T_R = -\sum_{i=1}^{N} \frac{1}{2fm_i} \left[\frac{\partial^2}{\partial r_i^2} + \frac{D-1-2\alpha}{r_i} \frac{\partial}{\partial r_i} \right] \tag{94}$$

$$\begin{aligned} T_\gamma = & -\sum_{i=1}^{N} \frac{1}{2fr_i^2 m_i} \left[\sum_{j \neq i}^{N} \sum_{k \neq i}^{N} (\gamma_{jk} - \gamma_{ij}\gamma_{ik}) \frac{\partial^2}{\partial \gamma_{ij} \partial \gamma_{ik}} \right. \\ & \left. -(D-1-2\alpha) \sum_{j \neq i}^{N} \gamma_{ij} \frac{\partial}{\partial \gamma_{ij}} \right] \end{aligned} \tag{95}$$

and

$$W = -\sum_{i=1}^{N} \frac{f}{2m_i} \chi^{-1} \Delta_i \chi + V' = \sum_{i=1}^{N} \frac{1}{2m_i h_i^2} + V' \tag{96}$$

where $V' = fV$ is the scaled potential energy of the system. In equations (94)-(97), the scaling factor, f, is given by

$$f \equiv \alpha(D - N - \alpha - 1) \tag{97}$$

In equation (97), h_i is defined by

$$h_i \equiv r_i \left(\frac{\Gamma}{\Gamma^{(i)}} \right)^{\frac{1}{2}} \tag{98}$$

where $\Gamma^{(i)}$ is the ith principal minor of Γ (i.e., the determinant of the matrix formed by deleting the ith row and ith column from the matrix of cosines). If the parameter α in the transformation function is chosen to be $\alpha = (D-1)/2$, then the first derivative terms disappear from T in the transformed Schrödinger equation. On the other hand, there is a considerable advantage in the alternative choice,

$$\alpha = \frac{D-3}{2} \tag{99}$$

since in that case, it follows from equations (95)-(98) that

$$T + W = fH_3 + \sum_{i=1}^{N} \frac{1}{2m_i h_i^2} \tag{100}$$

Where H_3 is the usual 3-dimensional Hamiltonian. Thus, with the second choice for α, existing quantum chemistry programs can easily be modified for use in dimensional scaling by adding the quasi-centrifugal potential.

In order to illustrate equations (94)-(98), we can apply them to the simple example of a two-electron atom or ion, in the approximation where the motion of the nucleus is neglected. In this approximation, we can treat the system as a two-body problem, and the scaled potential becomes:

$$V' = -\frac{Z}{r_1} - \frac{Z}{r_2} + \frac{1}{(r_1^2 + r_2^2 - 2r_1 r_2 \gamma_{12})^{\frac{1}{2}}} \tag{101}$$

where the origin of the coordinate system has been placed at the position of the nucleus. The Gramian determinant, Γ, for the two-electron system is given by

$$\Gamma = \begin{bmatrix} 1 & \gamma_{12} \\ \gamma_{12} & 1 \end{bmatrix} = 1 - \gamma_{12}^2 \tag{102}$$

and $\Gamma^{(1)} = \Gamma^{(2)} = 1$, so that

$$W = \frac{1}{2m_e(1-\gamma_{12}^2)} \left(\frac{1}{r_1^2} + \frac{1}{r_2^2} \right) + V' \tag{103}$$

It is interesting to ask whether the formalism of equations (94)-(98) can be applied to two particles of unequal masses, m_1 and m_2, and unequal charges, $Z_1 = -1$ and $Z_2 = Z$, in the case where there is no third particle at the origin of the coordinate system. In that case, W becomes

$$W = \frac{1}{2(1-\gamma_{12}^2)} \left(\frac{1}{m_1 r_1^2} + \frac{1}{m_2 r_2^2} \right) - \frac{Z}{(r_1^2 + r_2^2 - 2r_1 r_2 \gamma_{12})^{\frac{1}{2}}} \tag{104}$$

and the global minimum of W occurs when $r_1 = r_2 \to \infty$. If we let

$$r_1^2 + r_2^2 - 2r_1 r_2 \gamma_{12} = l^2 \tag{105}$$

then, in the limit where r_1 and r_2 are equal and very large, W can be written in the form:

$$W = \frac{1}{2\mu l^2} - \frac{Z}{l} \tag{106}$$

where μ is the reduced mass defined by equation (37). Comparing equation (107) with equation (47), we can see that they are identical.

7.Discussion

In the examples discussed above, we have seen that the Schrödinger equation for bound S-states of quantum mechanical systems can be solved exactly in the high-D limit. As D becomes large, the kinetic energy becomes progressively less important, and the distribution function becomes sharply localized at the minimum of the total effective potential, W. In the high-D ("pseudoclassical") limit, the relative positions of the particles in a system are thus sharply defined, regardless of whether the particles are nuclei or electrons. In quantum chemistry, it is customary to use the Born-Oppenheimer approximation, and to regard the nuclei in a molecule as having definite relative positions; but the electrons are regarded as delocalized. In dimensional scaling, the Born-Oppenheimer approximation is avoided, and the electrons and nuclei of a system are treated on equal footing. For low values of D, the wave functions of both electrons and nuclei are delocalized; and in the high-D limit, a definite structure, the Lewis structure, appears for both electrons and nuclei. For intermediate values of D, the vibrations about the Lewis structure are treated by finding the normal modes and using the corresponding harmonic oscillator wave functions as a basis set for the expansion of ϕ. Thus, in discussing electron correlation in a two-electron atom or ion, a vibrational normal mode analysis is used, similar to that which might be used for a triatomic molecule. This feature of the dimensional scaling approach closely parallels earlier studies of electron correlation by Herrick, Berry and others, who showed that the highly-correlated doubly-excited states of helium can be analysed in terms of the vibrational normal modes of a triatomic molecule. However, it should be noted that the equilibrium structure about which the system vibrates in the dimensional scaling picture is not structure for which $| \psi |^2$ is a maximum, but rather the structure at which the distribution function, $| \phi |^2$ is localized. This difference should be remembered when a comparison is made between studies using the dimensional scaling approach and the earlier studies of electron correlation.

Acknowledgements

The author would like to thank the Carlsberg Foundation for travel grants which allowed research visits to the laboratory of Professor Dudley Herschbach at Harvard University. He is also extremely grateful to Professor Herschbach and coworkers for much advice, warm hospitality, and encouragement.

References

1. Rosenthal, C.M. (1971), J. Chem. Phys. **55**, 2474.
2. Mlodinow, L.D. and Papanicolaou, N. (1980), Ann. Phys. (N.Y.) **128**, 314.
3. Herrick, D.R. and Stillinger, F.H. (1975), Phys. Rev. A **11**, 42.
4. Herrick, D.R. (1983), Adv. Chem. Phys. **52**, 1.
5. van der Merwe, P. du T., (1984), J. Chem. Phys. **81**, 5986; (1985), **82**, 5293.
6. van der Merwe, P. du T. (1987), Phys. Rev. A **36**, 3446; (1988), **38**, 1187.
7. Goscinski, O. and Mujica, V., (1986), Int. J. Quantum Chem. **29** 897.
8. Herschbach, D.R. (1986), J. Chem. Phys. **84**, 838-851.
9. Herschbach, D.R., (1987), J. Chem. Soc. Faraday Disc. **84**, 465.
10. Herschbach, D.R. (1988), Faraday Disc. Chem. Soc. **84**, 465-478.
11. Herschbach, D.R., Loeser, J.G. and Watson, D.K. (1988), Z. Phys. D **10**, 195-210.
12. Herschbach, D.R. (1989), Proc. Intl. Conf. Atomic Phys. **11**, 63-81.
13. Herschbach, D.R. (1989), Proc. Welsch Fd. Chem. Research **XXXII**, 95-116.
14. Herschbach, D.R., (1992), in Chemical Bonding; Structure and Dynamics, A. Zewail ed., Academic Press, New York.
15. Herschbach, D.R., (1993), Proc. Amer. Phil. Soc. **137**, 532.
16. Herschbach, D.R., Avery, J. and Goscinski, O. (eds.) (1993), Dimensional Scaling in Chemical Physics, Kluwer, Dordrecht, Netherlands.
17. Loeser, J.G. and Herschbach, D.R., (1985), J. Chem. Phys. **83**, 3444.
18. Loeser, J.G. and Herschbach, D.R. (1986), J. Chem. Phys. **84**, 3882-3892, 3893-3900.
19. Loeser, J.G. and Herschbach, D.R. (1987), J. Chem. Phys., **86**, 2114-2122, 3512-3521.
20. Loeser, J.G. (1987), J. Chem. Phys. **86**, 5635-5646.
21. Loeser, J.G. and Herschbach, D.R. (1988), J. Phys. Chem. **89**, 3444-3447.
22. Loeser, J.G., preprint, Harvard University Chemistry Department.
23. Loeser, J.G., Zhen, Z., Kais, S., and Herschbach, D.R., (1991), J. Phys. Chem. **95**, 4525.

24. Doren, D.J. and Herschbach, D.R. (1985), Chem. Phys. Letters **118**, 115-119.
25. Doren D.J., and Herschbach, D.R. (1986), J. Chem. Phys. **85**, 4557-4562.
26. Doren, D.J. and Herchbach, D.R. (1986), Phys. Rev. A, **34**, 2654-2664, 2665-2673.
27. Doren D.J., and Herschbach, D.R. (1987), J. Chem. Phys. **87**, 433-442.
28. Doren D.J., and Herschbach, D.R. (1988), J. Phys. Chem. **92**, 1816-1821.
29. Goodson D.Z., and Herschbach, D.R. (1987), Phys. Rev. Letters **58**, 1628-1631.
30. Goodson, D.Z. and Herschbach, D.R. (1987), J. Chem. Phys. **86**, 4997-5008.
31. Goodson, D.Z., Watson, D.K., Loeser, J.G., and Herschbach, D.R., (1991), Phys. Rev. A **44**, 97.
32. Goodson, D.Z., Morgan, J.D., III, and Herschbach, D.R., (1991), Phys. Rev. A **43**, 4617.
33. Frantz, D.D., Herschbach D.R. and Morgan, J.D. (1989), Phys. Rev. A **40**, 1175-1184.
34. Franz, D.J. and Herschbach, D.R. (1988), Chem. Phys. **126**, 59-74.
35. Kais, S., Herschbach D.R. and Levine, R.D. (1989), J. Chem. Phys. **91**, 7791-7796.
36. Kais, S., Morgan, J.D., III, and Herschbach, D.R., (1991), J. Chem. Phys. **95**, 9028.
37. Kais, S., Frantz, D.D., and Herschbach, D.R., (1992) Chem. Phys. **161**, 393.
38. Kais, S., Sung, S.M., and Herschbach, D.R., (1993), J. Chem. Phys. **99**, 5184.
39. Kais, S., and Herschbach, D.R., (1994) [preprint]
40. Avery, J., Goodson D.Z. and Herschbach, D.R. (1991) "Approximate Separation of the Hyperradius in the Many-Body Schrödinger Equation", Int. J. Quantum Chem. **39** 657-666.
41. Avery, J., Goodson D.Z. and Herschbach, D.R. (1991) "Dimensional Scaling and the Quantum Mechanical Many-Body Problem", Theor. Chem. Acta **81**, 1.

42. Avery, J. (1989), Hyperspherical Harmonics; Applications in Quantum Theory, Kluwer Academic Publishers, Dordrecht, Netherlands, Chapters 5 and 6.
43. Berry, R.S. and Krause, J.L. (1988), Adv. Chem. Phys. **70**, 35.
44. Berry, R.S. (1987), in Understanding Molecular Properties, J. Avery, J.P. Dahl and A.E. Hansen (eds.), Reidel, Dordrecht, Netherlands.
45. Ezra, G.S. and Berry, R.S. (1983), Phys. Rev. A **28**, 1974.
46. Burden, F.R., (1983) J. Phys. **B16**, 2289.
47. Feagin, J.M. and Briggs, J.S., (1986) Phys. Rev. Lett. **57**, 984.
48. Rost, J.M., Sung, S.M., Herschbach, D.R., and Briggs, J.S., (1992), Phys. Rev. A **46**
49. Rost, J.M. and Briggs, J.S., (1993) in Dimensional Scaling in Chemical Physics, D.R. Herschbach, J. Avery and O. Goscinski (eds.), Kluwer, Dordrecht, Netherlands.
50. Gantmacher, F.R. (1959), The Theory of Matrices, Vol. 1, Chelsea, New York, pp 246-258.

PROBING THE COLLECTIVE AND INDEPENDENT-PARTICLE CHARACTER OF ATOMIC ELECTRONS

R. STEPHEN BERRY
Department of Chemistry,The University of Chicago
5735 South Ellis Avenue,Chicago, Illinois 60637, U.S.A.

1. Introduction

1.1. HISTORICAL BACKGROUND

Niels Bohr and Arnold Sommerfeld developed their model of the hydrogen atom and other one-electron atoms based on the concepts of discrete, stationary states and quantized energies and angular momenta of these atoms. Every stationary state carried its own constants of motion, to each of which corresponded a quantum number. Transitions between stationary states corresponded to sudden changes in the values of constants of the motion and of the corresponding quantum numbers, and to the absorption or emission of radiation to maintain conservation of energy.

Bohr, with Henrik Kramers, tried to extend this model to the helium atom and larger atoms[1, 2]. So did many of their contemporaries[3-9]. Almost all of these efforts began with the assumption that each electron should have its own conserved energy and angular momentum, and its own corresponding quantum numbers. These assumptions became the constraints for a variety of mechanical models in which the two electrons of helium moved in orbits presumably stabilized by the attraction of the nucleus for the electrons and the repulsion of the electrons for each other. Only Irving Langmuir proposed a model, at that time, that did not preserve the individual angular momenta of each electron[6-8]. Rather, Langmuir said[6] "Bohr assumes that ... the angular momentum of every electron round the center of its orbit is $h/2\pi$,... . However it is an attractive hypothesis to assume that in the case of coupled electrons, the quantum theory is concerned not with the angular momentum possessed by one electron but rather with the angular momentum which, by being transferred from electron to electron, circulates in each of two directions about the nucleus." (An amusing historical footnote to this was the objection raised by the young van Vleck to Langmuir's model, on the grounds that it *did not* quantize the angular momenta of each electron, as Bohr had proposed[9, 10].) Then, beginning in 1925, came the quantum mechanics of Schrödinger, Heisenberg and Dirac, and its immediate, successful application to the hydrogen atom.

By 1928, quantum mechanics was well established and had become a powerful computational tool, especially because of the variational method with the proof that it

J. L. Calais and E. S. Kryachko (eds.), Structure and Dynamics of Atoms and Molecules: Conceptual Trends, 155–181.

always gives convergence *from above* to the lowest state of any system [11-13]. With only desk calculators, Hylleraas applied this approach in what became a long series of ever more accurate calculations of the ground state of the helium atom[11]. Perhaps just because of the accuracy that such calculations yield, interest faded, for many years, in any mechanistic interpretation of the structure of the helium atom or of more complex atoms. Only sporadically did works appear that treated mechanistic or structural aspects of the correlation problem[14, 15].

Then, triggered by the discovery of a series of unusually-shaped spectral lines of doubly-excited helium[16, 17], new interest sprang up in the mechanistic interpretation of such species. Despite their lying in the ionization continuum, broad spectral lines of doubly-excited helium atoms, He**, had been expected[18], but the observations did not fit the first predictions. Those predictions had been that two series of lines would appear; actually only one was strong enough to show in the first spectra. The interpretation appeared immediately[17]: configuration mixing, representing electron correlation, made one of the expected series extremely weak, while the other collected almost all of the oscillator strength.

Several attacks soon appeared, dealing with accurate computation[19-29] and with more interpretive orientations. Of the latter, some were based on the use of hyperspherical coordinates[30-34], usually chosen as $R=(r_1^2 + r_2^2)^{1/2}$, $\alpha=r_1/r_2$ and θ_{12}, the angle between the two electron-nucleus vectors. We shall discuss the hyperspherical coordinate approach briefly. Other early attacks were based on symmetry considerations, notably on the recognition that a) diagonalizing the electron-electron interaction operator e^2/r_{12} is very much like (but not identical to) diagonalizing the (square of the) vector obtained as the difference between the Runge-Lenz vectors of the two electrons, if they are each treated as hydrogenic[35, 36]. In classical terms, the Runge-Lenz vectors are essentially the semimajor axes of the Kepler elliptical orbits. Making the quantum operator corresponding to this length diagonal, so that it becomes at least approximately a constant of motion, corresponds to tying the two Kepler orbits so that they precess together, keeping the *length* of the vector between their semimajor axes constant. This approach was one of the threads that leads to the interpretation we shall pursue in most detail.

1.2. ALTERNATIVE QUANTIZATIONS AND CONSTANTS OF MOTION

This concept, generalized, eventually has become a powerful organizing concept of the quantum mechanics of complex systems: that the constants of motion and corresponding quantum numbers are characteristic of specific systems and states. There is no simple, universal criterion justifying identification of constants and quantum numbers with the real component particles of an atom, molecule or larger system. In more positive terms, this means that a central part of the analysis of any physical system is the identification of what quantities are strictly conserved and what quantities are approximately conserved. The strictly conserved quantities correspond to true fundamental symmetries of the system and its Hamiltonian. The approximately conserved quantities correspond to approximate symmetries of the real system and usually to exact symmetries of model systems and model Hamiltonians. Energy is an exactly conserved quantity because of the time invariance of any isolated system and its Hamiltonian. Total momentum is an exactly conserved quantity for any system in field-free space because of the the translational invariance of the system and its Hamiltonian. Total angular momentum is an exactly conserved quantity for any system isolated in a field-free space because of the angular invariance of the system and its Hamiltonian.

Orbital and spin angular momenta are approximately conserved quantities for light atoms because the exchange of angular momentum between them is weak: we can neglect the spin-orbit coupling of the electrons in light atoms in many situations. We express this by using approximate Hamiltonians in which spin-orbit coupling is simply omitted. Another familiar example of approximate quantization and approximately conserved quantities is the separation of normal modes of vibration of a molecule. The approximation here is that the vibrations are perfectly harmonic, which takes the potential energy of the Hamiltonian to be strictly quadratic in displacements of the atoms from positions of equilibrium. Such a Hamiltonian can always be separated by transformation to a sum of Hamiltonians of independent, noninteracting harmonic oscillators, the normal modes of vibration. This means that the energy of each normal mode is a constant of the motion, with its own quantum number, and there are as many constants of motion as there are oscillatory degrees of freedom. This model has been extremely effective for describing the lowest vibrational states of most molecules; only those nonrigid molecules whose atomic nuclei undergo large-amplitude motions fail to fit closely to the harmonic approximation. However states above the very lowest can display enough anharmonicity to make the harmonic model, separability and the corresponding approximate constancy of motion all fail. The water molecule is an example: the lowest quantum states of the stretching modes of its O-H bonds fit the harmonic model, and are well described as symmetric and antisymmetric stretching modes, with both O-H bonds stretching and contracting in synchrony in the former, and with one stretching while the other contracts synchronously in the other. However with three or more quanta in these bonds, the harmonic picture is a far less accurate representation of many of the states than is a local-mode picture, in which, at any instant, the quanta can be found localized in one O-H bond or the other, with only occasional transfer of the excitation between the two bonds[37-40]. Eventually, this exchange must take place so that the two O-H bonds are indistinguishable on a sufficiently long time scale, but for times of order several vibrational periods, the molecule appears to have its quanta localized, so that if the numbers of quanta in the two bonds are unequal, the bonds appear inequivalent.

Still one other example of approximate quantization is the Hartree-Fock model. This is, in effect, a quantum mechanical formulation of the Bohr-Sommerfeld quantum theory, insofar as it is based on the assumption that each electron has its own well-defined energy and angular momentum. Only the azimuthal quantum numbers, i.e. the orientations of the individual angular momenta, are spoiled in this picture. The Hamiltonian for the Hartree-Fock system has a potential energy that represents the electron-electron interaction by the mean field of that interaction, as felt by each electron. This means that the Hamiltonian is separable into a sum of one-electron Hamiltonians, each with its own eigenvalues and eigenfunctions. This model is therefore also called an "independent-particle" model. The

Hartree-Fock approximation is accurate enough to give well over ninety percent of the binding energy of the electrons in most atoms, yet it fails to represent many of the properties of atoms, including their capacity to form stable negative ions and chemical bonds. Configuration mixing is the standard way to correct this model for the effects of electron-electron correlation, in either a variational or perturbational formalism. Robert Mulliken made the remark, "A little configuration mixing goes a long way," and indeed this is so, as the later discussion will show.

This analysis will examine some of the possible ways that two-electron atoms and quasi-two-electron atoms--the valence electrons of the alkaline earth atoms--are approximately quantized and how this quantization can be determined. Most of the discussion will focus on the choices among the Hartree-Fock and two collective models. The latter are based on quantization in which the constants of motion--exact constants in the simplified models, but presumably approximate constants in the real atoms--involve motion of both electrons.

1.3 THE TOOLS OF THE ANALYSIS

Several means have been employed to give insight into the quantization of two-electron and many-electron atoms, and into the extent and consequences of electron correlation. Many of these involve solving Schrödinger equations based on exact and model Hamiltonians and comparing the energy levels E_j, their intervals $(E_i - E_j)$ (corresponding to spectral line frequencies) and sometimes the eigenfunctions ψ_j or their absolute squares $|\psi_j|^2$. A powerful tool in the early stages of the analysis[36, 41-46] and perhaps a very important one for ultimate understanding of these issues is the application of symmetry-based interpretations of the energy spectra and states of the atoms. Another has been the conditional probability distribution obtained by integrating $|\psi_j|^2$ over irrelevant variables and setting one variable equal to a significant value, such as its most probable; this allows us to study graphic representations of the spatial and momentum distribution of electron probability density[14, 47-54]. Still another such tool is the overlap $S_{j\alpha}$ (or its absolute square), of an accurate wavefunction ψ_j with an approximation $\psi_{j\alpha}$, based on the specific model α and its Hamiltonian H_α[55].

Apart from the intervals between energy levels, all of these diagnostic tools are theoretical constructs. However much we may employ these abstract devices, it is of course absolutely necessary to use other diagnostic tools that are observable, to determine the validity of any interpretation we make. But even here, testing the validity of a model or approximation depends on comparing what we observe with what the model says we should observe. This means that we are obligated to evaluate the expectation values of each of the observables we measure, within each model we test. Observables that serve as probes of electron correlation and the nature of multi-electron quantization include the intensities of spectral lines[56, 57], atomic quadrupole moments[58], possibly the specific mass contribution to the isotope shift of spectral lines[59] (a measure of the mean of the inner product of pairs of electron momentum vectors, $\langle \mathbf{p_1} \cdot \mathbf{p_2} \rangle$, probably useful only for He and other 2-electron atoms[60]), and angular correlations from (γ,2e) and (e,3e) processes, i.e. double-ionization processes initiated, respectively, by radiation[61-65] and by electron impact[66-78]. As yet, most of these probes of electron correlation and quantization have hardly been exploited. The vigor of theoretical studies will, one can hope, stimulate their investigation.

2. Symmetry–Based Interpretations

2.1. INITIAL EXPLORATIONS

Wulfman[35, 79, 80], Novaro and Freyre[81] and Herrick and Sinanoglu[36, 41] first suggested that the helium atom, particularly in its doubly-excited states, could be treated in terms of the hydrogen-like symmetry each electron would have in the absence of the other. This symmetry, the symmetry of the hydrogen atom consisting of two inequivalent, point-particles bound by their Coulomb attraction, is that of the rotations of a 4-dimensional sphere[82-84]. The symmetry group of this system is the orthogonal group O(4), and it has one constant of motion more than O(3), the corresponding rotation group in three dimensions. In addition to the orbital angular momentum with quantum number l, and energy of the electron with quantum number n, the length of the Runge-Lenz vector **B**, essentially the semimajor axis of the classical Kepler orbit of the electron, is the new constant of motion. This extra symmetry, characteristic specifically of the Coulomb potential, is what raises the degeneracy of the states of the hydrogen atom to n^2, rather than only 2l+1.

Hence a natural approach to the doubly-excited states of the helium atom would be based on the direct product of the O(4) groups of the two electrons, $O(4)_1 x O(4)_2$. The states of the helium atom would then correspond approximately to bases of the irreducible representations of a suitable subgroup of $O(4)_1 x O(4)_2$. A particularly attractive choice of subgroup is a new O(4) group: $O(4) \subset O(4)_1 x O(4)_2$. There are, however, several choices of the new O(4) group, corresponding to different choices of constants of motion. These are combinations of the constants of motion of the individual electrons. The choice selected by Wulfman and by Herrick and Sinanoglu takes as constants of motion the total orbital angular momentum, the vector sum $\mathbf{l_1} + \mathbf{l_2}$ of the angular momenta of the two electrons, and the length of the *difference* **B** of the two Runge-Lenz vectors, $|\mathbf{B_1} - \mathbf{B_2}|$. Wulfman and Kumei made this selection on the basis that it corresponds approximately to diagonalizing the electron-electron repulsion, e^2/r_{12}; Herrick and Sinanoglu found that this choice, diagonalizing $|\mathbf{B}|^2$, gave the best agreement of limited-basis computations with experiments and with far more elaborate and accurate computations of the doubly-excited states of helium. Specifically, they used only the doubly-excited, Hartree-Fock-type states in which both electrons have the same principal quantum numbers, $n_1 = n_2$, which they called the DESB or "doubly excited symmetry basis." The lowest-energy states of each manifold corresponding to a specific n are given remarkably well by this very approximate procedure, both in terms of the energies, which Herrick and Sinanoglu studied, and in terms of the wave functions, examined by Rehmus et al.[48, 50]. However the higher members were, in many cases, very poorly represented, as frequently happens in variational calculations carried out by expansion in a fixed basis set. More accurate calculations, with larger basis sets, corrected this[51, 52].

An interesting aside regarding the history of these ideas is the extent to which the idea of using $O(4) \subset O(4)_1 x O(4)_2$ was stimulated by the notion that the two electrons of doubly-excited helium would be far from each other and therefore very hydrogen-like and not strongly correlated. This stimulus is totally irrelevant to the development of the group theoretical interpretation--fortunately, because it turns out to be a misconception. The electrons in doubly-excited helium are in fact extremely strongly correlated, far more correlated that the electrons of helium in its ground or singly-excited states. In fact, of all atoms in the periodic table, the electrons of helium are the least correlated of the valence shell electrons in their ground state. The physical basis for the increased correlation in the excited states is the low kinetic energies of the electrons in those states, making the electron-electron scattering far more important than in the ground state, particularly when the two electrons begin to approach each other.

Quantization based on $O(4) \subset O(4)_1 x O(4)_2$ is more general than just the representations that involve strong electron-electron correlations. Nikitin and Ostrovsky showed that the same formal broken symmetry obtains in the case in which the two electrons have very different principal quantum numbers[85, 86]. Here, however, the independent-particle picture is very much the appropriate one, because the two electrons have very different momenta and very different spatial distributions in this situation.

2.2. MULTIPLETS AND SUPERMULTIPLETS

The next step, and almost the whole picture as it now stands, came with the analysis of Kellman, Herrick and Poliak[42-46]. First, in their 1978 paper, Kellman and Herrick found that subsets--multiplets--of the states of each manifold obtained with the DESB calculations correspond strikingly to rotor series, even more so than the rotor series encountered in nuclear structure[87]. Then they found that entire sets of states constructed from the basis sets with $n_1 = n_2$ have patterns of energies and quantum numbers that correspond to states of a linear, ABA system like a linear triatomic molecule, with very light A-particles and a much heavier B-particle. These are *supermultiplets,* in the terminology used for classifying groups of levels sharing some common quantum numbers. In particular, the intra-shell manifolds correspond to states of the linear, light-heavy-light three-body system with quanta in the modes of rotation and doubly-degenerate bending. The lowest state of the manifold with $n_1 = n_2 = 2$, for example, corresponding to the $2s^2\ ^1S^e$ state of doubly-excited helium, plays the role of the ground state. The next state on the energy scale, the $2s2p\ ^3P^o$ state, is the first excited rotational state; the next is the $2s2p\ 1P^o$ state, which is one of the pair of states with one quantum in the bending-vibration mode. These would be degenerate if there were no Coriolis interaction, but its occurrence splits the energies of these two states. The other member of the pair of bending states is the $2p^2$ $^1P^e$ state. In the manifold of states with n = 2, there are only two others, the $2p^2$ $^1D^e$, corresponding to the state with two quanta of rotation and no excitation of any bending, and the highest, the $2p^2$ $^1S^e$ state, corresponding to two quanta in the bending mode, one in each of the states. A sketch of the symmetry-based pattern and the accurately-calculated levels for the manifold with $n_1 = n_2 = 3$ are shown in Figure 1. Herrick, Kellman and Poliak also speculated on the possibility of states of helium corresponding to stretching modes, but these lay outside their symmetry structures.

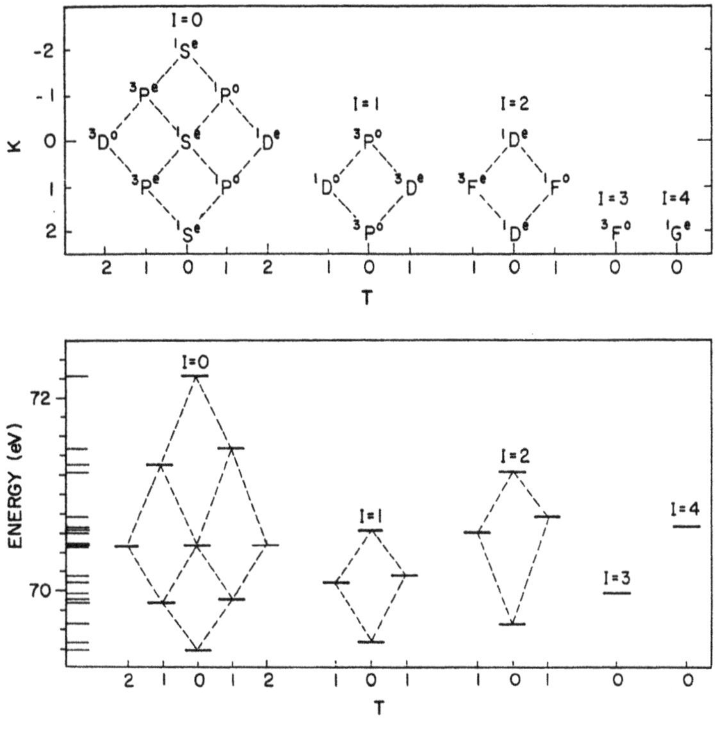

Figure 1. The pattern of energy levels (upper figure) based on the Herrick-Kellman supermultiplet pattern and the accurately calculated energy levels of doubly-excited helium with $n_1 = n_2 = 3$. The quantum numbers I, K and T correspond, respectively, to the rigid-rotor rotational quantum number, the number of bending quanta (but numbered up and down from the central value of the submanifold with I=0) and the number of units of angular momentum associated with the degenerate bending mode, angular momentum along the figure axis. These can be put in correspondence with the number of bending quanta, v, and the quantum number L of total angular momentum (independent of spin), i.e. into correspondence with the quantum numbers familiar in the spectroscopy of triatomic molecules, as follows: $I = L - T$, $K = L_{maximum} - l_{maximum} = L_{max} - (n-1)$.

A connection was made by Watanabe and Lin[88], between the supermultiplet approach of Herrick and Kellman and the representation of doubly-excited helium in terms of hyperspherical coordinates. A precursor to this appeared in a rediscovery of the quantization according to the quantum numbers I, K and T, with the addition of another very approximate two-valued quantum number A, that indicates an approximate odd-even kind of symmetry[89, 90].

One further advance in the symmetry-based approach has just appeared. While the analysis just described provides a persuasive classification and a useful analogy to the linear triatomic molecule, it still left open the real physics associated with the collective O(4) symmetry of the helium atom. It also left untouched the question of possible extensions to systems of more than two electrons. In fact the only extensions to three or more electrons have been made in the context of hyperspherical coordinates[91] and in models based on freezing the electrons onto spheres of fixed radii[92, 93], much as had been done previously with two electrons[94-96]. The new advance by Kellman[97] appears to be a significant step toward finding a strong, approximate symmetry for the two-electron and potentially several-electron Hamiltonians, based on unitary rather than rotational or orthogonal symmetry. The rotational symmetry emerges as a subgroup of the higher, unitary symmetry. If this approach is successful, it may lead to the identification of new, approximate collective quantum numbers for many-electron atoms, and to their association with particular kinds of collective modes of motion.

3. Interpretations Based on Wave Functions

3.1. GRAPHIC REPRESENTATIONS OF HELIUM

The two-electron system, with its six degrees of freedom, seems at first too complicated to lend itself to graphical representation, for example in terms of $\Psi(\mathbf{r}_1,\mathbf{r}_2)$ or $|\Psi(\mathbf{r}_1,\mathbf{r}_2)|^2$. However this is too pessimistic a view. First, for purposes of interpreting correlation and quantization, the spatial orientation of the system of two electrons and a nucleus is irrelevant, so that one can immediately eliminate three of the six variables by integrating $|\Psi(\mathbf{r}_1,\mathbf{r}_2)|^2$ over the three Euler angles that specify that orientation. At that stage, one has a choice of variables for representing the system in internal coordinates. The two most frequent choices are hyperspherical coordinates, usually taken to be $R \equiv (r_1{}^2 + r_2{}^2)^{1/2}$, $\alpha \equiv \arctan(r_1/r_2)$, and $\theta_{12} \equiv \arccos[(\mathbf{r}_1 \cdot \mathbf{r}_2)/r_1 r_2]$, and simply r_1, r_2 and θ_{12}. Both are useful, but the former tend to lock computations and interpretations to concepts in which R and α are the relevant coordinates, which is not always the case as we shall see. Hence the following discussion is based largely on the representation of the three-variable reduced form of $|\Psi(\mathbf{r}_1,\mathbf{r}_2)|^2$, namely the density ρ in terms of r_1, r_2 and θ_{12}, $\rho(r_1,r_2,\theta_{12})$.

This quantity is still too complicated to graph, even as a graph in three dimensions. One more reduction is required. We so this by constructing the *conditional probabiliy distribution* $\rho(r_2,\theta_{12}; r_1=a)$, with a specific choice or set of choices for the value a, such as the most probable value of r_1. This enables us to construct three-dimensional graphs of $\rho(r_2,\theta_{12}; r_1=a)$ or to generate entire sequences of images in the form of animations and thus make time a surrogate for the r_1 axis. In this way, the full behavior of $\rho(r_1,r_2,\theta_{12})$ can be represented graphically. Here, however, we are limited to discrete images, such as those shown in Fig.2. Note in this figure how similar are the three rotor states, in this internal-coordinate representation, and how alike are the two bending-mode states. The two distributions furthest to the right are those of the 2s3s $^3S^e$ and 2s3s $^1S^e$ states, corresponding respectively to the first excited antisymmetric and symmetric stretching modes. These, which will be discussed further, do not fit into the supermultiplet pattern but are expected on the basis of the analogy with the triatomic ABA molecule.

The conclusion we can draw from these distributions, completely consistent with the symmetry-based supermultiplet interpretation, is that the states of doubly excited helium, at least those in which the two electrons have the same or very similar quantum numbers, are well described by the same kind of collective quantization that describes the rotations and vibrations of a linear ABA triatomic molecule. When we look later at quantitative measures of validity of models, we will turn to the question of whether collective quantization gives a better description of this and other systems than does the traditional independent-particle model stemming from a mean-field, Hartree-Fock approach.

Before closing this section, we should mention the graphic representation of the ground state of the helium atom and of its isoelectronic counterpart, the H^- ion. These have been discussed in reviews of the subject[98, 99], as well as in the original articles based on very approximate[47, 100, 101] and accurate[49] wave functions. Because in these states the electrons are frequently near the nucleus, their kinetic energies are large relative to the potential energy of their repulsion, except when the electrons are very close together. Consequently electron correlation is not strong, especially in the ground state of helium, and its quantization is primarily independent-particle-like. The hydride ion is more an intermediate case; electron correlation is far more important than in He, but much of this correlation is radial, rather than angular. As a result, H^- does not seem to fit well into the supermultiplet picture, but since it has only a single bound state, there is little motivation to use a symmetry-based model to describe it.

3.2. GRAPHICAL REPRESENTATIONS FOR OTHER ATOMS: ALKALINE EARTHS

The doubly excited helium atom is not the only example in which electron correlation and collective quantization have been invoked. The role that correlation might play in alkaline earth atoms was pointed out by Greene and O'Mahony[102-104]. Krause[53] computed well-converged wave functions for the ground and low-lying excited states of the valence electrons of these atoms, using frozen-core effective potentials[105, 106]. Soon thereafter, he computed comparable functions for the alkali negative ions, also two-valence-electron atoms[54]. The wave functions for the alkaline earth atoms were refined further by

Hunter[55]. In fact the two valence electrons of the alkaline earth atoms, Be, Mg, Ca, Sr and Ba, are at least as interesting as the electrons of doubly excited helium, perhaps much more so because these atoms are far easier to study experimentally than the transient species we denote as He**. Moreover the *ground states* of the alkaline earth atoms display the same

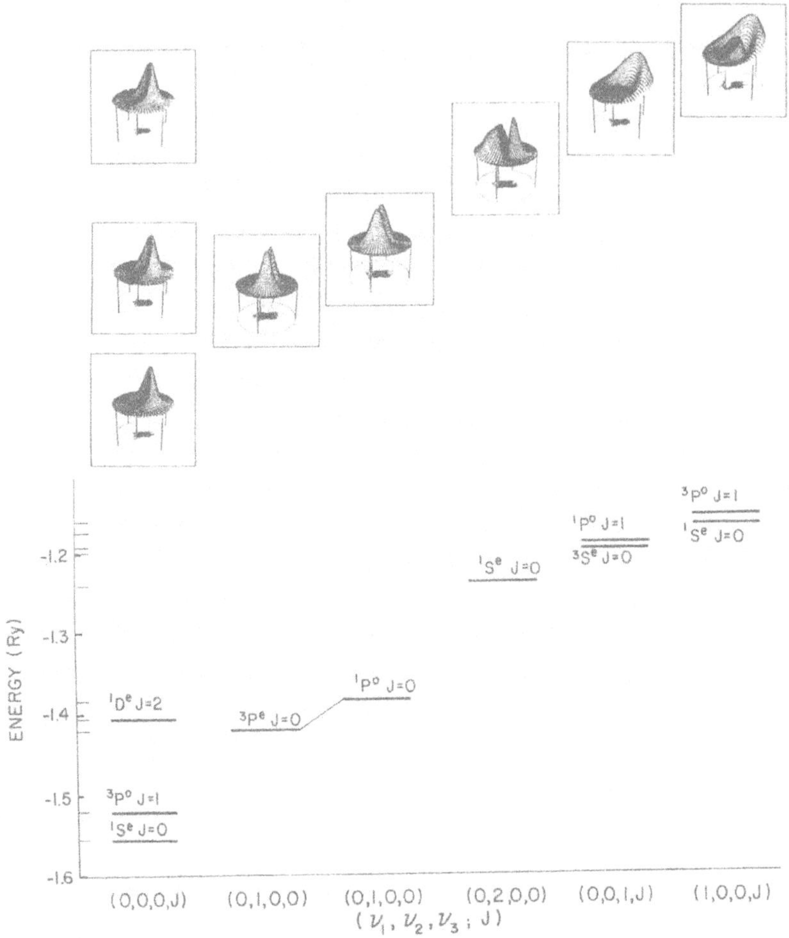

Figure 2. Conditional probability distributions and energy levels for the doubly excited states of helium with $n_1 = n_2 = 2$. The distributions are shown in cylindrical coordinates, with the angle $\theta_{12}=0$ indicated by the leftmost vertical bar in each figure. In fact these distributions are all multiplied by the factor r_2^2 of the Jacobian, so that the concentration of probability density close to the nucleus is suppressed, making the angular distributions easier to see.

extreme angular correlation that the $2s^2\ {}^1S^e$ state of He** shows. The same holds for the entire pattern of low-lying excited states of these atoms: they are all strongly correlated, analogous to their He** counterparts, but lie in the energy range of bound states, rather that of the ionization continuum, where all states eventually autoionize.

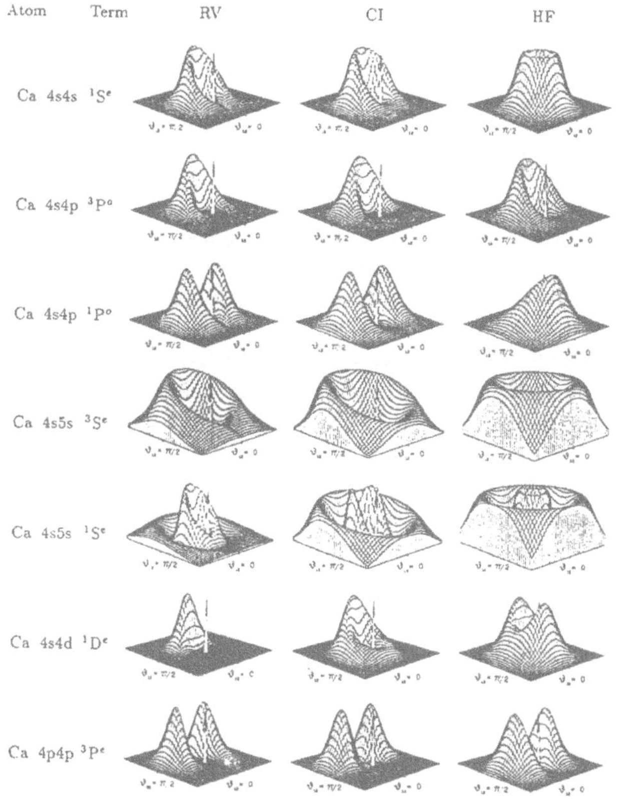

Figure 3. Conditional probability distributions for the ground and low-lying excited states of Ca. The central column of distributions are derived from well converged Sturmian configuration interaction expansions. The left column of distributions are based on single-term rotor-vibrator functions with harmonic bending modes and symmetric and antisymmetric combinations of localized Morse stretching modes. The right column of distributions is based on Hartree-Fock independent-particle wave functions.

The alkaline earth atoms offer a striking opportunity to compare the conditional probability distributions based on well-converged wave functions with those based on simple, independent-particle and rotor-vibrator models. Figure 3 illustrates such a comparison for states of the calcium atom, all with conditional probability distributions for the "fixed" electron taken at its most probable distance from the nucleus. While some states are represented about as well by both the independent-particle and collective, rotor-vibrator models, none is represented better in these graphs by the independent-particle model and several have collective, rotor-vibrator representations that are much more like the well converged functions than are the independent-particle, Hartree-Fock approximate functions. In short, this qualitative criterion implies that the collective model and collective quantization appears more realistic and therefore in some sense more accurate to describe the valence electrons of the alkaline earths than an independent-particle model and individual-electron quantization.

The role of configuration interaction in representing angular correlation is important and important to understand. Only a small fraction of the $np^2\ {}^1S^e$ configuration need be added to the $ns^2\ {}^1S^e$ principal configuration of the ground state to introduce so much angular correlation that the two electrons are virtually prohibited from being on the same side of the nucleus. Figure 4 shows two simplified models of the ground ${}^1S^e$ state of Be composed of only those two configurations. The upper figure is based on a function containing 98% of the $2s^2$ configuration and the lower figure, on a function that is 90% $2s^2$, in the sense that the squares of the coefficients of the $2s^2$ functions are 0.98 and 0.90, respectively.

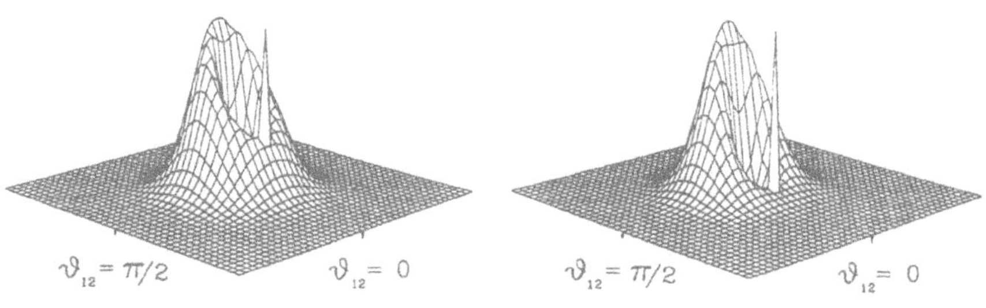

Figure 4. Illustrations of the effect on angular correlation of a small amount of configuration mixing. The left figure is a conditional probability distribution for a wave function of the ground state of the valence electrons of Be composed 98% of $2s^2$ and 2% of $2p^2$. The right figure is based on a function that is 90% $2s^2$ and 10% $2p^2$. The spikes indicate the position of the "fixed" electron in the conditional probability distribution.

3.3. OVERLAP CRITERIA

The next stage of comparing independent-particle and collective, rotor-vibrator quantization is a comparison of the overlaps, or more properly, the squares of the overlaps of the approximate functions with well-converged wave functions[55, 57]. The results of this comparison are, in some instances, a little surprising insofar as some independent-particle wave functions have overlaps with accurate functions that are larger than the overlaps of the collective, rotor-vibrator functions, even when the graphs such as those in Fig.3 indicate that the angular correlation is better represented by the rotor-vibrator model. Nonetheless the rotor-vibrator model comes off as somewhat more reliable, overall, than the independent-particle model, although not nearly so clearly so as it does according to the qualitative criterion of visual similarity. Table 1 contains these overlaps.

Table 1. Squared overlaps between configuration interaction (CI), rotor-vibrator (RV) and Hartree-Fock (HF) wave functions. The CI-RV overlaps are taken from Hunter[55] and the CI-HF and HF-RV overlaps, from Batka[57].

Atom	configuration	term	$\lvert(\Psi_{CI}\lvert\Psi_{RV})\rvert^2$	$\lvert(\Psi_{CI}\lvert\Psi_{HF})\rvert^2$	$\lvert(\Psi_{HF}\lvert\Psi_{RV})\rvert^2$
Be	2s2s	$^1S^e$	0.9966	0.8948	0.8957
Be	2s2p	$^3P^o$	0.9836	0.9869	0.9793
Be	2s2p	$^1P^o$	0.9108	0.9173	0.9105
Be	2s3s	$^3S^e$	0.9587	0.9719	0.9331
Be	2s3s	$^1S^e$	0.9102	0.9545	0.9213
Be	2p2p	$^1D^e$	0.8542	0.7157	0.6434
Be	2p2p	$^3P^e$	0.9856	0.9872	0.9858
Mg	3s3s	$^1S^e$	0.9973	0.9255	0.9275
Mg	3s3p	$^3P^o$	0.9505	0.9840	0.9385
Mg	3s3p	$^1P^o$	0.7970	0.9288	0.7952
Mg	3s4s	$^3S^e$	0.9648	0.9782	0.9385
Mg	3s4s	$^1S^e$	0.9241	0.9625	0.9211
Mg	3s3d	$^1D^e$	0.6457	0.6902	0.1573
Mg	3p3p	$^3P^e$	0.9928	0.9791	0.9783
Ca	4s4s	$^1S^e$	0.9963	0.9177	0.9216
Ca	4s4p	$^3P^o$	0.9551	0.9651	0.9221
Ca	4s3d	$^1D^e$	0.5236	0.8541	0.2624
Ca	4s4p	$^1P^o$	0.8649	0.8458	0.7198
Ca	4s5s	$^3S^e$	0.9578	0.9594	0.9232
Ca	4s5s	$^1S^e$	0.9233	0.9247	0.8919
Ca	4p4p	$^3P^e$	0.9783	0.8784	0.8915

Table 1, cont.

Sr	5s5s	$^1S^e$	0.9944	0.9244	0.9309
Sr	5s5p	$^3P^o$	0.9629	0.9267	0.8874
Sr	5s4d	$^1D^e$	0.7063	0.8324	0.3634
Sr	5s5p	$^1P^o$	0.9163	0.6837	0.6120
Sr	5s6s	$^3S^e$	0.9684	0.7153	0.7271
Sr	5p5p	$^3P^e$	0.9119	0.5732	0.6257
Sr	5s6s	$^1S^e$	0.7668	0.5879	0.7349
Ba	6s6s	$^1S^e$	0.9896	0.9200	0.9333
Ba	6s5d	$^1D^e$	0.5365	0.8813	0.3320
Ba	6s6p	$^3P^o$	0.9411	0.8877	0.8541
Ba	6s6p	$^1P^o$	0.8329	0.5319	0.5094
Ba	5d5d	$^3P^o$	0.7109	0.8339	0.4157
Ba	5d5d	$^1S^e$	(no convergence)	0.7552	(no convergence)
Ba	6s7s	$^3S^e$	0.9727	0.9232	0.9056
Ba	6s7s	$^1S^e$	0.9362	0.6678	0.7101
He	2p2p	$^3P^e$	0.9880	0.9870	0.9856
He	2p2p	$^1D^e$	0.8159	0.9204	0.6759

3.4. OSCILLATOR STRENGTH CRITERIA

The next quantitative criterion brings us to observable quantities, the values of oscillator strengths, as evaluated from the three kinds of wave functions--CI, RV and HF--and from experimental measurements. These are shown in Fig.5 for five allowed transitions. In the first four transitions, respectively the ground $^1S^e \rightarrow {}^1P^o$, $^1P^o \rightarrow {}^1D^e$, $^1P^o \rightarrow {}^1S^e$ and $^3P^o \rightarrow {}^3P^e$, the Hartree-Fock results are significantly further from both the accurately computed values and the experimental values, where they are available. For the fifth kind of transition, $^3P^o \rightarrow {}^3S^e$, the Hartree-Fock is closest to the accurate calculations. In all but that last case, the accurate calculations and the experimental results agree quite satisfactorily.

3.5. QUADRUPOLE MOMENTS

The quadrupole moment of a charge distribution offers a measure of its nonspherical character. In nuclear physics, quadrupole moments have been standard probes of shapes, but few atomic quadrupole moments have been measured, and among these, the only excited states to be observed have been two Rydberg states of helium[107, 108], and the metastable $^3P_2$ states of Ne, Ar, Kr and Xe[109]. Previously, Angel, Sandars and Woodgate had measured the quadrupole moment of aluminum atoms in their $^2P_{3/2}$ ground state[110].

Although the charge distribution in, for example, the ground states of the alkaline earth atoms is nonspherical in its own internal, "inertial axis" coordinates, such nonspherical character is not an observable. The ground states of the alkaline earth atoms are all ^1S_e states, which simply have quadrupole moments of zero. However many excited states of doubly excited helium and of the more experimentally accessible alkaline earth atoms do have nonvanishing quadrupole moments, which could be measured. Values of these quadrupole moments have been computed with well-converged Sturmian CI functions, to indicate approximately what experimental values should be expected[58]. Values have also been computed, as part of the same study, from unrestricted Hartree-Fock functions (for some of the states) and from rotor-vibrator functions. Typical expectation values of Q_{zz} = - $-(e/2)\ \Sigma_i\ (3z_i^2 - r_i^2)$ are shown in Table 2. In most cases, the three wave functions yield similar values for the quadrupole moments. The resonance ^1P_e excited state of Ca is an exception, for which the three values differ considerably. The rotor-vibrator values of some of the states of Ba are also far from the values predicted by the well-converged functions.

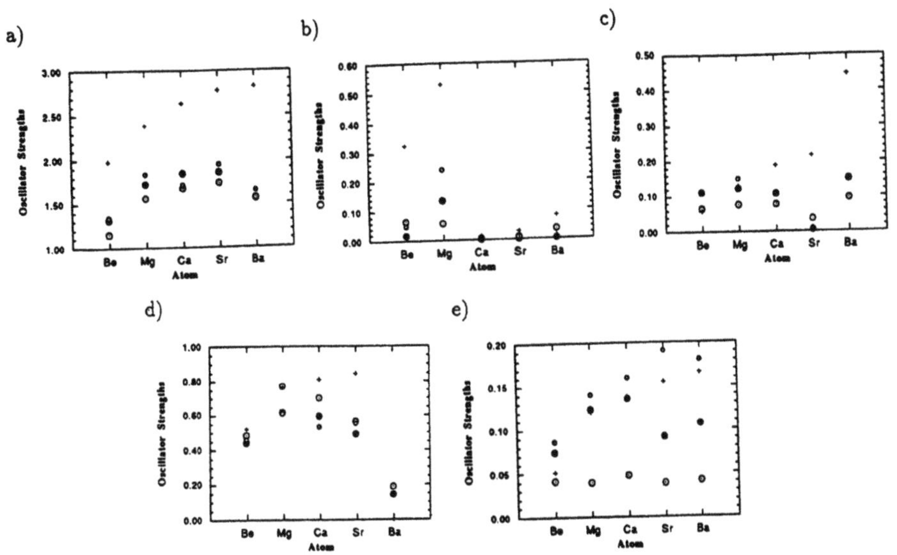

Figure 5. Graphic summary of the oscillator strengths f for five allowed transitions of the alkaline earth atoms: a) $^1S^e \rightarrow {}^1P^o$; b) $^1P^o \rightarrow {}^1D^e$; c) $^1P^o \rightarrow {}^1S^e$; d) $^3P^o \rightarrow {}^3P^e$, and e) $^3P^o \rightarrow {}^3S^e$. The small circles represent f_{CI}, the large circles represent f_{RV}, the crosses represent f_{HF}, and the solid circles represent the experimental values, where they are available. Results are taken from Batka[57].

While the quadrupole moments reflect the effects of correlation, it is not particularly easy to infer from simple models what these should be, or even what their signs should be. For example one might naively suppose that the pairs of "partner" states of the bending vibration would at least have quaddrupole moments of the same sign, but this is not the case. The nsnp $^1P^o$ states have positive quadrupole moments and the npnp $^3P^e$ states have negative quadrupole moments, corresponding to prolate charge distributions, while the positive quadrupole moments indicate oblate distributions.

Table 2. Some typical values of atomic quadrupole moments based on Sturmian CI, on unrestricted Hartree-Fock and on rotor-vibrator wave functions. Values are in units of ea_0^2, the electron charge times the Bohr radius, squared.

Atom	State	Q_{zz}(CI)	Q_{zz}(HF)	Q_{zz}(RV)
He	2s2p $^3P_2$	2.622		3.026
	2p2p $^1D_2$	5.015		3.871
	2s2p $^1P_1$	0.6929		1.656
	2p2p $^3P_2$	-2.357		-2.365
Be	2s2p $^3P_2$	2.140	1.833	1.919
	2p2p $^1D_2$	5.795		4.219
	2s2p $^1P_1$	2.783		1.538
	2p2p $^3P_2$	-2.073	-2.042	-2.076
Mg	3s3p $^3P_2$	3.972		3.212
	3s3d $^1D_2$	14.74		13.78
	3s3p $^1P_1$	5.174		2.199
	3p3p $^3P_2$	-3.622	-3.428	-3.616
Ca	4s4p $^3P_2$	6.121	6.630	5.324
	4s3d $^1D_2$	1.727		5.676
	4s4p $^1P_1$	4.919	6.630	1.988
	4p4p $^3P_2$	-5.215	-5.642	-5.236
Sr	5s5p $^3P_2$	6.419	6.080	6.008
	5s4d $^1D_2$	4.617		8.749
	5s5p $^1P_1$	2.873		1.366
	5p5p $^3P_2$	-4.744		-4.879
Ba	6s6p $^3P_2$	8.792		8.636
	6s5d $^1D_2$	2.015		9.537
	6s6p $^1P_1$	2.860		0.4626
	5d5d $^3P_2$	-4.136		-4.332

3.6. ANGULAR CORRELATIONS IN DOUBLE IONIZATION

A promising direction is the measurement and prediction of the angular correlation of the electrons in a two-electron ionization process. This may be accomplished either by electromagnetic radiation, in what is called a (γ,2e) process, or by electron impact, in what is called an (e,,3e) process. If, in an (e,3e) process, the incident electron has quite high energy, and the two electrons knocked out leave with energies high compared with their binding energy, then the electron-electron interactions in the final state are likely (but not guaranteed) to be small, so that the process measures predominantly a property of the initial state. In particular, it measures the distribution of angles θ_{12} between the two electron-nucleus vectors. This is an oversimplification, because there can always be some effect of the electron-electron interaction in the final stae. Nevertheless with interpretation, these two approaches are probably the most direct ways to study electron correlation that we know.

Glassgold and Ialongo[111] pointed out in 1968 that the angular correlation of electrons ejected in a fast process, such as the (e,3e) process, would be a particularly direct way to observe correlation effects, but the technology was not yet ready for such a sophisticated problem. Theoretical calculations were apppearing[112, 113], but the experiments were, at that time, simply too difficult to do. The mechanism of the double ionization had just been worked out[114]--and was confirmed by a later analysis[115]--to be predominantly a correlation effect that shows itself when one electron is suddenly knocked out, leaving the atom in a nonstationary superposition of new stationary states of the now-ionized system, a superposition that includes continuum states that correspond to a second ionization. This process, called "shake-off", dominates the double ionization when the incident electron has an energy of a few KeV and the ejected electrons carry kinetic energies of order a hundred or a few hundred eV.

Calculations continued to be done, largely for helium and other rare-gas atoms in their ground states[63, 64, 66-68, 73-77, 116-118] but also for alkaline earth atoms, for which correlation is much more important[65, 78]. Experiments with the corresponding double photoionization were carried out, first to study the threshold law [119, 120] whose theory had been pursued previously[121-124]. Then, angle- and energy resolved double photoionization studies were reported for the helium and argon atoms[61, 62]. Shortly before, the first experiments on angular correlations in the (e,3e) process appeared[69-72]. One set of experiments was done to study the simpler process abbreviated as (e,3–1e), i.e. only two of the three outgoing electrons are observed[125]. It seems that a great deal of information is lost in this simplification, at least regarding the effects of correlation[78].

Until this kind of study is applied to a system, such as an alkaline earth or conceivably even doubly excited helium, the question remains open of how powerful (e,3e) experiments will be to reveal effects of electron correlation and the nature of the quantization of two-electron and many-electron atoms. Such experiments are now underway with Mg atoms, but only the Auger double ionization process has been reported until now[126, 127]. Of course a natural way to carry out the analysis when we have reliable experimental results will, as with other methods, to compare not only the predictions of accurate theoretical calculations but of approximate models as well, models for which specific quantities are constants of the motion.

3.7. OTHER CORRELATIONS AND DISPERSIONS

There is one other class of probes we can consider, which is based on analysis of wave functions. This is the set of quantities in the category of expectation values of distributions, correlations and dispersions. As yet, these do not reveal as much as one might hope, but they are potentially useful and should be introduced in this context. The first is simply the angular distribution, p(q12), which was examined by Coulson and Neilson[15] and by Banyard and Ellis[100, 101]. In a sense, this is now superseded by the distributions in more than just a single variable, but such reduced distributions can still give insight into angular correlation.

The next to consider is the expectation value of the projection of the electron momenta onto each other, $\langle \mathbf{p}_1 \cdot \mathbf{p}_2 \rangle$, which is, in principle, measurable as the special mass effect of the isotope shift of atomic spectral lines. This is an attractive index to probe collective quantization because even its sign is a powerful indicator of the nature of correlation. In a rotationally excited state, insofar as the rigid rotor model is applicable, this quantity should be negative and moderately large because the electrons 1 and 2 should move along roughly antiparallel momentum vectors. Likewise, in a harmonic, symmetric stretch mode along the traditional normal mode for this motion, the projection should also be negative. For the antisymmetric stretch mode and for the bending modes, however, this quantity should be positive. This quantity was studied for doubly-excited helium, as well as for helium in its ground and singly-excited states, and the expectations of the model are indeed borne out[59]. In some respects, this is the most reassuring justification currently available for collective, molecule-like quantization in doubly excited helium.

The use of $\langle \mathbf{p}_1 \cdot \mathbf{p}_2 \rangle$ as an easily-interpreted index of correlation is perhaps limited to helium and helium-like ions. The reason is that inner-shell electrons contribute significantly to this quantity[60]. Even though their correlation is much less than that of valence electrons, their momenta are enough larger to compensate, so that their contributions to the expectation value are large enough to hide the contribution of the valence electrons. J. Morgan has suggested in private communication that, instead of $\langle \mathbf{p}_1 \cdot \mathbf{p}_2 \rangle$, one could use the pure angular quantity $\langle \mathbf{p}_1 \cdot \mathbf{p}_2 / |p_1|\,|p_2| \rangle$, which would very much reduce the inner-shell contributions.

Still another kind of probe that goes beyond the mean interelectronic angle (or, as with $\langle \mathbf{p}_1 \cdot \mathbf{p}_2 \rangle$ or $\langle \mathbf{p}_1 \cdot \mathbf{p}_2 / |p_1| |p_2| \rangle$, simply the expectation value of the cosine of the momentum-space angle $\theta^p{}_{12}$) is the mean deviation of that angle[57]. In general, Hartree-Fock functions have considerably larger values of the root-mean-square deviation $\Delta\theta_{12}{}^{rms}$ than do either the collective, rotor-vibrator or configuration interaction wave functions. This is of course consistent with the Hartree-Fock being a mean field representation, so that there is little to create any angular focusing of the wave function.

Analyses of the momentum correlations of the valence electrons of correlated systems will continue to be a useful tool for theoretical investigation, and perhaps experiments with electrons, analogous to the Coulomb explosion imaging experiments with molecules, will eventually enable us to observe these momentum correlations.

4. Interpretations Based on Classical and Semiclassical Models

4.1. THE BOHR-SOMMERFELD APPROACH

This discussion opened with a historical review of efforts to apply Bohr-Sommerfeld quantization to some suitable model of the helium atom. In the period before quantum mechanics developed, this effort was notably unsuccessful. However the difficulties were surmountable, and in 1980 Leopold and Percival succeeded in creating a satisfactory description of the helium atom in its ground state within the old quantum theory[128]. (Later these authors extended their work to treat the symmetric/antisymmetric symmetry of identical particles[129], in order to treat excited singlet and triplet states.) They achieved this by including zero-point motion, by using variational rather than perturbational methods wherever possible, and with the use of expansions for the interelectronic repulsion operator, angular momentum coupling and specific perturbation terms developed for this purpose[130]. They also followed the strict imposition of the Einstein-Brillouin-Keller prescription that the motion should correspond to motion on invariant tori (but not to truly periodic orbits, because the different modes have incommensurate natural periods). This requires that the quantization be half-integral, not integral, as all the treatments of the 1920's would have had it. The outcome can be summarized by a comparison of the total energy of the ground state and the ionization potential, from experiment and from several calculations including that of Leopold and Percival. The experimental values are -2.9037 atomic units and 24.58 eV, respectively. The best quantum mechanical calculations give results identical with these, to as many figures as we have quoted. The *variational* results from Leopold and Percival's semiclassical calculation are -2.8407 a.u. and 22.88 eV, respectively. Their corresponding results based on perturbation methods are -2.741 a.u. and 20.16 eV. The classical results derived in 1922 by van Vleck[9, 10] were -2.765 a.u. and 20.70 eV. Kramers, with a model of inclined orbits, obtained a total binding energy of -2.762 a.u. and an ionization energy of 20.60 eV[2]. The method is also applicable to excited states, as demonstrated in the 1980 paper, particularly after particle symmetry is accounted for.

4.2. ORBITS, PLANETARY ATOMS, LANGMUIR STATES

The development of the semiclassical description was actually preceded by Percival's analysis of highly excited states in which the electrons have angular momenta approaching the maximum allowed by the principal quantum numbers. The classical analogue of these states has the electrons going around the nucleus in nearly-circular orbits, never approaching the nucleus. Percival introduced the term "planetary atoms" to describe such states in his first publication on this subject[131]. Such states have very little overlap with the ground state and consequently are not easily produced by absorption of a single quantum. They are, however numerous and therefore may play important roles in situations in which atoms may be highly excited.

Quantitative methods for finding energies and classifying the planetary states emerged somewhat later. Richter and Wintgen found a means to regularize the three-body Coulomb system and then, with Tanner, implemented this to find the variety of categories of classical orbits of the helium-like system[132-134]. These include the Langmuir orbits[6-8, 135], which were shown to correspond to regions of stability[136], and were then computed by Müller, Burgdörfer and Noid[137, 138]. The quantum analogue of a Langmuir orbit has both electrons on the same side of the nucleus, corresponding to being at a turning point in a state in which the two electrons oscillate in synchrony, like two pendula on opposite sides of the nucleus.

The corresponding semiclassical explorations also proceeded successfully. Weidenmüller[139] showed how to incorporate the Pauli principle into the sum (trace) of periodic orbits as derived by Gutzwiller[140]. A particularly important advance was that of Ezra, Richter, Tanner and Wintgen[141], who studied the near-collinear states of helium, corresponding to the stretching vibrations in the collective, molecular picture. Arguments had been made for some time, based largely on the analyses done in hyperspherical coordinates, that the stationary states of such a system included states analogous to both the symmetric stretch and the antisymmetric stretch modes of a harmonic model of a molecule. The former would be the "ridge states" because they would have maximum probability on the ridge of the potential if it is represented in the hyperspherical coordinates described earlier. Ezra et al. quantized the dynamics of the model, which they showed is chaotic, and found evidence for the antisymmetric states but nothing at all like the symmetric stretch or "ridge" states. They concluded that such states do not exist. However the manifold developed as the states of the rotor-vibrator model virtually demands stretching-mode states that are totally symmetric with respect to the symmetry operations of the rotor-vibrator Hamiltonian. The answer is that they are indeed there, but do not look like the symmetric-stretch normal modes of a harmonic oscillator. Instead, they are the sums--symmetric combinations--of local-mode excitations, analogous to the local-mode states of the water molecule[37-39]. They had been identified as such, or at least the lowest member of the series had, the 2s3s $^1S^e$ state of He and its analogues in the alkaline earth atoms, in the work of Hunter[55]. Figure 6 shows conditional probability distributions for He in collinear geometry, for the 2s3s $^1S^e$ and 2s3s $^3S^e$ states, from well-converged Sturmian CI functions and from the three approximate functions: single-configuration, Hartree-Fock type functions(but based on hydrogenic orbitals), symmetric and antisymmetric combinations of local Morse oscillators with only one quantum in either one or the other of the local oscillators, and finally the symmetric and antisymmetric normal mode states.

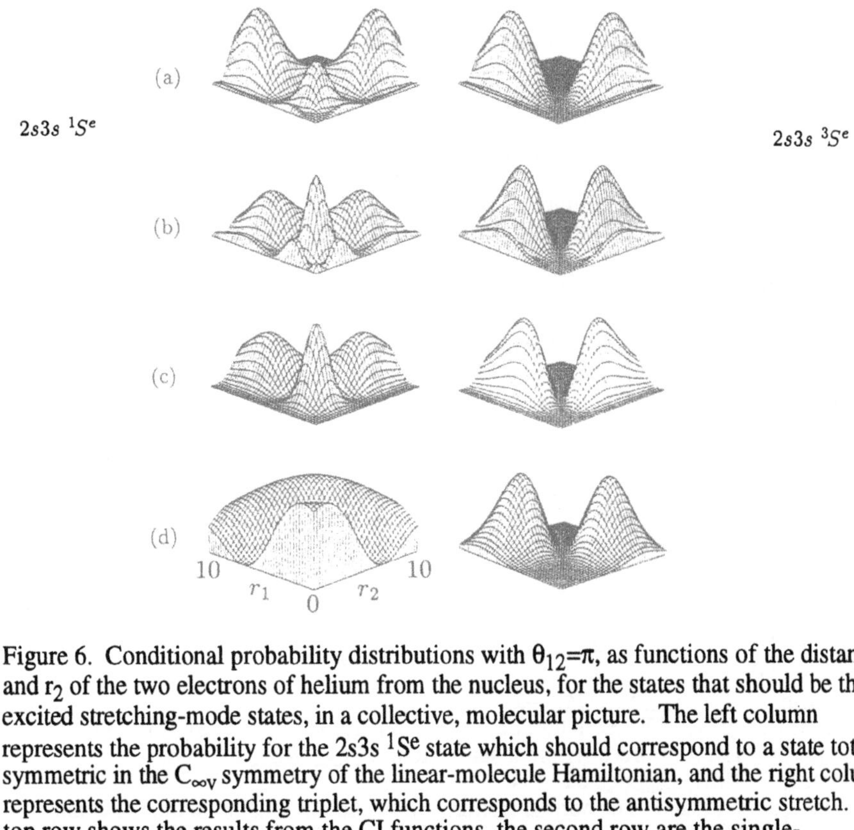

Figure 6. Conditional probability distributions with $\theta_{12}=\pi$, as functions of the distances r_1 and r_2 of the two electrons of helium from the nucleus, for the states that should be the first excited stretching-mode states, in a collective, molecular picture. The left column represents the probability for the 2s3s $^1S^e$ state which should correspond to a state totally symmetric in the $C_{\infty v}$ symmetry of the linear-molecule Hamiltonian, and the right column represents the corresponding triplet, which corresponds to the antisymmetric stretch. The top row shows the results from the CI functions, the second row are the single-configuration Hartree-Fock-like functions, the third row are the symmetric and antisymmetric combinations of Morse local-mode states and the fourth row are based on harmonic, normal-mode functions. Clearly all four of the triplet functions are similar to each other, but the normal-mode symmetric-stretch function gives results very different from the others, notably from either the well-converged CI function or the symmetric local-mode function.

While all the representations are similar for the triplet, the one which is clearly not at all like the accurate function is the harmonic, normal-mode symmetric stretch. This is borne out quantitatively for this state and the corresponding states of the alkaline earths, through the comparison of the mean values of $\cos\theta_{12}$ and of $\cos^2\theta_{12}$ for these states. While the former is very small, the latter is not, showing that the low value of the former is due to cancellation of significant contributions with opposite sign, not to orthogonality of the motion of the two electrons[142].

5. Summary and Conclusions

This discussion has begun with the early history of the attempts to find the nature of the quantization in simple, two-electron and few-electron atoms, then went into the means by which we can assess the extent of validity of one or another model, and then turned to the specific probes and findings. Finally, the discussion returned to a very brief survey of the progress that has been made in trying to solve the same classical and semiclassical forms of these problems that we used to introduce the Chapter. Many, perhaps most of the questions of what the best constants of motion are, and of how we can recognize them from experiments, remain unanswered. However we have far clearer ideas now than anyone had in the 1920's of how to get answers, and even more so of how rich and enlightening are the questions we have to ask along the way. The several-body problem is terribly, tantalizingly difficult, even in special cases such as the planetary and Langmuir orbits. Let us hope that this Chapter is a stimulus to some of its readers to help move us toward new answers.

References.

1. Bohr, N. (1976) Niels Bohr: Collected Works, J.R. Nielsen, ed., North-Holland, Amsterdam.
2. Kramers, H. A. (1923) "Über das Modell des Heliumatoms", Z. Physik **13**,
3. Landé, A. (1919) Phys. Z. **20**, 233.
4. Landé, A. (1920) Phys. Z. **21**, 114.
5. Kemble, E. C. (1921) Phil. Mag. **42**, 123.
6. Langmuir, I. (1921) "The Structure of the Helium Atom", Phys. Rev. **17**, 339.
7. Langmuir, I. (1921) "Forces within a static atom", Phys. Rev. **18**, 104.
8. Langmuir, I. (1921) Science **52**, 434.
9. van Vleck, J. H. (1922) "The normal Helium Atom and the relation to the Quantum Theory", Phil. Mag. **44**, 311.
10. van Vleck, J. H. (1922) "The Dilemma of the Helium Atom", Phil. Mag. **44**, 842-869.
11. Hylleraas, E. A. (1928) Z. Phys. **48**, 469.
12. Hylleraas, E. A. (1929) Z. Phys. **38**, 739.
13. Hylleraas, E. A. (1929) Z. Physik **54**, 347.
14. Dickens, P. G. and Linnett, J. W. (1957) Quart. Revs. **11**, 291.
15. Coulson, C. A. and Neilson, A. H. (1961) Proc. Phys. Soc. **78**, 831.
16. Madden, R. P. and Codling, K. (1963) Phys. Rev. Lett. **10**, 516.
17. Cooper, J. W., Fano, U. and Prats, F. (1963) "Classification of Two-Electron Excitation Levels of Helium", Phys. Rev. Lett. **10**, 518-521.

18. Fano, U. (1961) "Effect of Configuration Interaction on Intensities and Phase Shifts", Phys. Rev. **124**, 1866-1878.
19. Lipsky, L., Anania, R. and Conneely, M. J. (1977) At. Data Nucl. Data Tables **20**, 127.
20. Callaway, J. (1978) Phys. Lett. A **66**, 201.
21. Bhatia, A. K. and Temkin, A. (1964) Rev. Mod. Phys. **36**, 1050.
22. Bhatia, A. K., Temkin, A. and Perkins, J. F. (1967) "Hylleraas Variational Calculations of Autoionizing States", Phys. Rev. **153**, 177.
23. Bhatia, A. K. (1972) Phys. Rev. A **6**, 120.
24. Bhatia, A. K., Burke, P. G. and Temkin, A. (1973) Phys. Rev. A **8**, 21.
25. Bhatia, A. K. and Temkin, A. (1975) Phys. Rev. A **11**, 2018.
26. Bhatia, A. K. (1977) Phys. Rev. A **15**, 1315.
27. Ho, Y. K. (1979) "Autoionization states of helium isoelectronic sequence below the n=3 hydrogenic threshold", J. Phys. B **12**, 387-399.
28. Ho, Y. K. (1980) Phys. Lett. **79A**, 44.
29. Ho, Y. K. (1981) Phys. Rev. A **23**, 2137.
30. Macek, J. H. (1967) "Application of the Fock Expansion to Doubly Excited States of the Helium Atom", Phys. Rev. **160**, 170.
31. Macek, J. (1968) "Properties of autoionizing states of He", J. Phys. B **1**, 831-843.
32. Lin, C. D. (1974) "Correlation of excited electrons. The study of channels in hyperspherical coordinates", Phys. Rev. A **10**, 1986-2001.
33. Fano, U. (1976) Physics Today **29**, 32.
34. Lin, C. D. (1976) "Properties of resonance states in H-", Phys. Rev. A **14**, 30-35.
35. Wulfman, C. E. (1968) Phys. Lett. **26A**, 397.
36. Sinanoglu, O. and Herrick, D. R. (1975) "Group theoretic prediction of configuration mixing effects due to Coulomb repulsion in atoms with applications to doubly-excited spectra", J. Chem. Phys. **62**, 886-892.
37. Lawton, R. T. and Child, M. S. (1979) "Local mode vibrations of water", Mol. Phys. **37**, 1799-1807.
38. Lawton, R. T. and Child, M. s. (1980) "Excited stretching vibrations of water: the quantum mechanical picture", Mol. Phys. **40**, 773-792.
39. Lawton, R. T. and Child, M. S. (1981) "Local and normal stretching vibrational states of H_2O. Classical and semiclassical considerations", Mol. Phys. **44**, 709-723.
40. Child, M. S. and Lawton, R. T. (1981) Farad. Disc. **71**, 1.
41. Herrick, D. R. and Sinanoglu, O. (1975) "Comparison of doubly-excited helium energy levels, isoelectronic series, autoionization lifetimes and group-theoretical configuration-mixing predictions with large -configuration-interaction calculations and experimental spectra", Phys. Rev. A **11**, 97-110.
42. Herrick, D. R., Kellman, M. E. and Poliak, R. D. (1980) "Supermultiplet classification of higher intershell doubly excited states of H- and He", Phys. Rev. A **22**, 1517-1535.
43. Herrick, D. R. and Kellman, M. E. (1980) "Novel supermultiplet levels for doubly excited He", Phys. Rev. A **21**, 418-425.
44. Kellman, M. E. and Herrick, D. R. (1978) "Rotor-like spectra for some doubly excited two-electron states", J. Phys. B **11**, L755.
45. Kellman, M. E. and Herrick, D. R. (1980) "Ro-vibrational collective interpretation of supermultiplet classifications of intrashell levels of two-electron atoms", Phys. Rev. A **22**, 1536.
46. Herrick, D. R. (1982) Adv. Chem. Phys. **52**, 1.

47. Munschy, G. and Pluvinage, P. (1963) Rev. Mod. Phys. **35**, 494.
48. Rehmus, P., Kellman, M. E. and Berry, R. S. (1978) "Spatial correlation of atomic electrons: He**", Chem. Phys. **31**, 239-262.
49. Rehmus, P., Roothaan, C. C. J. and Berry, R. S. (1978) "Visualization of electron correlation in ground states of He and H^-", Chem. Phys. Lett. **58**, 321-325.
50. Rehmus, P. and Berry, R. S. (1979) "Visualization of electron correlation in a series of helium S states", Chem. Phys. **38**, 257-275.
51. Yuh, H.-J., Ezra, G. S., Rehmus, P. and Berry, R. S. (1981) "Electron correlation and Kellman-Herrick quantization in doubly excited helium", Phys. Rev. Lett. **47**, 497-500.
52. Ezra, G. S. and Berry, R. S. (1983) "Collective and independent-particle motion in doubly excited two-electron atoms", Phys. Rev. A **28**, 1974-1988.
53. Krause, J. L. and Berry, R. S. (1985) "Electron correlation in the ground and low-lying states of alkaline earth atoms", J. Chem. Phys. **83**, 5153-5162.
54. Krause, J. L. and Berry, R. S. (1986) "Electron Correlation in Alkali Negative Ions", Comments At. Mol. Phys. **18**, 91-106.
55. Hunter, J. E., III and Berry, R. S. (1987) "Projection of accurate configuration-interaction wave functions for He** and the alkaline-earth-metal atoms onto simple rotor-vibrator wave functions", Phys. Rev. A **36**, 3042-3053.
56. Hunter, J. E., III and Berry, R. S. (1987) "Oscillator Strengths for the Alkaline-Earth Atoms Using Rotor-Vibrator and Configuration-Interaction Wave Functions", Phys. Rev. Lett. **59**, 2959-2962.
57. Batka, J. J., Jr. and Berry, R. S. (1993) "Validity Criteria for Rotor-Vibrator and Independent-Particle Models of Atoms: Overlaps, Oscillator Strengths and Angular Deviations", J. Phys. Chem. **97**, 2435-2442.
58. Ceraulo, S. C. and Berry, R. S. (1991) "Quadrupole moments as measures of electron correlation in two-electron atoms", Phys. Rev. A **44**, 4145-4153.
59. Krause, J. L., Morgan, J. D., III and Berry, R. S. (1987) "Expectation values of $\mathbf{p}_1 \cdot \mathbf{p}_2$ as a measure of electron correlation in two-electron atoms", Phys. Rev. A **35**, 3189-3196.
60. Wen, J., Travis, J. C., Lucatorto, T. B., Johnson, B. C. and Clark, C. W. (1988) Phys. Rev. A **37**, 4207.
61. Schwarzkopf, O., Krässig, B., Elmiger, J. and Schmidt, V. (1993) "Energy- and Angle-Resolved Double Photoionization in Helium", Phys. Rev. Lett. **70**, 3008-3011.
62. Schwarzkopf, O., Krässig, B. and Schmidt, V. (1993) "Energy- and angle-resolved (γ,2e) experiments with synchrotron radiation: helium and argon", J. Physique IV **Colloq. C6, supplément au Volume 3**, 169-174.
63. Maulbetsch, F. and Briggs, J. S. (1993) "The angular distribution of equal-energy electrons following double photoionization", J. Phys. B **26**, L647-L652.
64. Maulbetsch, F. and Briggs, J. S. (1993) "Angular distribution of electrons following double photoionization", J. Phys. B **26**, 1679-1694.
65. Ceraulo, S. C., Stehman, R. M. and Berry, R. S. (1994) "Six-fold differential cross sections for atomic helium, magnesium and calcium in (γ,2e) experiments", Phys. Rev. A **49**, 1730-1744.
66. Tweed, R. J. (1973) "Double ionization by electron impact. II. Calculations of cross sections for H-, He and Li+", J. Phys. B **6**, 270-285.

67. Smirnov, Y. F., Pavlitchenkov, A. V., Levin, V. G. and Neudatchin, V. G. (1978) "A study of the two-electron Fourier amplitudes of atomic and molecular wave functions using the (γ,2e) and (e,3e) processes at high energies", J. Phys. B **11**, 3587-3602.
68. Neudatchin, V. G., Smirnov, Y. F., Pavlitchenko, A. V. and Levin, V. G. (1977) Phys. Lett. **64A**, 31.
69. Lahmam-Bennani, A., Dupré, C. and Duguet, A. (1989) Phys. Rev. Lett. **63**, 1582.
70. Lahmam-Bennani, A. (1991) "Recent developments and new trends in (e,2e) and (e,3e) studies", J. Phys. B **24**, 2401-2442.
71. Lahmam-Bennani, A., Ehrhardt, H., Dupré, C. and Duguet, A. (1991) "Identification of mechanisms of electron impact double ionizing collisions by e,(3-1)e experiments", J. Phys. B **24**, 3645-3653.
72. Lahmam-Bennani, A., Duguet, A., Grisogono, A. M. and Lecas, M. (1992) "(e,3e) absolute five-fold differential cross sections for double ionization of krypton", J. Phys. B **25**, 2873-2884.
73. Dal Cappello, C. and Le Rouzo, H. (1991) "Angular distributions in the double ionization of helium by electron impact", Phys. Rev. A **43**, 1395-1404.
74. Joulakian, B., Dal Cappello, C. and Brauner, M. (1992) "Double ionization of helium by fast electrons: use of correlated two electron wavefunctions", J. Phys. B **25**, 2863-2871.
75. Joulakian, B. and Dal Cappello, C. (1993) "Theoretical study of the optimal conditions for the measurement of the differential cross section of the double ionization of helium by fast electrons", Phys. Rev. A **47**, 3788-3795.
76. Berakdar, J. and Klar, H. (1993) "Structures in the cross section of double ionization of helium by the impact of fast electrons", J. Phys. B **26**, 4219-4235.
77. Berakdar, J. and Briggs, J. S. (1994) "The three-body Coulomb continuum problem", (in press)
78. Ceraulo, S. C., Stehman, R. M. and Berry, R. S. (1994) "Eight-fold differential cross sections for double ionization of helium, magnesium and calcium by electron impact", (submitted)
79. Wulfman, C. D. (1973) "Approximate dynamical symmetry of two-electron atoms", Chem. Phys. Lett. **23**, 370-372.
80. Wulfman, C. D. and Kumei, S. (1973) "A simple O(4,2) approximation for hydrogenic Coulomb integrals", Chem. Phys. Lett. **23**, 367-369.
81. Novaro, O. and Freyre, A. (1972) "O(4) and U(3) symmetry breaking in the second row of the periodic table", Mol. Phys. **20**, 861-871.
82. Fock, V. (1935) "Zur Theorie des Wasserstoffatoms", Z. Phys. **98**, 145-15.
83. Bargmann, V. (1934) "Zur Theorie des Wasserstoffatoms", Z. Phys. **99**, 576-582.
84. Englefield, M. J. (1972) Group Theory and the Coulomb Problem, Wiley-Interscience, New York.
85. Nikitin, S. I. and Ostrovsky, V. N. (1976) "On the classification of the doubly excited states of the two-electron atom", J. Phys. B **9**, 3141-3147.
86. Nikitin, S. I. and Ostrovsky, V. N. (1978) "The symmetry of the electron-electron interaction operator in the dipole approximation", J. Phys. B **11**, 1681-1693.
87. deShalit, A. and Feshbach, H. (1974) Theoretical Nuclear Physics, Wiley, New York.
88. Watanabe, S.-I. and Lin, C. D. (1986) "Demonstration of moleculelike modes of doubly excited states in hyperspherical coordinates", Phys. Rev. A **34**, 823-837.
89. Lin, C. D. (1984) "Classification of doubly excited states of two electron atoms", Phys. Rev. Lett. **52**, 1252.

90. Ezra, G. S. and Berry, R. S. (1984) "Comment on "Classification of doubly excited states of two electron atoms" by C. D. Lin", Phys. Rev. Lett. **52**, 1252.
91. Watanabe, S.-I. and Lin, C. D. (1987) "Classification of triply excited states from a molecular viewpoint", Phys. Rev. A **36**, 511-522.
92. Bao, C.-g. (1992) "Possible modes of angular motion in $^4S^o$ triply excited states", J. Phys. B **25**, 3725-3734.
93. Bao, C.-g. (1993) "The gentle collective internal oscillation in the ground state of four-valence-electron atoms", J. Phys. B **26**, 4671-4682.
94. Ezra, G. S. and Berry, R. S. (1982) "Correlation of two electrons on a sphere", Phys. Rev. A **25**, 1513-1527.
95. Ezra, G. S. and Berry, R. S. (1983) "Quantum states of two particles on concentric spheres", Phys. Rev. A **28**, 1989-2000.
96. Ojha, P. C. and Berry, R. S. (1987) "Angular correlation of two electrons on a sphere", Phys. Rev. A **36**, 1575-1585.
97. Kellman, M. E. (1994) "Origin of two-electron atomic supermultiplets in U(4) group embedding", Phys. Rev. Lett. **73?**, (in press).
98. Berry, R. S. (1986) "Collective and Planetary Motion in Atoms" in The Lesson of Quantum Theory, deBoer, J., Dal, E. and Ulfbeck, O., ed., Elsevier, Amsterday.
99. Berry, R. S. (1989) "How good is Niels Bohr's atomic model?", Contemp. Phys. **30**, 1-19.
100. Banyard, K. E. and Ellis, J. D. (1972) "A distribution function for angular correlation in He- and Be-like ions", Molec. Phys. **24**, 1291-1296.
101. Banyard, K. E. and Ellis, J. D. (1975) "Excited states of He: the behavior of interelectronic angular distribution functions", J. Phys. B **8**, 2311-2319.
102. Greene, C. H. (1981) "Doubly excited states of the alkaline earth atoms", Phys. Rev. A **23**, 661-678.
103. O'Mahony, P. F. (1985) "Electron correlations in atomic valence shells: magnesium and aluminum", Phys. Rev. A **32**, 908-916.
104. O'Mahony, P. F. and Greene, C. H. (1985) "Doubly excited states of beryllium and magnesium", Phys. Rev. A **31**, 250-259.
105. Barthelat, J. C., Durand, P. and Serafini, A. (1977) Mol. Phys. **33**, 159.
106. Bachelet, G. B., Hamann, D. R. and Schlüter, M. (1982) Phys. Rev. B **26**, 4199.
107. Miller, T. A. and Freund, R. S. (1971) Phys. Rev. A **4**, 81.
108. Miller, T. A. and Freund, R. S. (1972) Phys. Rev. A **5**, 5188.
109. Sandars, P. G. H. and Stewart, A. J. (1973) Phys. **3**, 429.
110. Angel, J. R. P., Sandars, P. G. H. and Woodgate, G. K. (1967) "Direct Measurement of an Atomic Quadrupole Moment", J. Chem. Phys. **47**, 1552-1553.
111. Glassgold, A. E. and Ialongo, G. (1968) "Angular Distributions of the Outgoing Electrons in Electronic Ionization", Phys. Rev. **175**, 151-159.
112. Byron, F. W., jr. and Joachain, C. J. (1966) "Importance of correlation effect in the ionization of helium by electron impact", Phys. Rev. Lett. **16**, 1139-1142.
113. Byron, F. W., jr. and Joachain, C. J. (1967) "Multiple Ionization Processes in Helium", Phys. Rev. **164**, 1.
114. Mittlelman, M. H. (1966) "Single and double ionization of He by electrons", Phys. Rev. Lett. **16**, 498-499.
115. McGuire, J. H. (1982) "Double Ionization of Helium by Protons and Electrons at High Velocities", Phys. Rev. Lett. **49**, 1153-1157.

116. Levin, V. G., Neudatchin, V. G., Pavlitchenkov, A. V. and Smirnov, Y. U. (1984) "A study of the electron correlations in the H_2 molecule using the double photoionization process (γ,2e)", J. Phys. B **17**, 1525.
117. Huetz, A., Selles, P., Waymel, D. and Mazeau, J. (1991) J. Phys. B **24**, 1917.
118. Dal Cappello, C., Langlois, J., Dal Cappello, M. C., Joulakian, B., Lahmam-Bennani, A., Duguet, A. and Tweed, R. (1992) Z. Phys. D **23**, 389.
119. King, G. C., Zubek, M., Rutter, P. M., Read, F. H., MacDowell, A. A., West, J. B. and Holland, D. M. P. (1988) J. Phys. B **21**, L403.
120. Kossmann, H. and Schmidt, V. (1988) Phys. Rev. Lett. **60**, 1266.
121. Wannier, G. H. (1953) "The Threshold Law for Single Ionization of Atoms and Ions by Electrons", Phys. Rev. **90**, 817-825.
122. Rau, A. R. P. (1971) Phys. Rev. A **4**, 207.
123. Peterkop, R. (1971) "WKB approximation and threshold law for electron-atom ionization", J. Phys. B **4**, 513-521.
124. Read, F. H. (1985) in Electron Impact Ionization, Märk, T. D. and Dunn, G. H., ed., Springer, Berlin.
125. Duguet, A. and Lahmam-Bennani, A. (1992) Z. Phys.. D **23**, 383.
126. Ford, M. J., Doering, J. P., Coplan, M. A., Cooper, J. W. and Moore, J. H. in *ICPEAC Satellite Meeting*, Paris.
127. Ford, M. J., Doering, J. P., Coplan, M. A., Cooper, J. W. and Moore, J. H. (1994) "(e,3e) Observation of the angular correlation between ejected and Auger electrons in the double ionization of magnesium", Phys. Rev. A (submitted).
128. Leopold, J. G. and Percival, I. C. (1980) "The semiclassical two-electron atom and the old quantum theory", J. Phys. B **13**, 1037-1047.
129. Leopold, J. G., Percival, I. C. and Richards, D. (1982) "Classical and semiclassical theory for the exchange symmetry of identical particles", J. Phys. A **15**, 805-824.
130. Leopold, J. G., Percival, I. C. and Tworkowski, A. S. (1980) "Semiclassical perturbation theory for energy levels of planetary atoms", J. Phys. B **13**, 1025.
131. Percival, I. C. (1977) "Planetary atoms", Proc. Roy. Soc. Lond. A **353**, 289-297.
132. Richter, K. and Wintgen, D. (1990) Phys. Rev. Lett. **65**, 1965.
133. Richter, K. and Wintgen, D. (1991) J. Phys. B **24**, L565-L571.
134. Richter, K., Tanner, G. and Wintgen, D. (1993) "Classical mechanics of two-electron atoms", Phys. Rev. A **48**, 4182-4196.
135. Langmuir, I. (1921) "The structure of the helium atom and the hydrogen molecule", Phys. Rev. **17**, 401(A).
136. Richter, K. and Wintgen, D. (1990) J. Phys. B **23**, L197.
137. Müller, J., Burgdörfer, J. and Noid, D. (1992) "Torus quantization of symmetrically excited helium", Phys. Rev. A **45**, 1471-1478.
138. Müller, J. and Burgdörfer, J. (1993) "Calculation of Langmuir States in Doubly Excited Helium", Phys. Rev. Lett. **70**, 2375-2378.
139. Weidenmüller, H. A. (1993) "Semiclassical periodic-orbit theory for identical particles", Phys. Rev. A **48**, 1819-1823.
140. Gutzwiller, M. C. (1990) Chaos in Classical and Quantum Mechanics, Springer, New York.
141. Ezra, G. S., Richter, K., Tanner, G. and Wintgen, D. (1991) "Semiclassical cycle expansion for the helium atom", J. Phys. B **24**, L413-L420.
142. Batka, J. J., Jr. and Berry, R. S. (1994) (in preparation)

ELECTRONIC STRUCTURE MODELS: COMPUTATIONS, CHEMICAL INSIGHTS AND APPROPRIATENESS

by

Mark A. Ratner
Department of Chemistry
Northwestern University
Evanston, Illinois 60208-3113
USA

ABSTRACT:

Electronic structure calculations lie at the very heart of modern theoretical chemistry. We discuss the nature of such calculations, stressing the importance of the central fact that these are model calculations: both *ab-initio* and semiempirical calculations use electronic structure models, in the form of a representation of the coulomb Hamiltonian over some finite basis set. Different basis sets, with different interaction integrals, correspond to different model Hamiltonians. In this sense, there is no fundamental difference between semiempirical and *ab-initio* calculations.

After a cursory overview of the nature of electronic structure calculations, we concentrate, for illustration, on the calculation of the nonlinear optical response of organic and organometallic chromophores. We find that appropriate semiempirical models do extremely well in calculating the frequency-dependent nonlinear response of a number of strong frequency-doubling chromophores, and that, when appropriate calculation models are chosen, semiempirical theories can lead to deep insight into the nature of the nonlinear optical response process.

J. L. Calais and E. S. Kryachko (eds.), Structure and Dynamics of Atoms and Molecules: Conceptual Trends, 183–212.

I. Introduction

Calculations of molecular electronic structure are at the center of theoretical chemistry: they provide direct, bias-free numerical guidance into the structure and response of molecules. These are issues at the heart of chemistry, and therefore electronic molecular calculations offer a deep and powerful way to interpret chemical phenomena in theoretical terms. Indeed, for some particular questions (such as the role of relativity in changing the observable properties of the hydrogen molecule, or structures of very unstable species) electronic structure methods are now superior to experiment. Moreover, the formal tools for calculating molecular electronic structure and response at quite accurate levels, especially for relatively small organic molecules, are now in place. Indeed, the development and application of electronic structure methods has probably been the central triumph of theoretical chemistry over the past half century.

On the other hand, often these computations result in a myriad of confusing numbers, such that while the computer may know the answer to the chemical question being posed, neither the computational chemist or the laboratory scientist can really understand its answer. While many formal tools for the calculation of molecular electronic structure are largely in place, the interpretive tools still lag fairly far behind; even a simple question such as how polar a particular bond may be, or how dominant the electrostatic component of hydrogen bomds is, remain difficult to answer. Development of interpretative tools for answering questions such as these is often much less precise than actual electronic structure methods. Such tools as Mulliken[1] or Lowdin[2] populations, natural orbital populations[3], bond polarity indices[4], reactivity indices[5], molecular hardness and softness[6], gradient functions for bonding[7], or total electrostatic potentials[8] are useful, and still require substantial insight and care before they can answer the questions the chemist poses to the computation.

In this contribution, I would like to stress that three levels of effort are really necessary before the results of electronic structure calculations can be useful in understanding chemical phenomena. These are:

(1) *Definition of a Model Hamiltonian.* Here, while theorists understand, experimentalists generally do not. There is a great deal of misinformation and confusion, particularly around the terms "*ab-initio*" calculation and "semiempirical electronic structure" study. Here, the notion (more common in the physics than the chemistry literature) of an electronic structure *model* is extremely helpful; it can be best understood[9] in the language of second quantization, and is the subject of Section II.

(2) *Level of Theory Involved in the Computed Functional Property.* Given model hamiltonians, wave functions and response properties can be calculated at various levels of approximation; starting with an SCF analysis, this nearly always involves discussion of how strongly correlated the ground state wave function is, and what the nature of the excited state or response computation may be. In calculation of response[10,11], the state of the art is less advanced than in ground state property calculation, especially since there is no convenient variational theorem to aid in honing the accuracy of such computations. This is discussed in Section III.

(3) *Interpretative Schemes.* Given the necessarily complicated wave function that results from use of an accurate electronic structure model and a reasonable degree of correlation, reduction of the numerical result to a conceptual framework is a very difficult challenge for the theorist. In general, theorists have been less active in this area

(particularly over the last few decades) than in the actual computation. While a book would be required to deal with this problem properly, we present in Section IV some particular examples, based on nonlinear optical response.

None of these three issues is an easy one, and is probably fair to say that progress has been greatest in the first, impressive in the second, and seriously lacking in the third. Within the past 20 years, density functional methods have become popular and important in the understanding of molecular electronic structure. While most of the remarks in this paper can be transferred to a density functional format, there are slight differences in the mechanics. We therefore will limit our remarks to calculations based on the use of self consistent fields (Hartree-Fock) techniques.

II. Model Hamiltonians

The Schrödinger equation, with the Pauli principle, defines the states and energy levels of a molecule. If, in addition, we assume the Born-Oppenheimer separability, then for any given choice of geometry one can solve for the electronic wave function of a molecule.*

The electronic structure Hamiltonian, for fixed nuclear coordinate, can be the written as[9,12]

$$H = \sum_{\mu}\sum_{ij} \beta_{ij} a^{+}_{i\mu} a_{j\mu} + \frac{1}{2}\sum_{\mu\nu}\sum_{ijk\ell} \langle ij|k\ell\rangle a^{+}_{i\mu} a^{+}_{\ell\nu} a_{\ell\nu} a_{k\mu} \qquad (1)$$

Here the Greek letters label spin, the Latin letters label spatial basis functions. The creation and destruction operators follow the usual anticommutator relationships,

$$[a_{i\mu}, a^{+}_{j\nu}]_{+} = \delta_{\mu\nu} S_{ij} \qquad (2)$$

where S_{ij} is the overlap of basis functions u_i and u_j, and $\delta_{\mu\nu}$ is the Kronecker symbol. The operator $a_{i\nu}$ destroys an electron in basis function. This Hamiltonian is *completely described* given the matrix elements β_{ij} and $\langle ij|kl\rangle$, as well as the overlap integral S_{ij}. *Any* specification of these parameters, then, along with a list of functions (the range over which the Latin subscripts run), completely describes the electron structure Hamiltonian.

This view is not new; indeed, it has been clear since the first work in which second quantization esd used to describe molecular electronic structure[9,13] that *all* electronic structure Hamiltonians were indeed model CS; even numerical basis calculations, such are commonly used for atoms, use a model Hamiltonian and a basis (albeit, in this case the basis is numerical).

Characteristic *ab-initio* calculations, then, are defined in terms of basis functions of well defined analytic type (or occasionally numerical type), whose matrix elements are

---

*We will limit our discussions to nonrelativistic situations, since the great majority of computations reported to date do not directly address the issues of relativity.

calculated directly from their forms; this involves a six dimensional integral for the two electron terms (ij|kl), and three dimensional integrals for the overlap integral S_{ij} and the tunneling matrix elements β_{ij}. In semiempirical models, these parameters are chosen differently. For instance, in the Huckel model, the summation is limited to a single π-electron p-type basis function on each center, S_{ij} is chosen as the Kronecker symbol, all two electron interactions are neglected, and β_{ij} is chosen according to a particular recipe. All of the semiempirical models (PPP, CNDO, INDO, Fenske-Hall, Hubbard, extended Hubbard...) correspond simply to different arbitrary choices of the basis sets and the appropriate matrix elements.

In this context, it is quite clear that semiempirical models are in no sense more approximate than *ab-initio* models: both correspond to a particular choice of basis set, and particular choice of the interaction matrix elements among the members of that set. In semiempirical models, one generally fixes the values of the integral with an eye both to simplicity and to accuracy for a particular set of properties within a particular class of molecules; to assure the latter, the matrix elements are often chosen by comparison with experiment for a particular set of sample species.[14]

The important point here is that all electronic structure descriptions of wave function type correspond to particular model Hamiltonians. The great advantage of *ab-initio* methods is that they are systematically improvable, in the sense that the basis sets can be extended, and as they are extended, one generally expects (in the absence of peculiarities like overcompleteness and near linear dependence) a systematic improvement in the wave function, and wave function related properties including energies and response. Nevertheless, it is quite clear from the second quantization description that there is no fundamental difference between semiempirical models of whatever sort and *ab-initio* basis function models, both are Hamiltonians.

III. Response Properties, Wave Functions, and Levels of Theory

Just as a great deal of research work has clarified, at least for ground state calculations of *ab-initio* type, how specific improvements in the basis set can result in specific improvements in ground state properties, so for simple static properties of the ground state, a good deal has been learned about how much electronic correlation (improvement to the simple single determinant Hartree-Fock structure) is necessary[15]. In terms of the usual "Pauling diagram", in which basis set is plotted along the abscissa and level of correlation along the ordinate, ground state calculations for particular quantities are well understood (at least for first row molecules, and in particular for small organics). The situation with respect to dynamical response properties is more complicated.

If one wishes to compute, for example, a simple linear response approximation to the absorption line shape, one must analyze[16,17] the Fourier transform of a dipole correlation function, as in equation 3.

$$I(\omega) = A\int_{\infty}^{\infty} <0|\mu(t)\mu(0)|0> e^{i\omega t}\, dt \qquad (3)$$

Here A is a proportionality constant, and $\mu(t)$ is the time dependent dipole operator. The expectation value is computed over the ground state (the calculation is done at zero temperature). As written, this computation involves simply a dynamical evolution on the ground electronic state: the dipole moment must be computed as a function of time, and then its scalar product with the dipole moment function at time zero is averaged over the ground state. It is entirely equivalent, however, to rewrite this in the spectral resolution form as[9a,18,19]

$$I(\omega) = A\hbar\sum_{s} \frac{<0|\mu|S><S|\mu|0>}{E_s - E_0 - \hbar\omega + i\eta} \qquad (4)$$

where η is an infinitesimal factor added to insure causality.

Both equation 3 and equation 4 are exact within linear response: if the ground state wavefunction denoted |0> is calculated exactly, and either the time evolution dynamics calculated exactly (equation 3) or the complete sum-over-exact-excited- states properly summed (equation 4), then the calculated line shape will be exact. Spectral resolutions of this kind exist for higher order responses as well; for example, one can derive an exact expression for the first hyperpolarizability, based on the use of time dependent perturbation theory. As presented by Orr and Ward[20,21], this reads

$$\beta_{ijk}(-\omega_3;\omega_1,\omega_2) = \hbar^{-2}\sum_{p}\left\{\sum_{mn}{}' (\mu_i)_{gm}(\hat{\mu}_j)_{mn}(\mu_k)_{ng} / (\omega_{gm} - \omega_3)(\omega_{gn} - \omega_1)\right\}. \qquad (5)$$

Here the notation is that the property β_{ijk} is the hyperpolarizability component formally describing the response of a molecular electronic system to an applied electromagnetic field. In turn, this property is defined by[22,23,24,25,26]

$$P_i = \mu_i + \sum_{i\geq j} \alpha_{ij} Fj + \sum_{i\leq j\leq k} \beta_{ijk} F_j F_k + \cdots \qquad (6)$$

where P_i, μ_i, α_{ij}, F_j and β_{ijk} are respectively the electronic polarization of a molecule along the ith axis, the permanent dipole moment along that axis, the polarizability tensor, the applied electric field, and the first hyperpolarizability tensor. Since the fields can be time (or frequency) dependent, so can the polarizations. The notation $\beta(-\omega_3;\ \omega_1,\ \omega_2)$ means that the hyperpolarizability tensor is computed at a frequency $\omega_3 3$ which is the sum of ω_1 and ω_2. On the right side of equation 5, the first sum is over all permutations of the pairs $(\mu_i, -\omega_3)$, (μ_j, ω_1), (μ_k, ω_2). The second sum is over all electronic states of the molecule, and the subscripts g, m and n denote respectively ground state and different molecular excited states. The matrix elements are defined with respect to the ground state, such that, by convention[21,25]

$$(\mu_i)_{mn} = <m|\mu|n> \qquad (7a)$$

$$(\hat{\mu}_i)_{mn} = (\mu_i)_{mn} - \delta_{mn}\mu_{gg} \qquad (7b)$$

The essential difference between equations 4 and 5 is that the line shape, that arises from photon mixing of the ground state with a singly excited state, involves a sum only over singly excited states; similarly, the spectral resolution of the first hyperpolarizability arises formally, due to three photon mixing between the ground state and excited states. It therefore has two energy denominators and three matrix elements and in the numerator. Similar formulas can be given for higher hyperpolarizabilities[21,27]; in general these will involve matrix elements in the numerator and products of energy denominators. Once again, these spectral-resolutions are exact, if the ground states, matrix elements, excited states and frequencies are calculated exactly.

The practical part of chemistry does not generally deal with exact calculations. For any real molecule, the current state-of-the-art requires selection both of the one electron basis set and of a multi-electron basis set; on the Pauling type diagram of Fig. 1, the one electron basis is the abscissa, and the multi-electron basis the ordinate. However, in calculating response properties, a third issue arises, involving calculation of the dynamics itself. In the time dependent response of equation 3, one must actually calculate the time dependence of an operator. In the equivalent spectral-resolution form of equations 4 or 5, one is required to do a summation over all excited states. In this formulation, then, one has to consider the one electron basis, the nature of the correlations in the ground state, and the nature of the excitations and excited states permitted in the calculation. Calculation of the static response properties (where all frequencies in equations 3, 4 and 5 vanish) is substantially simplified, by noting that equation 6 is in the form of a Taylor series. Therefore, it is clear that one can use the formal definitions

$$\alpha_{ij} = \frac{\partial P_i}{\partial F_j}\Big|_{F=0} \tag{8a}$$

$$\beta_{jk} = \frac{\partial^2 P_i}{\partial F_j\, \partial F_k}\Big|_{F=0} \tag{8b}$$

to calculate the static response properties.

These derivatives can be performed either using finite field methods, giving rise to so-called FF approximations[28], or using analytic derivatives (calculation of analytic derivatives at correlated and uncorrelated levels has been one of the most significant modern advances in the calculation of response[15,29]). Modern *ab-initio* techniques, with correlation included at various levels including coupled cluster[30], MCSCF[31], singles and doubles configuration interaction, or various levels of perturbation theory (MP2, MP3, MP4) have indeed been reported, and reviewed[23,26,31,32], for static polarizabilities and hyperpolarizabilities. Our focus here will be on the more general question of dynamical response.

The simplest approach to the problem of dynamical response within a given model Hamiltonian is simply to take the ground state as the Hartree-Fock state, and to calculate all excited states that enter. Often this summation is truncated (see Sec. 4); this is irrelevant for the general discussion. It is then clear that the only non-vanishing contributions to the excited states in either equation 4 or equation 5 correspond to single excitations from the groundstate since the dipole operators are of one-electron type. This means that the excited states to be summed are limited in number, being essentially $N_{occ}\,N_{unocc}$, where N_{occ} and N_{unocc} are respectively the number of occupied and unoccupied Hartree-Fock orbitals. Such a calculation is generally referred to as a sum over states, or SOS, calculation in the literature. It is exactly equivalent[23] to a random phase approximation (RPA)[9a,33] and in the static limit it is also equivalent to doing either finite field or analytic derivative calculations of the polarization as calculated in the ground state.* Again, in the static limit, if one chooses for instance an MCSCF ground state, then the sum-over-states formula will give an MC-RPA approximation, which will be precisely equivalent to the MCSCF analytical derivative, or MCSCF finite field approximation, calculation of the response[23].

When frequency dependence is required, different schemes have been used. Generally, estimates are made: for example, one can arrive at an estimate for the hyperpolarizability as[34]

$$\beta^{best}(-\omega_3;\omega_2,\omega_1) = \beta^{MP2}(-\omega_3;\omega_2,\omega_1)\cdot\left[\frac{\beta^{best}(0;0,0)}{\beta^{MP2}(0;0,0)}\right] \tag{9}$$

The assumption (unproven and almost certainly untrue in detail) behind this approximation is that the corrections due to electronic correlation of the ground state and to frequency dependence are to some extent separable: equation 9 means that one obtains a best estimate for frequency-dependent hyperpolarizability by taking the most correlated available calculation of the frequency dependent property (in this case, for example, done at the MP2 level) and multiplies it by the ratio of the static response between the best possible static calculation and the MP2 static calculation. Results using corrections of this kind appear in literature, as do results in which the correction is assumed to be additative, as in equation 10, rather than multiplicative[34].

$$\beta^{best}(-\omega_3;\omega_2,\omega_1) = \beta^{MP2}(-\omega_3;\omega_2,\omega_1) + [\beta^{best}(0;0,0) - \beta^{MP2}(0;0,0)] \tag{10}$$

Shelton and Rice[34] discuss these approximations, and comment on situations in which one or the other is to be preferred. It is important to note here that if one chooses to work within highly correlated description of the static problem, one must often make approximations of this kind to calculate frequency-dependent responses. There are a number of techniques for generating frequency dependent response without approximations of this kind; these include higher order polarizability propagator approximations[19], analyses based on the use of perturbation theory[31], and approaches using general response formulations[26]. In fact, calculations using these techniques are just beginning to appear in the *ab-initio* literature, in connection with the calculation of hyperpolarizabilities or high order response properties.

---

*In fact, the usual analysis involves calculation of the change in the energy in the presence of the field, rather than changing the dipole moment in the presence of the field. For electronic structure approximations that satisfy the Hellmann-Feynman approximation, such as the SCF approximation itself, the MCSCF approximation, and full CI, these two analytic derivatives will be the same. For other methods, such as perturbation theory in which the Hellmann-Feynman theory does not hold, the more usual analysis is based on the approximation to derivatives of the energy.

From the point of view of convenience, utilizing the spectal resolutions of equations 4 or 5 is in fact fairly easy; one simply needs to approximate the ground state and the excited state, approximate the energies, and choose which terms enter into the sum. While the efficiency of such calculations is often very unattractive, such sum-over-states methods are attractively simple, and generally applicable. Perhaps more important, they also can be useful for interpretative purposes: one can then discuss exactly which excited states dominate the response at a particular level[25,35].

For accurate response computations within the *ab-initio* context, actual sum-over-states calculation is extremeluy inefficient, and therefore very rare. Instead, the sums are replaced by equivalent formulations in which linearl equations are solved. In this way, both linear and higher-order response calculations have been reported, starting from correlated ground states and corresponding to sums over millions of configurations.

In calculations of vibrational properties, the actual time dependent propagator of equation 3 is very generally used; wavepacket methods and exact propagation techniques have made calculations of this kind quite common in such areas as vibrational spectroscopy, unimolecular decay, electron transfer, proton transfer reactive collision and dynamical problems[36]. In the electronic structure literature, explicit time dependence calculated via propagation methods are difficult, because of the deBroglie wavelength of the electron; few such general calculations are yet available[36].

In a very incomplete way, we've just summarized the current state of approaches to calculation of electronic response properties, in the Born-Oppenheimer limit. The viewpoint we would like to stress is that electronic structure calculation of a response property requires a number of choices: one must choose a model Hamiltonian (equivalent to choice of the one particle basis set and particular allowed interactions), a level of correlation for the ground state, and a level of treatment of the dynamics and excited states. Because the electronic excited states generally involve expansion of electronic clouds, the one particle basis necessary for calculating them and can often differ substantially from that required for ground state calculations. Thus, for example, accurate *ab-initio* calculations of hyperpolarizability responses requires, at a minimum, a basis that includes diffuse functions and polarization functions, to describe the structure of the expanded electronic cloud[23,24,31,26]. An attractive alternative approach, then, is to choose a semiempirical model Hamiltonian in which a minimum basis is still used, but the matrix elements themselves are scaled, semiempirically, to describe excitations in the sample set of molecules. This was a very important historical step in the analysis of optical spectra: early approximate calculations on the optical spectra of conjugated molecules required[37] changes in the matrix elements used (in modern language, these were minimal basis semiempirical calculations) to get any reasonable agreement with experiment. Similarly, the approximations involved in this simple Huckel model of hydrocarbons are only mutually consistent if a semiempirical value is chosen for the Huckel resonance integral, β; indeed, it has been stressed for many years[38] that different experiments, operating on different time scales and accessing different states of the molecule, require different semiempirical patametrization.

Development by a number of workers, particularly Pople[39] and Dewar[40], of semiempirical electronic structure schemes starting nearly four decades ago again is based on use of semiempirical parametrization within the model Hamiltonian, to deal partially with effects such as basis set incompleteness and changes in electron localization that are not included directly in the model.

In the calculation of such properties as electronic hyperpolarizabilities, the use of semiempirical models is now suspect in several quarters. The view we take here is that such suspicions are *entirely inappropriate*: well-chosen semiempirical models can provide in-depth understanding both of magnitudes of hyperpolarizability response and of the sources, in the electronic and molecular structure, of those response properties.[25,41-48] It is of course necessary to be entirely consistent in the use of the semiempirical model, and appropriate semiempirical models must be chosen. Given an appropriate electron semiempirical Hamiltonian, with appropriate treatment of correlation of the ground state and appropriate dynamical analysis, convincing, accurate and useful semiempirical calculations of electronic structure response (such as hyperpolarizability) can be obtained. We proceed to some illustrations of this, in the particular case of electronic hyperpolarizability.

IV. Example: The Frequency Dependent Hyperpolarizabilities of Push Pull Substituted Aromatics

A. *Model Hamiltonian*

Hyperpolarizabilities of molecules, either isolated or in a model solvent, are of current substantial interest for possible technical applications in information processing.[49-55] From a fundamental point of view, calculation of hyperpolarizabilities is challenging and complicated, because it involves interaction with more than one photon. Extensive computational study of hyperpolarizabilities has occurred over the past decade or so;[24,25,31,32] we overview here a few insights that have been gained from our own work[25,56-66] on the first hyperpolarizability response. We will concentrate on the particular category of push pull organic pi systems; we also make some comments on applications to other molecules, and the nature of what has been learned, and on how these results compare both with experiment and with *ab-initio* calculation.

The calculations are all based on the use of two slightly different semiempirical models. These models are the PPP model[9,67] of pi electron systems, and the INDO/S model Hamiltonian for all valence electron systems, parametrized by Zerner with the idea of fitting spectra and ground state properties of organometallic species as well as organics.[68] The specification of these Hamiltonians is given elsewhere;[9,14] in the PPP model, the sums in equation 1 are limited to one p – pi basis function on each carbon or heterocycle atom. In the INDO model, the sum runs over a minimum Slater orbital valence basis set (one orbital on hydrogen, four orbitals on carbon or nitrogen or oxygen); the interaction integrals in equation 1 are chosen as described in the literature.[13,68]

It is to be noted that both of these Hamiltonians are semiempirical, and that the appropriate interactions are chosen, by fitting to some selected experimental data, to reproduce particular properties. Both the PPP and INDO/S models have proven very effective for calculating linear optical response (absorption spectra).[14,69] Since the spectral representation of equation 5 shows that the hyperpolarizability involves exactly the same kinds of molecular properties (dipole matrix elements, dipole moments, excitation energies to singly excited states) that are involved in the optical spectrum, we suspected that these model Hamiltonians would also be appropriate for computing the first hyperpolarizability.

B. Ground State

We chose the ground state to be the Hartree-Fock state, so that

$$|0\rangle = |g\rangle = |HF\rangle . \quad (11)$$

Once again, we point out that this choice of the uncorrelated ground state is also used in calculating optical spectra using these model Hamiltonians.

For small systems, it is not necessary, of course, to choose the Hartree-Fock ground state. Indeed, elegant work by Soos and others[70-72] has been done that reports the hyperpolarizability response for PPP models using full CI representations both in the ground state and in the excited states; that is, these are the exact results within the model Hamiltonian. Important conclusions follow from this work: as Table 1 shows, when different levels of correlation are included, very substantial changes in the calculated nonlinear response properties follow; indeed, even the sign changes. I do *not* view this as a criticism of the use of semiempirical models: the semiempirical model Hamiltonian is parametrized and chosen to describe particular kinds of response, and it is probably inconsistent to overcorrelate the states (that is, to go beyond the Hartree-Fock description of the ground state) if these same parameters are chosen for the model Hamiltonian. That is: within a given choice of semiempirical model Hamiltonian, particular response properties can be calculated accurately using a given level of theory (with respect to treatment of correlations and dynamical response). It is inconsistent to overbalance these descriptions - that is, choice of the semiempirical model Hamiltonian parameters to some extent takes into account correlation and screening effects in the ground state. If correlation is then added by going beyond the Hartree-Fock description, one risks inconsistent description, and artifactual result. The full CI work shows explicitly how important it is to be consistent in the choice of model Hamiltonian and treatment of correlation and response. This is not a new insight - indeed, in their description of the Huckel model, the well-known textbook[38] by Murrell, Kettle and Tedder explicitly points out that different semiempirical parametrizations will give different response properties — as they should!

C. Nature of the Excitations

To calculate the hyperpolarizability from the spectral representation of equation 5, it is necessary, having defined the Hamiltonian model and the ground state choice, to describe the excited states. Since the dipole operators are one-electron operators, the excited states $|m\rangle$ (or $|n'\rangle$) and $|n\rangle$ should be chosen as single electron excited states — that is, as monoexcited configurations, in which an electron is promoted from an occupied to an empty molecular orbital. Diagrammatically, this can be sketched in Fig. 1..

Two sorts of excited configurations will in fact appear in the sum of eqn. 5. Two level terms of the sort of Fig. 2 are the simplest. Three level terms of the sort of Fig. 3 will also enter.[25] We will make more comments on these below.

The computation, then, is straightforward: one chooses the semiempirical model Hamiltonian and the appropriate geometry, calculates the Hartree-Fock ground state, calculates the singly excited configurations, computes the dipole matrix elements and dipole moment differences and energies of the excited configurations, and does the summation in equation 5. Truncation of the monoexcited configuration interaction (MECI) sum, therefore limiting the number of configurations is often done, to simplify calculation.[25] We conclude this section with a discussion of the results of these semiempirical calculations, the nature of this sum-over-states convergence, comparison with experiment, insight into frequency-dependent response, specific mechanistic conclusions involving specific molecules, and some comments on appropriateness.

D. Some Selected Results

Comparison of calculated and observed nonlinear optical response is a bit more difficult than that for simple properties such as ionization energy, bond length or frequency of maximum absorption. The usual experimental measurement is made in solution, using the EFISH technique;[50] comparison with T=0 vapor phase computational work is then slightly suspect. Nevertheless, these calculations really scale rather nicely by comparison with experiments; Fig. 4 shows the calculated versus measured response for a series of organic and organometallic chromophores. The variation of the slope (smaller) from unity is largely due to solvent polarization effects.[66,74-77] The qualitative behavior is that excited stated dipoles are generally larger than ground state dipole moments (for a positive β), so that the energies of the excited state are lowered more by solvent than those of the ground state. This makes the excited state easier to reach in the liquid than in the vapor, and (as is clear from the spectral-resolution of equation 5) this will increase the magnitude of β. For weak solvation (small dielectric constant) these simple corrections in fact make the slope very close to unity (Fig. 4b).[66]

The qualitative behavior of Fig. 4 is clear: there is a good mapping between those chromophores that experimentally show large nonlinear optical responses and those that are computed to have large nonlinear optical responses.

2. Convergence of the sum-over-states expression: a legitimate concern arises in truncating the expansion of equation 5: it is clear that this expansion converges (in fact, comparable expansions for properties like spin-spin coupling constants do not converge very well, and to describe completely the electronic polarization in those cases, truncation of the sum is a dubious technique). In the case of nonlinear optical response, however, we and others have shown that the convergence is in fact rather good:[25] Fig. 5 shows characteristic convergence both for positive and negative β response. (Negative β response simply means, as is again clear from the spectral resolution, that the excited state dipole moment is generally smaller than the ground state dipole moment.) The characteristic feature that the first, strongly allowed transition gives a result that is roughly twice the converged result has been found in a large number of SOS (sum-over-states, a notation that is used to specify calculations using the spectral resolution explicitly). Generally, terms involving three level structure such as those in Fig. 3 come in with opposite signs to the two level structures of Fig. 2, the two level structures include the strongest allowed, usually homo-lumo, mixing.

3. Comparison with experiment: Table 2 shows some results for a series of donor/acceptor chromophores using the PPP model. The calculations are done at two relevant experimental frequencies; the overall scaling between experiment and theory is excellent. Similar results are found for other categories of molecules, including

heterocycles, phthalocyanines,[78] octopolar species,[79] polyaniline oligomers,[80] heterocycles,[81] and other systems.

4. Comparison with *ab-initio* calculations: the simplest comparison with *ab-initio* calculations is made at the zero frequency limit, where there is no problem, in the *ab-initio* context, in dealing with the dispersion. There is a bit of a difficulty in comparing the semiempirical results with the *ab-initio* results because the categories of molecules for which the two techniques are well suited show little commonality. *Ab-initio* techniques, because of large computational demands that arise from the necessity for extensive basis sets and proper treatment of correlation, are most accurate and most useful for small molecules, such as HCl, H_20, CH_2O, CH_2F_2, *etc.*[31,32,76] *Ab-initio* calculations have been done for a few larger molecules.[75,76] Semi-empirical calculations, on the other hand, clearly do a very poor job of describing molecules in which the dominant behavior arises from Rydberg type excitations, from transitions that involve extensive Duschinsky rotation, from strongly correlated ground states, or generally in situations in which nonlinear response is weak. We have seen this in calculations on distorted benzenes,[65] where the response is very small (benzene itself is centrosymmetric, and has no β response): under those situations, ZINDO/S type calculations show very poor convergence and dubious results. On the other hand, the semiempirical models are very good for calculation of strong non-linear optical response, especially that characterized by strong charge transfer transitions as, as dominant in the first hyperpolarizability (for the second hyperpolarizability, where some transitions involving charge transfer are still important, semiempirical methods, again, prove to be quite good).[46,64]

There is also an issue of the conventions that are used to define nonlinear response: these are exhaustively discussed in several places in the literature.[32,76,64,82] A comparison of calculations can be done on p-nitroaniline using both the simple PPP model and an *ab-initio* MCSCF calculation,[75] employing the full pi valence space of orbitals and complete active space correlation (CASSCF). The basis set includes double ζ plus polarization functions; a total of 180 basis functions were used. As Table 3 demonstrates, the semiempirical result is in better agreement with experiment than is the *ab-initio* study.

These results are simply indicative: other comparable calculations on other extended pi type systems are beginning to appear in an *ab-initio* context.[76] The net upshot is that, as in the calculation of optical spectra, the demands on an *ab-initio* calculation to calculate properly the optical properties is extended, polarized pi-type systems are substantial: both correlation and basis set effects must be dealt with carefully, and the appropriate screening of the pi manifold must be properly described (this requires both correlation in the sigma manifold and extended basis sets).[31,32,76] All of these are, qualitatively, taken care of in the semiempirical calculations: it is therefore not surprising that semiempirical calculations do a much more efficient, and much better, job of describing the optical excitations in these systems.

5. Frequency dependence: the frequency-dependent results, again, follow straightforwardly from the semiempirical theory. Figure 6 shows the calculated frequency dependence of two of the vector components of β for all the different nonlinear optical response properties; the divergences arise because the optical excitation of aniline occurs at roughly twice of the divergence frequency (close to 5 eV). In fact,

frequency dependence can be calculated directly using the spectral resolution, but broadening (damping) terms, that arise from vibronic interactions should also be included.[9,84] These will change the line shapes from those calculated ignoring these vibronic effects. Figure 7 shows the computed frequency dependent response of dimethylaminonitrostilbene. The sharp, spiky structure calculated ignoring vibronic effect will certainly become smooth, as it will be experimentally, when vibronic effects are included.

6. Sensitivity to parameterization: semiempirical models require the definition of matrix elements, and one valid and often-voiced criticism is that the results of the calculation depend on the choice of the model Hamiltonian; as we have already stressed, different choices for the parameters simply correspond to different model Hamiltonians, and some model Hamiltonians will be better than others. Nevertheless, one would hope that relatively small changes in parameterization should not give substantial variances in the predictions. As Table 4 shows, in fact this is true: changing, for example, the ionization potential by as much as 13 volts on the nitrogen atom, or the electron affinity by a factor of 5 on the carbon, change the predicted β_{vec} by less than a factor of 2 in each case. Therefore, within a fairly broad range, small changes in the parameterizations of this semiempirical model will not effect the predicted nonlinear response.

7. Some chemistry: Identification of the donor. A most simplified numerical estimate for the first hyperpolarizability response, developed long ago by Oudar and Chemla,[85] limits the sum in equation 5 to a single excited state. The resulting two-level model gives the hyperpolarizability as

$$\beta_{vec} \cong \frac{\hbar\omega_{gn} f_{gn} \Delta\mu}{[(\hbar\omega_{gn})^2 - (2\hbar\omega)^2][(\hbar\omega_{gn})^2 - (\hbar\omega)^2]} \tag{12}$$

here, $\Delta\mu$ is the change in dipole moment from the ground to the excited state, and f is the oscillator strength. It is clear from this two level formula that chromophores involving large changes in dipole elements, relatively low-lying excited states, and strongly-allowed transitions will have large nonlinear responses. One category of π-electron chromophores that obviously fits this description is the push-pull (or donor acceptor) conjugated systems such as the stilbenes of structure A or the benzenes of structure B. We and others have shown that increasing the donor strength, or the acceptor strength,[43,45,47,48] of the donors and acceptors in A and B results in increasing hyperpolarizability. The molecular orbitals involved in such transitions are sketched qualitatively in Fig. 8. Notice that the accepting orbital, the lumo, is centered almost entirely on the nitro group (when nitro is chosen as the acceptor species). The donor, on the other hand, is not localized in the para position of the benzenoid ring, but has contributions both from benzene rings and the ethylene bridge. This suggests that the addition of donor substituents in these positions would in fact increase the nonlinear response, while acceptors in any of these positions tend to reduce response. Table 5, taken from INDO/S calculations, substantiates this: notice that the effective donor site is indeed largely the entire pi electron system, while the acceptor is the isolated nitro group. Of course, the farther the donor is from the nitro group, the larger one expects the change in dipole moment to be, and therefore the larger the non-linear response. Nevertheless, the overall response is increased by placing donor substituents anywhere on the bridge. This sort of insight, that suggests new synthetic strategies, is typical of what can be learned from semiempirical calculations on well-defined systems.

As extensively documented elsewhere, semiempirical calculations (and, indeed, *ab-initio* calculations) have been very useful in understanding the nonlinear optical response of isolated chromophores.[35] Among other issues, the responses of organometallic species[62,63] (including its relative smallness in almost all cases), the important geometric variations of a nonlinear response,[61] dependence upon solvation,[66,74-77] the role of non-bonded interactions,[80] frequency dependence and the interpretation of the relative response magnitudes of a number of molecules have all been clarified using appropriate calculational advances. Again, the operative word here is appropriate: if the right questions are asked, relatively straightforward calculations on well-defined model systems can be used to derive real understanding even of a complicated response property like the hyperpolarizability of a molecule.

All model calculations (that is, all electronic structure calculations) and all analytic schemes for analysis of those calculations have limitations. Semiempirical models, in particular, do have problems: the most important of these is the fact that they are not systematically improvable (although attempts have been made to do so). Semiempirical model calculations remain a useful tool for the chemist. If the purpose of theoretical chemistry is to understand chemistry, then this understanding can, in many cases, be very effectively based on appropriate semiempirical calculations.

Acknowledgments. I am grateful to David Kanis, Albert Israel, John Pople, Dequan Li, Seth Marder, Santo di Bella and Tobin Marks for helpful discussions concerning hyperpolarizabilities, and to Jan Linderberg for teaching me what a model Hamiltonian is. This work was supported by the NSF-MRC program through the Northwestern Materials Research Center (grant # DMR9120521) and by the AFOSR (#90-0071).

Table 1. Comparison of β Components for Aniline ($C_6H_6NH_2$) at 1064 nm ($\hbar\omega$ = 1.17 eV) as Calculated with PPP (Full CI and MECI) and CNDO (MECI)[a]

	$PPP^{Exact,b}$	$PPP^{MECI,c}$	$CNDO^{MECI,d}$
β_{xxx}	0.46	1.71	1.22
β_{xyy}	0.04	−0.32	0.25
β_{yxy}, β_{yyx}	0.10	−0.27	0.43
β_x	0.54	1.42	1.59

a. All NLO data are in units of 10^{-30} cm^5 esu^{-1}, λ = 1064 nm ($\hbar\omega$ = 1.17 eV).
b. From ref. 25.
c. From ref. 25.
d. From ref. 25.

Table 2. Comparison of PPP β_{vec} Values with Reported Experimental Results. From Ref. 25.

Molecule	$\beta_{vec}^{(PPP)a}$	$\beta_{vec}^{(exp)}$	$\hbar\omega$, eV
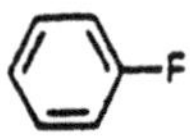	1.02	1.06	1.17
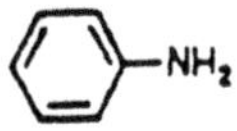	1.84	0.79-2.46	1.17
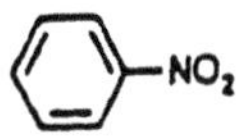	4.55	1.97-4.6	1.17
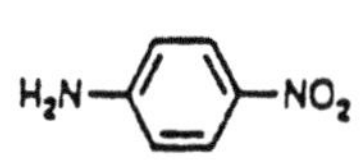	34.4	16.2-47.7	1.17
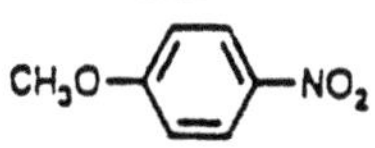	11.7	14.3-17.5	0.656
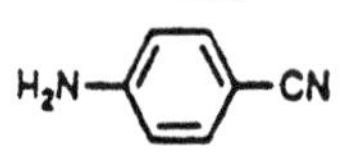	13.26	13.4	0.656
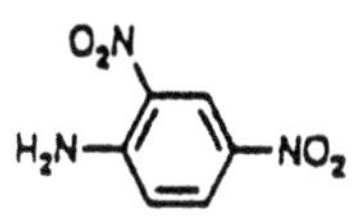	29.3	21	1.17
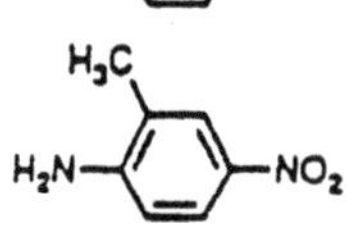	36.3	16-42	1.17
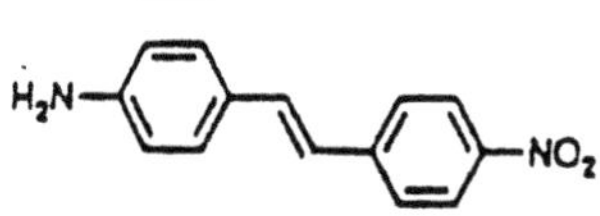	298.1	225-295	1.17
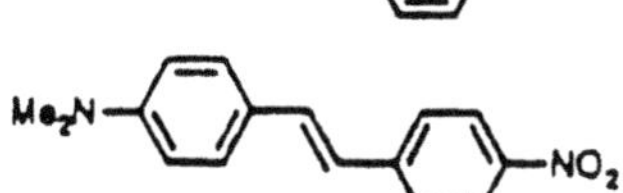	451.9	450	1.17
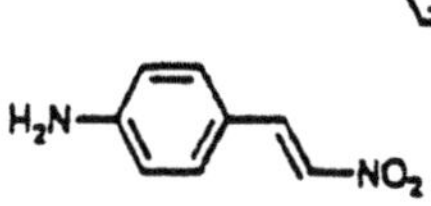	213.1	180-260	1.17
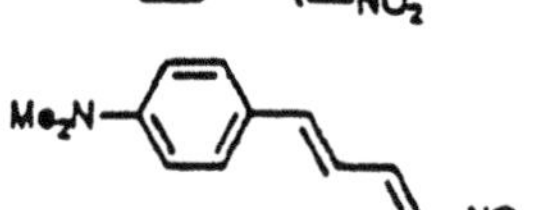	466.8	470-790	1.17

a. All NLO data are in units of 10^{-30} cm^5 esu^{-1}.

Table 3. Computed β_{zzz} (-2ω; ω,ω) for p-nitroaniline at ℏω = 1.17 eV[a]

Method	Value
PPP[b]/SOS	34.4
CASSCF-*ab initio*[c]	8.20
Expt.[c]	50.7

(a) This comparison is a bit problematic: the computations are in vacuo, the experiment in solution; also the issue of conventions and standards is complicated — see, e.g., Refs. 76, 82, and 83.

(b) D. Li, M. A. Ratner and T. J. Marks, *J. Am. Chem. Soc.* **110,** 1707 (1988).

(c) K. V. Mikkelsen, Y. Luo, H. Ågren and P. Jørgensen, *J. Chem. Phys.* **100,** 8240 (1994).

(d) Units of 10^{-30} cm^5/esu. The experiment measures β_{vec}, whic is very close to $\boldsymbol{\beta}_{zzz}$ for this strongly axial molecule.

Table 4. The Sensitivity of PPP-SOS-Derived β_{vec} of *para*-Nitroaniline with Choice of Carbon and Nitrogen Parmeterization

H_2N NO_2

Carbon Parameterization

Vary C_{IP}; C_{EA} = 0.03

IP	8.93	10.04	11.16	12.28
β_{vec}	45.2	39.4	35.9	31.9

Vary C_{EA}; $C_{IP} \backslash 11.16$

EA	−2.24	−0.43	0.03	0.23	0.53
βvec	49.8	36.6	35.9	35.7	35.8

Nitrogen Parameterization

Vary N_{IP}^{nitro}, N^{amino}; N_{EA} = 8.97

$IP_{nitro-N}$	20.58	23.16	25.73	28.30	33.88
$IP_{amino-N}$	17.18	19.32	21.47	23.62	25.76
β_{vec}	29.5	33.6	35.9	39.1	51.9

Vary N_{EA}; N_{IP}^{nitro} = 25.73, N_{IP}^{amino} = 21.47

EA	7.18	3.07	8.97	9.87
β_{vec}	49.8	43.0	35.9	30.8

a. All NLO data are in units of 10^{-30} cm^5 esu^{-1}; β_{vec} calculated at λ = 1064 nm ($\hbar\omega$ = 1.17 eV).

b. Excerpts of Table taken from Reference 25.

c. EA refers to electron affinity (in eV) and IP refers to ionization potential (in eV).

TABLE 5. Calculated response of substituted stilbenes, showing the effects of substitution by various donor and acceptor substituents on aminonitrostilbene.

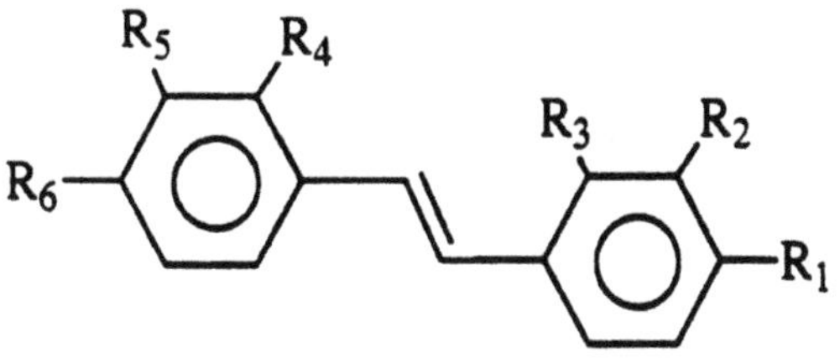

A

Calculated				
β_{vec} (0.65eV) $R_1 = NO_2$, $R_6 = NH_2$				(units of $10^{-30}cm^5/esu$)
R_2	R_3	R_4	R_5	β_{vec}
H	H	H	H	46
H	NH_2	H	H	52
H	H	NO_2	H	28
H	H	NH_2	H	55

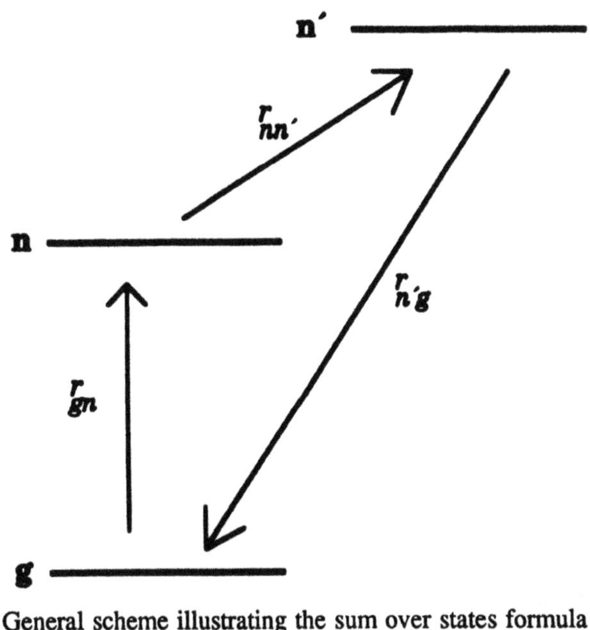

Figure 1: General scheme illustrating the sum over states formula (Equation 5) as applied to the calculation of molecular hyperpolarizabilities. Notice that both n and n' are excited states; which can mix with the ground state only by single electron (single excitation) terms.

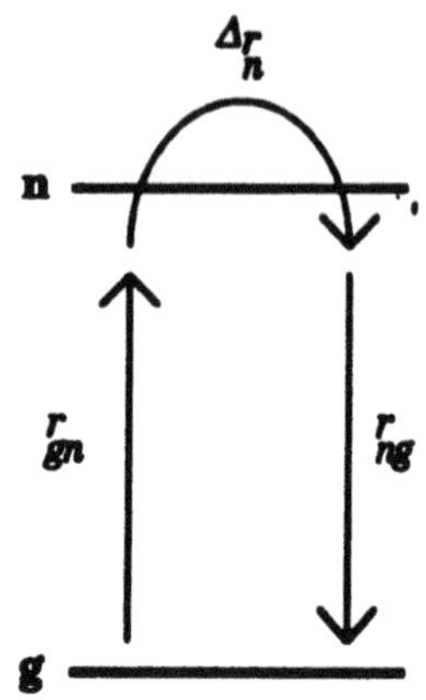

Two-Level Terms

Figure 2: The so-called two-level terms, that form the basis for the two-level formula approximating the first molecular hyperpolarizability. Notice that in this the n and n' excited states in Figure 1 have been collapsed into a single state.

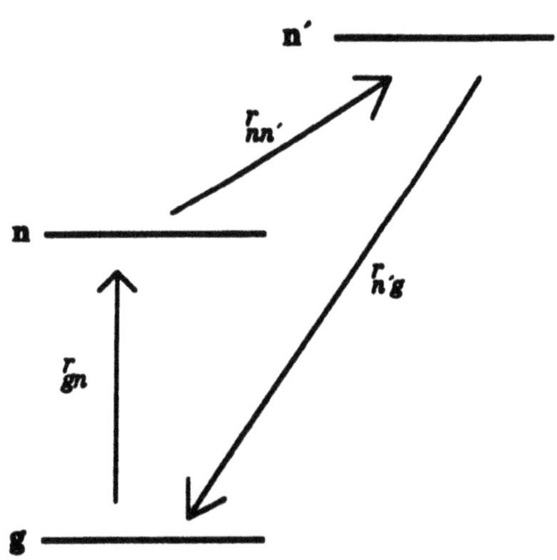

Three-Level Terms

Figure 3: The so-called three-level terms, that add to the two-level terms to give the total molecular hyperpolarizability, using the sum over states (spectral representation) formula. Note here that the intermediate states n and n' must differ.

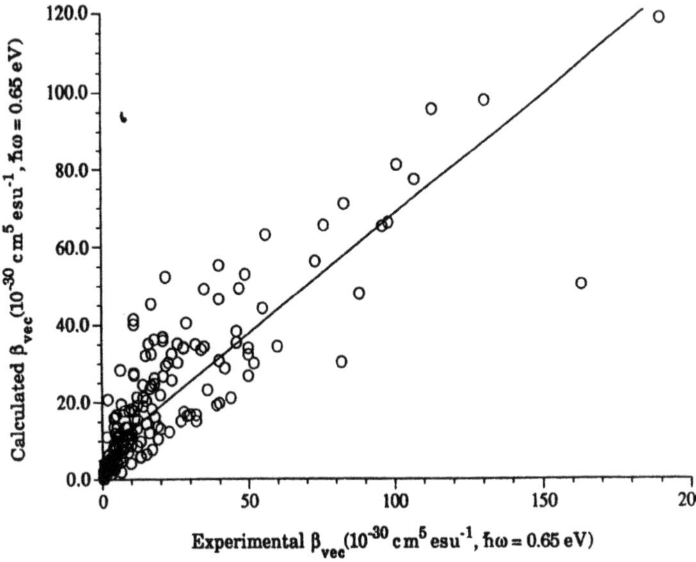

Figure 4a: Comparison of calculated and computed (INDO/S) calculations for 203 π-type organic chromophores. The value of β_{vec} is plotted. Note that the slope is roughly .8 (from reference 25).

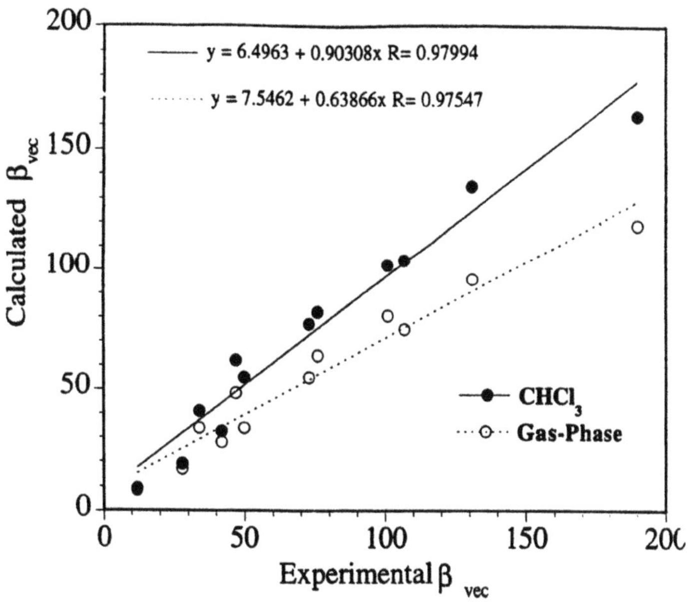

Figure 4b: As in Figure 4a, but for the calculation done using a modification of the energies in the spectral representation of Equation 5, caused by polarization due to solvent. The comparison is made with measurements in chloroform (from reference 66).

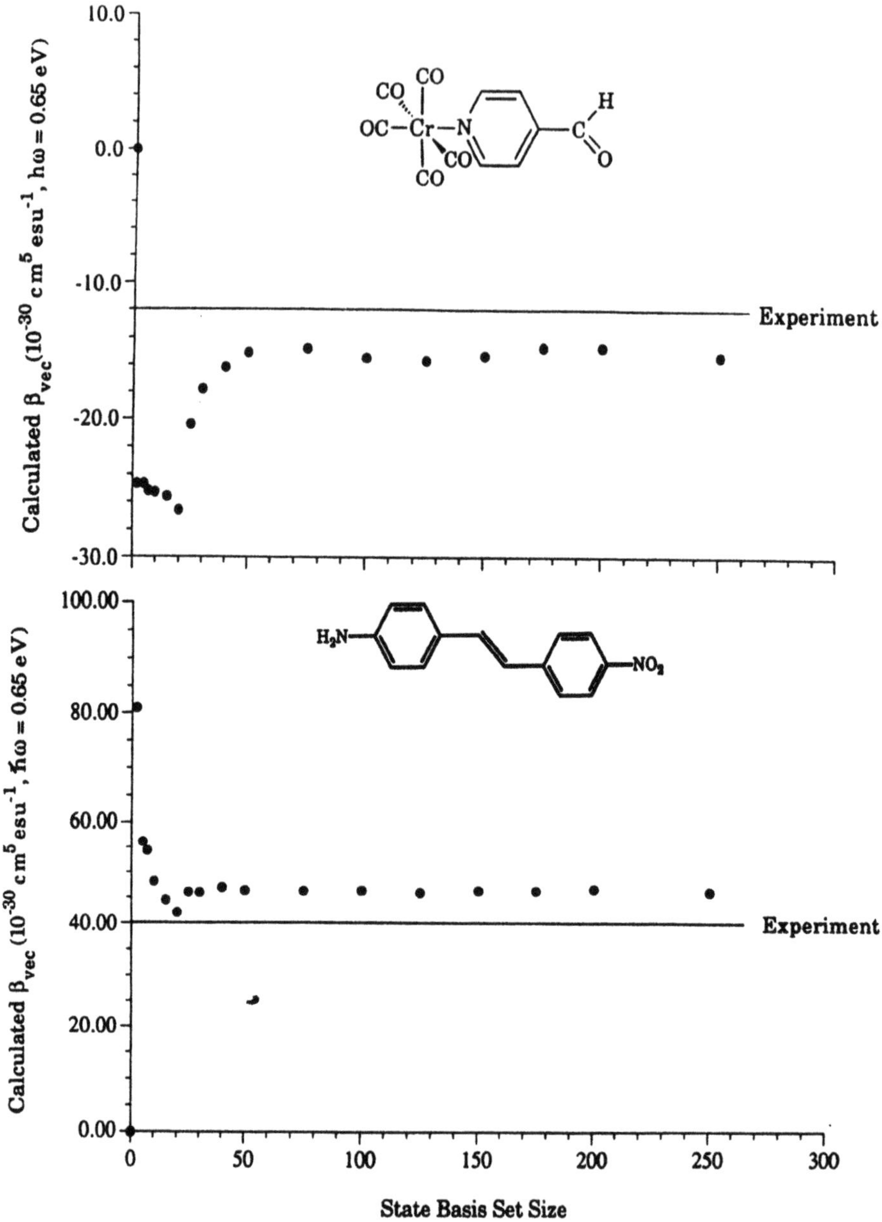

Figure 5: Illustrating the convergence of the sum-over-states formula of Equation 5, for nonlinear optical response of characteristic organic and organometallic species (from reference 25).

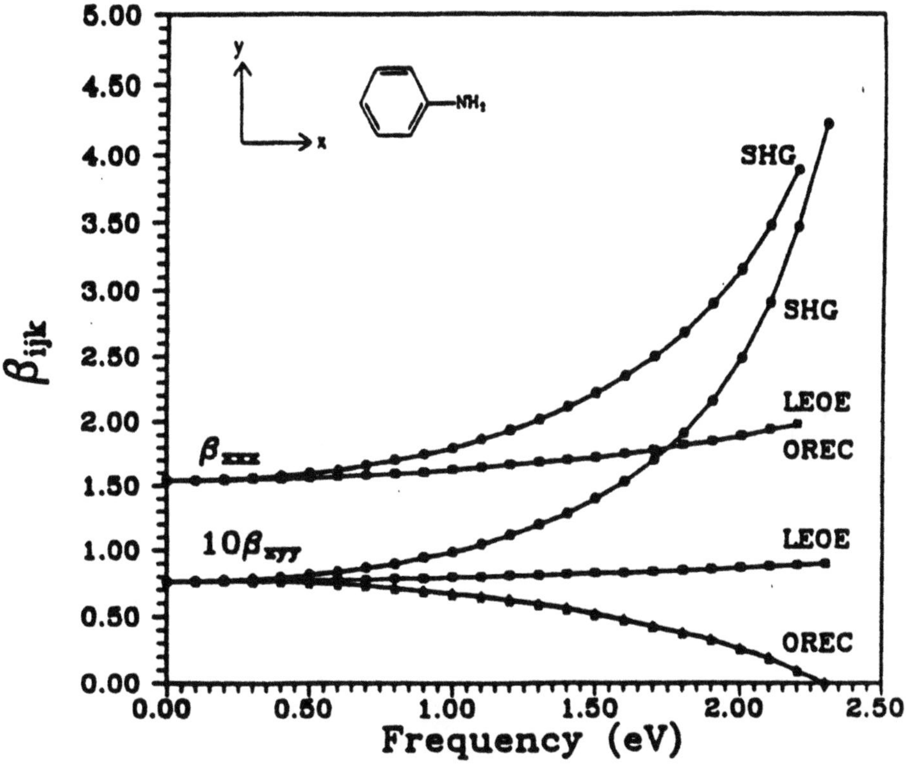

Figure 6: Dispersion (frequency dependence) of various second-order nonlinear optical properties, calculated using the PPP model Hamiltonian and the sum over states formula (from reference 25). The notations SHG, LEOE and OREC refer, respectively, to second harmonic generation, linear electrooptic effect and optical rectification. Calculations are for aniline (from Reference 25).

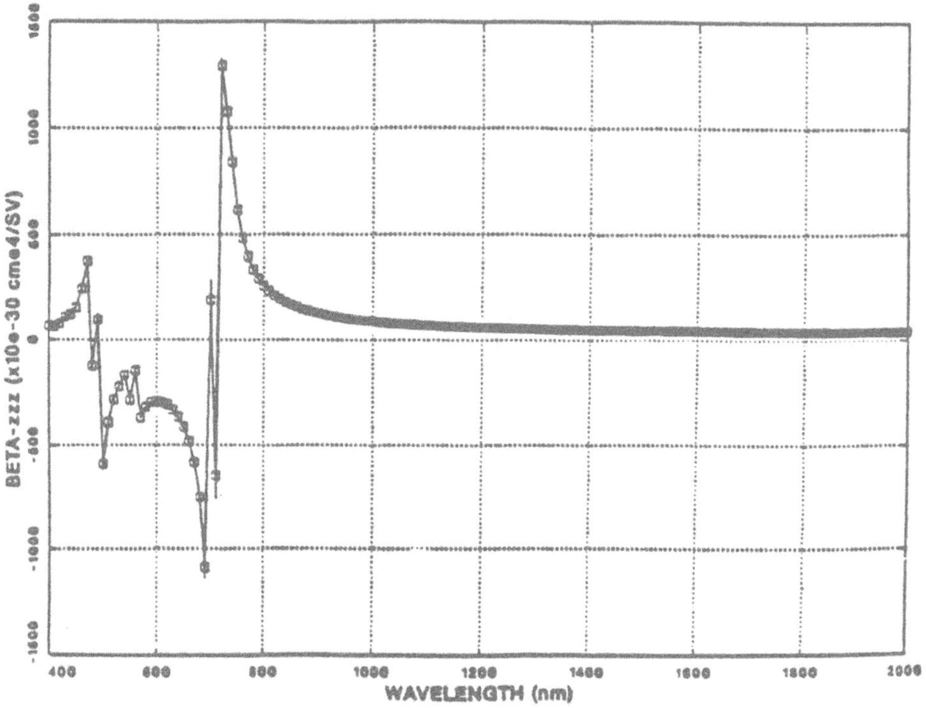

Figure 7: The calculated frequency dependence of the β_{ZZZ} tensor (Z is the charge transfer direction) in the dimethylaminonitrostilbene molecule. This sharp structure occurs in the electronic calculation, and should be smoothed by vibronic effects. (reference 25)

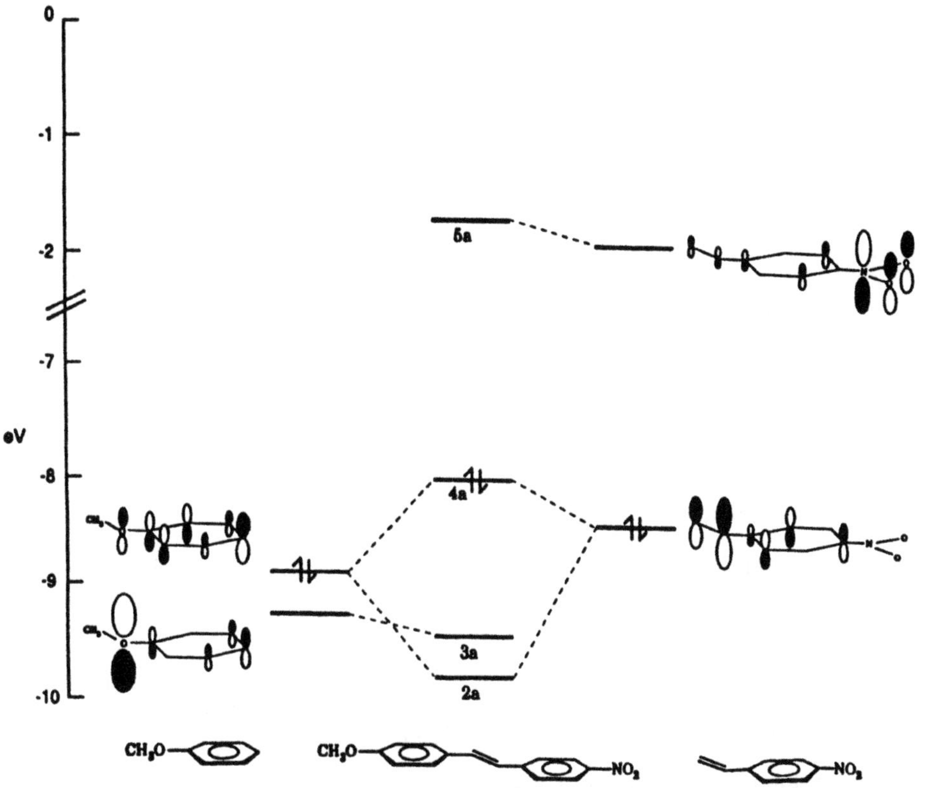

Figure 8: The frontier molecular orbitals for 4-methoxynitrostilbene (from reference 25). Notice that the LUMO is almost entirely centered on the nitro group, whereas the higher occupied orbitals are delocalized over the acceptor and the stilbene structure. Thus the acceptor group is almost entirely localized, whereas the donor is spread over the entire pi system of the molecule.

References

1. R.S. Mulliken, *J. Chem. Phys.* **23**, 1433 (1955).

2. A Szabo and N.S. Ostlund, *Modern Quantum Chemistry* (McGraw-Hill, New York, 1989) p. 204.

3. A.E. Reed & F. Weinhold, *J. Chem. Phys.* **83**, 1737 (1985).

4. D.P. Armstrong, P. Perkins and J.J.P. Stewart, *J. Chem. Soc.* Dalton 1973, 838.

5. Compare the discussion and references in M. Sola, J. Mestres, R. Carbo and M. Duran, *J. Am. Chem. Soc.* **116**, 5909 (1994).

6. R.G. Pearson, *J. Am. Chem. Soc.* **85**, 3533 (1963); K. Fukui & H. Fujimoto, *Bull. Chem. Soc. Jap.*, **42**, 3399 (1969); R.G. Pearson, *Accts. Chem. Res.* **26**, 250 (1993); R.G. Parr and Z. Zhou, ibid., p. 225.

7. R.F.W. Bader, *Accts. Chem. Res.* **18**, 9 (1985).

8. P.C. Hiberty, *Int. J. Quant. Chem.* **19**, 259 (1981); See also the discussion and references in E.D. Jemmis, G. Subramanian, I.J. Srivastava and S.R. Gadre, *J. Phys. Chem.* **98**, 6445 (1994).

9. J.Linderberg, Y. Ohrn, *Propagators in Quantum Chemistry*; Academic Press: London, 1974; G.C. Schatz & M.A. Ratner, *2Quantum Mechanics in Chemistry*; Prentice-Hall: Englewood Cliffs, NJ, 1993, Chapter 6.

10. H. Sekino and R.J. Bartlett, *J. Chem Phys.*, 85:976, 1986; S.P. Karna, P.N. Prasad, and M. Dupuis, *J. Chem. Phys.*, 94:1171, 1991; J.E. Rice, R.D. Amos, S.M. Colwell, N.C. Handy, and J. Sanz, *J. Chem. Phys.*, 93:8828, 1990; H.J. Aa. Jensen, H. Koch, P. Jørgensen, and J. Olsen, *Chem. Phys.*, 119:297, 1988; P. J. Jørgensen, H.J. Aa. Jensen, and J. Olsen, *J. Chem. Phys.*, 89:3654, 1988; H. Hettema, H.J.Aa. Jensen, P. Jørgensen, and J. Olsen, *J. Chem. Phys.*, 97:1174, 1992; O. Vahtras, H.Ågren, P. Jørgensen, H.J.Aa. Jensen, T. Helgaker, and J. Olsen, *J. Chem. Phys.*, 97:9178, 1992; K.V. Mikkelsen, E. Dalgaard, and P. Swanstrøm, *J. Phys. Chem.*, 79:587, 1987.

11. E. Dalgaard, *Phys. Rev. A*, 26:42, 1982; E. Dalgaard and P. Jørgensen, *J. Chem. Phys.*, 69:3833, 1978.

12. G. Khanarian, Ed. *Proc. SPIE-Int. Soc. Opt. Eng.* **1990**, 1147 (*Nonlinear Optical Properties of Organic Molecules II*).

13. J. Linderberg & Y. Öhrn, *J. Chem. Phys.* **49**, 716 (1968).

14. J.N. Murrell; A.J. Harget, *Semi-Empirical Self-Consistent-Field Molecular Orbital Theory of Molecules;* Wiley: London, 1972; J.A. Segal, ed., *Semi-Empirical Methods of Electronic Structure Calculation*, Plenum, New York, 1977.

15. W.J. Hehre, L. Random, P.v.R.Schleyer, J.A. Pople, *Ab-Initio Molecular Orbital Theory*; Wiley: New York, 1986; Chapter 2.

16. R.G. Gordon, *Adv. Mag. Res.* 3, 1 (1968).

17. G.C. Schatz & M.A. Ratner, *Quantum Mechanics in Chemistry*; Prentice-Hall: Englewood Cliffs, NJ, 1993, Chapter 9.

18. D.N. Zubarev, editor, *Nonlinear Statistical Thermodynamics*. Plenum, New York, 1974.

19. J. Oddershede, *Adv. Chem. Phys.* **69**, 201 (1987); W.A. Parkinson & J. Oddershede, *J. Chem. Phys.* **94**, 7251 (1991).

20. E. Dalgaard, *Phys. Rev. A* 1982, 26, 42-52; D.L. Yeager, P. Jørgensen, *Chem. Phys. Lett.* 1979, 65, 77-80.

21. S.L. Lalama, A.F. Garito, *Phys. Rev.* **20**, 1979, 1179-1194; G.C. Schatz & M.A. Ratner, *Quantum Mechanics in Chemistry*; Prentice-Hall: Englewood Cliffs, NJ, 1993, Chapter 11; Y.R. Shen, *The Principles of Nonlinear Optics*; Wiley: New York, 1984.

22. N.D. Prasad, D.J. Williams, *Introduction to Nonlinear Optical Effects in Molecules and Polymers;* Wiley: New York, 1991.

23. D.R. Shelton & J. Rice, *Chem. Revs.*, **94** 3(1994).

24. *Int. J. Quantum Chem.* **43**, 1 (1992): Special issue on Molecular Nonlinear Optics.

25. G.R. Meredith, *J. Mater. Educ.*, 1987, **9**, 719-750.

26. Y. Luo, H. Ågren, P. Jørgensen and K.V. Mikkelsen, *Adv. Quantum Chem*, in press.

27. N. Bloembergen, H. Lotem, R.T. Lynch, Ind. J. Pure Appl. Phys. 16, 151 (1978).

28. A. Schweig, *Chem. Phys. Lett.*, 1967, I, 195-199; J. Zyss, *J. Chem. Phys.*, 1979, 70, 3333-3340.

29. P. Pulay, *Mol. Phys.*, 1969, 17, 197-207; C.E. Dykstra, P.G. Jasien, *Chem. Phys. Lett.*, **109**, 1984, 388-393.

30. See, for example: R.J. Bartlett, *Ann. Rev. Phys. Chem.*, **32**, 1981, 359-401.

31. H. Sekino, R.Bartlett, *Int. J. Quantum Chem.*, **43**, 1992, 119-134.

32. J.E. Rice,N.C. Handy, *Int. J. Quantum Chem.*, **43**, 1992, 91-118.

33. H. Sekino and R.J. Bartlett, *J. Chem. Phys.*, **58**, 3022 (1993).

34. H. Sekino and R.J. Bartlett, *J. Chem. Phys.*, **94**, 3665 (1991); see also Ref. 32.

35. S.P. Karna and M. Dupuis, *Chem. Phys. Lett.*, **171**, 201 (1990); F. Meyers, J.L. Brédas, J. Zyss, *J. Am. Chem. Soc.*, **114**, 1992 (2914-2921); F. Meyers, J. Zyss, J.L. Brédas *Synth. Met.*, **43**, 1991, 3191-3194; F. Meyers, J.L. Brédas, *Int. J. Quantum Chem.*, **42**, 1992, 1595-1614.

36. R. Kosloff, *Ann. Revs. Phys. Chem.*, in press; R. Kosloff, *J. Phys. Chem.*, **92**, 2087 (1988).

37. M. Goeppert-Mayer, A.L.Sklar, *J. Chem. Phys.*, **6**, 645 (1936).

38. J.N. Murrell, S.F.A. Kettle and J.M. Tedder, *Valence Theory* (Wiley, London, 1965).

39. J.A. Pople and D.N. Beveridge, Approximate Moleuclar Orbital Theory (McGraw-Hill, New York, 1970).

40. M.J.S. Dewar, The Molecular Orbital Theory of Organic Chemistry (McGraw-Hill, New York, 1969); Faraday Disc. 62, 197 (1977).

41. A. Ulman, C.S. Willand, W. Kohler, D.R. Robello, D.J. Williams, L. Handley, *J. Am. Chem. Soc.*, **112**, 1990, 7083-7090; A. Ulman, *J. Phys. Chem.*, **92**, 1988, 2385-2390.

42. C.W. Dirk, R.J. Twieg, G. Wagniere, *J. Am. Chem. Soc.* **108**, 1986, 5387-5395.

43. H.A. Kurtz, J.J.P. Stewart, K.M. Dieter, *J. Comput. Chem.*, **11**, 1990, 82-87.

44. J.A. Morrell, A.C. Albrecht, *Chem. Phys.Lett.*, **64**, 1979, 46-50; J.A. Morrell, A.C. Albrecht, K.H. Levin, *J. Chem. Phys.*, **71**, 1979, 5063-5068; C.S. Willand, A.C. Albrecht, *Opt. Commun.*, **57**, 1986, 146-152.

45. J.O. Morley, *Int. J. Quantum Chem.*, **46**, 1993, 19-26; J.O. Morley, D.A. Pugh, *J. Chem.Soc., Faraday Trans.2* **87**, 1991, 3021-3025; J.O. Morley, V.J. Docherty, D. Pugh, *J. Chem. Soc., Perkin Trans. 2* 1987, 1361-1363; V.J. Docherty, D. Pugh, J.O. Morley, *J. Chem. Soc., Faraday Trans. 2*, **81**, 1985, 1179-1192; S. Ramasesha, I.D.L. Albert, *Phys. Rev. B: Condens. Matter*, **42**, 1990, 8587-8594; S. Ramasesha, I.D.L.Albert, *Chem. Phys. Lett.*, **196**, 1992, 287-293.

46. B.M. Pierce, Proc. S.P.I. 9, 971, 25 (1988); C.P. de Melo, R. Silbey, *J. Chem. Phys.*, **88**, 1988, 2567-2571.

47. D. Li, T.J. Marks, M.A. Ratner, *Chem. Phys. Lett.*, **131**, 1986, 370-375; D. Li, T.J. Marks, M.A. Ratner, *J. Am. Chem. Soc.*, **110**, 1988, 1707-1715; D. Li, M.A. Ratner, T.J. Marks, *J. Phys. Chem.*, **96**, 1992, 4325-4336; D. Li, T.J. Marks, M.A. Ratner, *Mat. Res. Soc. Symp. Proc.* **109**, 1988, 1707-1715.

48. S.R. Marder, C.B. Gormann, F. Meyers, J.W. Perry, G. Bourhill, J.L. Bredas and B.M.Pierce, *Science*, **632**, 265 (1994); S.R. Marder, D.N. Beratan and L.T. Cheng, Ibid. 252, 193 (1991).

49. D.J. Williams, Ed. *Proc. SPIE-Int. Soc. Opt. Eng.* 1992, 1775 (Nonlinear Optical Properties of Organic Molecules V); D.J. Williams, Ed. *Nonlinear Optical Properties of Organic Molecules and Polymeric Materials*; ACS Symposium Series 233, American Chemical Society: Washington, DC 1984.

50. R.W. Boyd, *Nonlinear Optics*, Academic Press: New York, 1992; P.N. Prasad, D.J. Williams *Introduction to Nonlinear Optical Effects in Molecules and Polymers*; Wiley: New York, 1991.

51. S.R. Marder, J.E. Sohn, G.D. Stucky, Eds. *Materials for Nonlinear Optics: Chemical Perspectives*, ACS Symposium Series 455, American Chemical Society: Washington, DC, 1991.

52. J.L. Brédas, R.J.Silbey, Eds. *Conjugated Polymers: The Novel Science and Technology of Highly Conducting and Nonlinear Optically Active Materials*, Kluwer: Dordrecht. Neth., 1991.

53. G. Khanarian, Ed. *Proc. SPIE-Int Soc. Opt. Eng.* 1990, 1337 (Nonlinear Optical Properties of Organic Molecules III).

54. J. Messier, F. Kajar, P. Prasad, D. Ulrich, Eds. *Nonlinear Optical Effects in Organic Polymers*, Kluwer Academic Pubishers: Dordrecht, 1989.

55. D.S. Chemla, J. Zyss, Eds. *Nonlinear Optical Properties of Organic Molecules and Crystals*, Academic Press: New York, 1987; Vols. 1 and 2.

56. D. Li, T.J. Marks and M.A. Ratner, *Mat. Res. Soc. Symp. Proc.*, **135**, 665-672 (1989).

57. D. Kanis, T.J. Marks, M.A. Ratner, *J. Am. Chem. Soc.*, **112**, 8203--8204 (1990).

58. S. DiBella, M.R. Ratner, and T.J. Marks, *J. Am. Chem. Soc.*, **114**, 5842-5849 (1992).

59. D.R. Kanis, T.J. Marks, and M.A. Ratner, *Nonlinear Optics*, **6**, 317-335 (1994).

60. S. DiBella, I.L. Fragalá, M.A. Ratner and T.J. Marks, *J. Am. Chem. Soc.*, **115** (2), 682-686 (1993).

61. D. Kanis, T.J. Marks and M.A. Ratner, *Int. Jour. of Quantum Chem.*, **43**, 61-82 (1992).

62. D. Kanis, M.A. Ratner, T.J. Marks, *Chem. Mater.*, **3**, 19-22 (1991).

63. D. Kanis, M.A. Ratner and T.J. Marks, *J. Am. Chem. Soc.*, **114**, 10338-10357, (1992).

64. D. Li, T.J. Marks and M.A. Ratner, *J. Phys. Chem.*, **96**, 4325-4336 (1992).

65. D.R. Kanis, J.S. Wong, T.J. Marks, M.A. Ratner, H. Zabrodsky, S. Keinan and D. Avnir, *J. Am. Chem. Soc.*, submitted.

66. S. DiBella, T.J.Marks, and M.A. Ratner, *J. Am. Chem. Soc.*, **116**, 4440-4445 (1994).

67. R. Pariser, R.G. Parr, *J. Chem. Phys.*, **21**, (1953) 466-471; J.A. Pople, *Trans. Faraday Soc.*, **49**, 1953, 1375-1385.

68. J. Ridley, M.C. Zerner, *Theoret. Chim. Acta (Berlin)*, **32**, 1973, 111-134; A.D. Bacon, M.C. Zerner, *Theoret. Chim. Acta (Berlin)* **53**, (1979), 21-54.

69. J. Linderberg, E.W. Thulstrup, *J. Chem. Phys.*, **49**, 710 (1968).

70. Z.G. Soos and S. Ramasesha, *J. Chem. Phys.*, **90**, 1067 (1989).

71. I.D.L. Albert, S. Ramasesha, and P.K. Das, *Phys. Rev.*, **B43**, 7013 (1991).

72. S. Ramasesha and I.D.L. Albert, *Phys. Rev.*, **B42**, 8587 (1990).

73. A.D.L. Israel, work in progress.

74. J. Yu and M.C. Zerner, *J. Phys. Chem.*, in press.

75. K.V.Mikkelsen, Y. Luo, H. Ågren, and P. Jørgensen, *J. Chem. Phys.*, 100: 8240, 1994.

76. H. Hettema, H.J. Aa. Jensen, P. Jørgensen, and J.Olsen, *J. Chem. Phys.*, 97: 1174, 1992.

77. A. Willetts and J.E. Rice, *J. Chem. Phys.*, 99:426, 1993.

78. D.Li, T.J.Marks and M.A. Ratner, *J. Am. Chem. Soc.*, **110**, 1988, 1707-1715.

79. M. Joffre, D. Yaron, R.J. Silbey, J. Zyss, *J. Chem. Phys.*, **97**, 1992, 5607-5615.

80. T.R.M. Sales, C.P. deMelo, M.C. Dossantos, *Synth. Met.*, **43**, 1991, 3751-3754.

81. K.J. Atkins C.L. Honeybourne, *Spec. Publ.-R. Soc. Chem.*, **91**, 1991 (Org. Mater. Nonlinear Opt 2), 121-127.

82. A. Willets, J. Rice, D. Burland, D. Shelton, *J. Chem. Phys.*, **97**, 7590 (1992).

83. B.F. Levine, C.G. Bethea, *J. Chem. Phys.*, **63**, 1975, 2666-2682.

84. Y.R.Shen, *The Principles of Nonlinear Optics*; Wiley: New York, 1984.

85. J.L.Oudar, D.S. Chemla, *J. Chem. Phys.*, **67**, 1977, 446-457; D.S. Chemla, J.L.Oudar, J. Jerphagnon, *J. Phys. Rev. B*, **12**, 1975, 4534-4546.

THE WORK FORMALISM: A NEW THEORY OF ELECTRONIC STRUCTURE

VIRAHT SAHNI
Department of Physics,
Brooklyn College of the City University of New York
Brooklyn, New York 11210, USA

1. Introduction

The electronic properties of a system are described by its stationary state wavefunction Ψ, which is the eigenfunction of the time-independent Schrödinger [1] equation. The determination of the wavefunction Ψ, however, is complicated by the presence in the Hamiltonian of the ***nonlocal*** operator representing Coulomb repulsion between the electrons. Thus, although electron correlations due to the Pauli exclusion principle are accounted for by the requirement of antisymmetry of the wavefunction, the explicit representation of Coulomb repulsion in the wavefunction is unknown. As a consequence the wavefunction must be approximated. The best wavefunctions are obtained by application of the variational principle for the energy. As the representation of electron correlations in the approximate wavefunction is improved, as in going from the Hartree to Hartree-Fock to the Configuration-Interaction approximations, the energy too is improved. In each case a rigorous upper bound to the exact non-relativistic value is obtained, the bound improving with the wavefunction.

An alternate approach to the determination of electronic structure is Hohenberg-Kohn-Sham [2] density functional theory [3] (DFT). According to this theory, which is also derived via the variational principle, every observable (and hence the energy) can be written as a unique functional of the electronic density $\rho(\boldsymbol{r})$. In the corresponding Kohn-Sham theory differential equation, which is equivalent to the Euler equation for the density, the operator representing ***all*** electron correlations including the correlation contribution to the kinetic energy is ***local***. This then simplifies the calculations for the determination of the density. However, as a result of the application of the variational principle, the local operator (potential) has a strictly mathematical

J. L. Calais and E. S. Kryachko (eds.), Structure and Dynamics of Atoms and Molecules: Conceptual Trends, 213–259.

definition as a functional derivative. Furthermore, it is the functional derivative of an as yet unknown "exchange-correlation" energy functional which represents Pauli and Coulomb correlations as well as the correlation-kinetic-energy contribution. In Kohn-Sham theory it is this energy functional that is approximated. However, in approximating the energy functional, as for example in the Local Density [3] or Gradient [3] or Generalized Gradient [4,5] Expansion approximations, the rigor of the Hohenberg-Kohn theorems is lost, and the bounds to the energy obtained no longer rigorous. Substantial progress [4,5] in the development of accurate density-functional theory exchange, correlation and exchange-correlation energy functionals as well as of the properties of the exact functionals has, however been made, and the approximate functionals employed extensively.

In this chapter we describe the work formalism of electronic structure due to Harbola and Sahni. [6,7] The work formalism, which is founded in Schrödinger theory, is derived by arguments that are entirely physical and based on Coulomb's law. It leads ***in principle*** to the system wavefunction Ψ via a Sturm-Liouville differential equation. The relationship of the work formalism to Schrödinger theory and the system wavefunction is through the pair-correlation ***density*** which constitutes the fundamental quantity in the formalism. It represents the ***quantum-mechanical source charge*** distribution which gives rise via Coulomb's law to both the electron interaction energy as well as a ***local*** potential representing electron correlations. The wavefunction is then determined by solution of the resulting Sturm-Liouville equation in which the potential depends self-consistently on the wavefunction.

The work formalism also overlaps with Kohn-Sham density-functional theory in that Pauli and Coulomb correlations between the electrons are represented by a local potential. The theory thus possesses all the attributes attendant to having a multiplicative operator representing electron correlations in the differential equation governing the system. Furthermore, whereas in Kohn-Sham theory the corresponding local potential is derived variationally to be a functional derivative, the potential in the work formalism is obtained directly via Coulomb's law. Thus, the work formalism also provides a physical interpretation for the potential of Kohn-Sham theory.

Approximations within the work formalism are derived by approximating the pair-correlation density or source charge distribution. One way [8] to achieve this is to approximate the wavefunction. In this manner approximations within the context of Schrödinger theory are derived. These approximations and those of Kohn-Sham theory at the same level of electron correlations assumed, can be reinterpreted [8] via the perspective of the work formalism. For example, when only correlations due to the Pauli exclusion principle are considered, both Hartree-Fock theory [9] as well as exchange-only Kohn-Sham theory [10] can be accurately described in terms of the same quantum-mechanical source charge distribution. Thus, the work formalism provides a ***unifying*** physical interpretation of these approximation schemes of Schrödinger and Kohn-Sham theory. Another way [8] to approximate the pair-correlation density is via an expansion in gradients of the density about the uniform electron gas result. With this expansion, approximations within Kohn-Sham theory such as the Kohn-Sham Hartree [8] and Local Density [3] approximations can be ***rigorously*** rederived [8,11] through the work formalism. Furthermore, the identification of the corresponding source charge distributions in turn provides a deeper understanding of electron correlations in these approximations than that achieved by Kohn-Sham theory. Higher order terms in the expansion of the pair-correlation density also lead [8,12] to a physical explanation for the singular nature of the Gradient Expansion approximations for exchange and correlation.

We begin section 2 with a general description of the work formalism, and then derive various approximations of the formalism obtained by approximating the pair-correlation density via the wavefunction. These approximations are the work formalism Hartree, Hartree-Fock and Configuration-Interaction schemes. This then allows a reinterpretation of the corresponding approximations of Schrödinger theory. We conclude this section by a discussion of the ***universal*** nature of the asymptotic structure of the exchange-correlation potential of the work formalism. For ***all*** nonuniform density systems, the asymptotic structure is a consequence ***only*** of the Pauli exclusion principle, and as such exactly determinable. In section 3 we compare the work formalism to Kohn-Sham and Slater [13] theories, and thereby provide physical insights into the local many-body potentials of these theories. We also rederive via the work formalism the equations of the extensively employed Local Density approximation of Kohn-Sham theory, and demonstrate the existence of correlations within this approximation not previously known to be present. This derivation also explains the limitations of the Slater theory Local Density approximation [13] and $X\alpha$ method [14]. The work formalism has thus far been applied to few-electron atomic and many-electron metal surface nonuniform density systems. To demonstrate the accuracy of the formalism we present in section 4 the results of application to atoms. We begin by a discussion of the structure of the Fermi and

Coulomb hole source charge distributions, and the exchange and correlation potentials to which they give rise to respectively. We also give the results of fully-self-consistent calculations within the Pauli-correlated approximation for properties such as the ground-state energy, Kato-Steiner [15] cusp condition, electronic shell structure and core-valence separation charges, ionization potentials, electron affinities and transition energies. Finally in section 5 we provide concluding remarks and discuss directions for future work.

2. The Work Formalism

2.1. GENERAL FRAMEWORK

The Schrödinger equation for a system of N electrons in an external potential $v_{ext}(r)$ is

$$\left[-\frac{1}{2} \sum_i \nabla_i^2 + \sum_i v_{ext}(r_i) + \frac{1}{2} \sum_{i,j}{}' \frac{1}{|r_i - r_j|} \right] \Psi = E \Psi \quad , \tag{1}$$

where H is the Hamiltonian, Ψ the normalized system wavefunction and $E = \langle \Psi | H | \Psi \rangle$ is the energy. As noted above, the fundamental quantity in the work formalism is the pair-correlation ***density*** g(r, r′) whose structure also describes the effects of Pauli and Coulomb correlations. It is defined as the expectation

$$g(r,r') = \langle \Psi | \sum_{i,j}{}' \delta(r_i - r)\, \delta(r_j - r') \,|\, \Psi \rangle / \rho(r) \quad , \tag{2}$$

where in turn the electronic density $\rho(r)$ is the expectation

$$\rho(r) = \langle \Psi | \sum_i \delta(r_i - r) \,|\, \Psi \rangle \quad . \tag{3}$$

The pair-correlation density is thus the density at r' for an electron at r. The total charge of the pair-correlation density for arbitrary electron position is then $\int g(r, r')dr' = N\text{-}1$.

The physics underlying the work formalism is based on the fact that the pair-correlation density is a ***dynamic*** (nonlocal) charge distribution which changes as a function of electron position. The dynamic nature of this distribution must therefore be incorporated in the construction of the local potential in which the electrons move. Thus, the force field $\mathcal{E}(r)$ due to this charge, which by Coulomb's law is

$$\mathscr{E}(r) = \int \frac{g(r,r')\,(r - r')}{|r - r'|^3}\, dr' \quad , \tag{4}$$

must first be obtained. The potential $W(r)$ is then the work done on an electron to bring it from infinity to r against the force of this field:

$$W(r) = -\int_{\infty}^{r} \mathscr{E}(r') \cdot dl' \quad . \tag{5}$$

Since the potential $W(r)$ which accounts for both Pauli and Coulomb correlations is local, the effective potential of the electrons is $[v_{ext}(r) + W(r)]$, and consequently the differential equation governing the system is

$$\left[-\frac{1}{2} \nabla^2 + v_{ext}(r) + W(r) \right] \phi_i(r) = \epsilon_i\, \phi_i(r) \quad . \tag{6}$$

This is a Sturm-Liouville differential equation whose solutions form a complete set. Thus, the wavefunction Ψ can, ***in principle*** be obtained [16] as an infinite linear combination of N-electron Slater determinants $\Phi_i\{\phi_i\}$ of the spin-orbitals $\phi_i(r)$ corresponding to the occupied and virtual states of the system:

$$\Psi = \sum_i B_i \Phi_i\{\phi_i\} \quad , \tag{7}$$

where the B_i are appropriately chosen coefficients. Note that since the spin-orbitals $\phi_i(r)$ are generated self-consistently from the pair-correlation density, the effects of Pauli and Coulomb correlations are ***implicitly*** incorporated in their structure. The energy E is then the expectation value of the Hamiltonian. The electron interaction energy E_{ee} can also be written in terms of the pair-correlation density $g(r, r')$ as the energy of interaction between it and the electronic density $\rho(r)$, so that

$$E_{ee} = \langle\Psi| \frac{1}{2} \sum_{i,j}{}' \frac{1}{|r_i - r_j|} |\Psi\rangle = \frac{1}{2} \int\int \frac{\rho(r)\, g(r,r')}{|r - r'|}\, dr\, dr' \quad . \tag{8}$$

Thus, the pair-correlation density constitutes a ***source charge*** distribution which gives rise via Coulomb's law to ***both*** the local potential representing the correlations between electrons as well as the electron correlation energy. Furthermore, by definition there is no self-interaction in this source charge distribution, and thus within the work formalism.

The equations of the work formalism Eqs. (4)-(8) can be put in a more recognizable form, and one in which the relationship to Kohn-Sham theory is more explicit, by rewriting the pair-correlation density for an electron at **r** as the sum of the density $\rho(\mathbf{r}')$ and the Fermi-Coulomb hole charge $\rho_{xc}(r, r')$ at $\mathbf{r}'$:

$$g(r,r') = \rho(r') + \rho_{xc}(r,r') \quad . \tag{9}$$

Thus, the Fermi-Coulomb hole satisfies the charge conservation constraint $\int \rho_{xc}(r, r')dr' = -1$. The force field $\mathcal{E}(r)$ can then be written as the sum of the Hartree $\mathcal{E}_H(r)$ and exchange-correlation $\mathcal{E}_{xc}(r)$ fields where

$$\mathcal{E}_H(r) = \int \frac{\rho(r')(r-r')}{|r-r'|^3} dr' \quad \text{and} \quad \mathcal{E}_{xc}(r) = \int \frac{\rho_{xc}(r,r')(r-r')}{|r-r'|^3} dr' \quad . \tag{10}$$

The potential $W(r)$ is now the sum of the potentials $W_H(r)$ and $W_{xc}(r)$ where

$$W_H(r) = -\int_{\infty}^{r} \mathcal{E}_H(r') \cdot dl' \quad \text{and} \quad W_{xc}(r) = -\int_{\infty}^{r} \mathcal{E}_{xc}(r') \cdot dl' \quad , \tag{11}$$

are respectively the work done on an electron against the Hartree and exchange-correlation fields. However, since the electronic density is a static charge distribution whose structure is independent of the electron position, the Hartree field can be written as $\mathcal{E}_H(r) = -\nabla W_H(r)$ with

$$W_H(r) = \int \frac{\rho(r')}{|r-r'|} dr' \quad , \tag{12}$$

which then defines the scalar potential $W_H(r)$. Further, this potential $W_H(r)$ is path independent since $\nabla \times \mathcal{E}_H(r) = 0$. With the assumption that the curl of the exchange-correlation field vanishes, i.e., $\nabla \times \mathcal{E}_{xc}(r) = 0$, so that the work $W_{xc}(r)$ is path-independent, the system differential equation is now

$$\left[-\frac{1}{2}\nabla^2 + v_{ext}(r) + \int \frac{\rho(r')}{|r-r'|} dr' + W_{xc}(r) \right] \phi_i(r) = \epsilon_i \phi_i(r) \quad . \tag{13}$$

The corresponding electron interaction energy E_{ee} is then the sum of the Coulomb self-energy E_H and the exchange-correlation energy E_{xc} where

$$E_H = \frac{1}{2}\int\int \frac{\rho(r)\,\rho(r')}{|r-r'|}\,dr\,dr' \quad \text{and} \quad E_{xc} = \frac{1}{2}\int\int \frac{\rho(r)\,\rho_{xc}(r,r')}{|r-r'|}\,dr\,dr' \quad , \tag{14}$$

and where the latter is the energy of interaction between the density $\rho(r)$ and the Fermi-Coulomb hole charge $\rho_{xc}(r, r')$.

The assumption of path-independence of the work $W_{xc}(r)$ is rigorously valid for symmetrical density systems such as spherically symmetric atoms and jellium metal surfaces and clusters, and non-spherically symmetric density systems in the central field approximation. There is as yet no general proof of the path-independence of $W_{xc}(r)$ for systems of arbitrary symmetry in which the external potential is ***physically realistic***. Let us thus assume [7,17] that there are systems for which the curl of the force field $\mathcal{E}_{xc}(r)$ does not vanish. Now according to the Helmholtz theorem, the most general vector field has both a non-zero divergence and a non-zero curl, and can be derived from the negative gradient of a scalar potential and the curl of a vector potential. Writing the electric field $\mathcal{E}_{xc}(r)$ as a sum of its irrotational $\mathcal{E}^{I}_{xc}(r)$ and solenoidal $\mathcal{E}^{S}_{xc}(r)$ components, the statement of Helmholtz's theorem in this case is

$$\begin{aligned}\mathcal{E}_{xc}(r) &= \mathcal{E}^{I}_{xc}(r) + \mathcal{E}^{S}_{xc}(r) \\ &= -\nabla \int \frac{\nabla' \cdot \mathcal{E}_{xc}(r')}{4\pi|r-r'|}\,dr' + \nabla \times \int \frac{\nabla' \times \mathcal{E}_{xc}(r')}{4\pi|r-r'|}\,dr' \quad .\end{aligned} \tag{15}$$

A local path-independent effective exchange-correlation potential $W^{eff}_{xc}(r)$ may then be obtained [6,7] from the irrotational component of the force field, the solenoidal component being neglected. The irrotational component and effective potential then arise from a scalar ***static*** effective Fermi-Coulomb hole charge $\rho^{eff}_{xc}(r)$ which is

$$\rho^{eff}_{xc}(r) = \frac{1}{4\pi}\nabla \cdot \mathcal{E}_{xc}(r) \quad , \tag{16}$$

so that

$$\mathcal{E}^{I}_{xc}(r) = -\nabla\, W^{eff}_{xc}(r) = \int \frac{\rho^{eff}_{xc}(r')\,(r-r')}{|r-r'|^3}\,dr' \quad , \tag{17}$$

and

$$W_{xc}^{eff}(r) = \int \frac{\rho_{xc}^{eff}(r')}{|r-r'|} dr' \quad . \tag{18}$$

Since the total charge of the Fermi-Coulomb hole is unity, so is the total charge of the effective source: $\int \rho_{xc}^{eff}(r)dr = -1$. The potential $W_{xc}^{eff}(r)$ is then substituted for $W_{xc}(r)$ in the differential equation Eq. (13) and determined self-consistently. The solenoidal component of the field $\mathcal{E}_{xc}^{S}(r)$, which is neglected, is then due to an exchange-correlation vector vortex source $J_{xc}(r)$ given as

$$J_{xc}(r) = \frac{1}{4\pi} \nabla \times \mathcal{E}_{xc}(r) \quad , \tag{19}$$

so that

$$\mathcal{E}_{xc}^{S}(r) = \nabla \times \int \frac{J_{xc}(r')}{|r - r'|} dr' = \int J_{xc}(r') \times \frac{(r-r')}{|r-r'|^3} dr' \quad . \tag{20}$$

It is evident that for systems (or regions of space) for which the vortex source vanishes, the potentials $W_{xc}^{eff}(r)$ and $W_{xc}(r)$ are equivalent and path-independent. Thus, another way of obtaining the potential $W_{xc}(r)$ for such systems or regions (other than determining the line integral of Eq. (11)) is via a static effective charge distribution given by the divergence of the field due to the Fermi-Coulomb hole.

Prior to discussing the important aspect of the asymptotic structure of the exchange-correlation potential $W_{xc}(r)$, we describe approximations to the formalism achieved through a hierarchy of approximate wavefunctions. In this manner it is possible to also provide a physical interpretation to the corresponding approximation schemes of Schrödinger theory in terms of a quantum-mechanical source charge distribution.

2.2. THE WORK FORMALISM HARTREE APPROXIMATION

With the wavefunction assumed to be a Hartree product of orthonormal spin-orbitals $\phi_i(r)$:

$$\Psi(r_1, \ldots r_N) = \prod_i \phi_i(r_i) \quad ; \quad \langle \phi_i | \phi_j \rangle = \delta_{ij} \quad , \tag{21}$$

the pair-correlation density of Eq. (2) is

$$g_h(r,r') = \sum_i \phi_i^*(r)\phi_i(r)\rho_i(r')/\rho(r) \quad , \tag{22}$$

where

$$\rho_i(r') = \sum_{\substack{j \\ j \neq i}} \phi_j^*(r')\phi_j(r') \quad , \tag{23}$$

is the density due to all but the ith electron or equivalently the ***orbital-dependent-density***. Thus, the pair-correlation density $g_h(r, r')$ is the weighted average of the density seen by an electron in state i, weighted by the probability of occupation of the state. The differential equation of the work formalism is then

$$\left[-\frac{1}{2}\nabla^2 + v_{ext}(r) + W_h(r)\right]\phi_i(r) = \epsilon_i\phi_i(r) \quad , \tag{24}$$

where $W_h(r)$ is the work done in the force field $\mathcal{E}_h(r)$ of the pair-correlation density $g_h(r, r')$:

$$W_h = -\int_{\infty}^{r} \mathcal{E}_h(r') \cdot dl' \quad \text{where} \quad \mathcal{E}_h(r) = \int \frac{g_h(r,r')(r-r')}{|r-r'|^3} dr' \quad . \tag{25}$$

The electron interaction energy E_{ee} which is the energy of interaction between the electronic and pair-correlation densities is

$$\begin{aligned} E_{ee} &= \frac{1}{2}\int\int \frac{\rho(r)g_h(r,r')}{|r-r'|} dr dr' \\ &= \frac{1}{2}\sum_{i,j}{}' \int\int \frac{|\phi_i(r|^2 |\phi_j(r')|^2}{|r-r'|} dr dr' \quad . \end{aligned} \tag{26}$$

Equations 24-26 constitute the Hartree approximation within the work formalism, the quantum-mechanical source charge giving rise to the electron interaction potential which generates the orbitals and the electron interaction energy being the pair-correlation density $g_h(r, r')$.

The differential equation Eq. (24) is not the same as that of Hartree theory [18] since it is a Sturm-Liouville equation in which all the electrons move in the same local potential $W_h(r)$. In Hartree theory each electron is assumed to move in a potential due to the charge distribution of all the other electrons. The ***orbital-dependent-potential*** differential equation originally proposed by Hartree is therefore

$$\left[-\frac{1}{2}\nabla^2 + v_{ext}(r) + v_{h,i}(r)\right]\phi_i(r) = \epsilon_i\phi_i(r) \quad , \tag{27}$$

where the orbital-dependent-potential

$$v_{h,i}(r) = \int \frac{\rho_i(r')}{|r-r'|}dr' \quad , \tag{28}$$

is due to the orbital-dependent-density $\rho_i(r')$. The expression for the electron interaction energy in Hartree theory is the same as that of Eq. (26). Since the Hartree differential equation can also be rigorously derived by application of the variational principle for the energy, the total energy obtained by the orbitals of the work formalism differential equation constitute a rigorous upper bound to that of Hartree theory.

The orbital-dependent-potential $v_{h,i}(r)$ of Hartree theory can also be interpreted in the context of the work formalism. The potential $v_{h,i}(r)$ is the work done to move an electron in the force field of the orbital-dependent-density $\rho_i(r')$. Since this is a static charge distribution, the curl of its force field vanishes, and the work done can be written as in Eq. (28). In the work formalism on the other hand, the local potential $W_h(r)$ is the work done in the force field of the ***weighted average*** of the orbital-dependent density.

2.3. THE WORK FORMALISM HARTREE-FOCK APPROXIMATION

The next approximation we consider is the Pauli-correlated approximation in which the wavefunction and thereby the pair-correlation density is improved by assuming it to be a Slater determinant of spin-orbitals $\phi_i(r)$:

$$\Psi(r_1, \ldots r_N) = \text{Slater det}\,\{\phi_i(r)\} \quad ; \quad \langle\phi_i|\phi_j\rangle = \delta_{ij} \quad . \tag{29}$$

On substitution of this wavefunction into Eq. (2), the resulting pair-correlation density is

$$g_x(r,r') = \rho(r') + \rho_x(r,r') \quad , \tag{30}$$

where $\rho_x(\boldsymbol{r},\boldsymbol{r}')$ is the Fermi hole charge distribution at $\boldsymbol{r}'$ for an electron at $\boldsymbol{r}$. The Fermi hole, which is the reduction in density about an electron due to the Pauli exclusion principle, is defined in terms of the ***idempotent*** Dirac density matrix $\gamma(\boldsymbol{r},\boldsymbol{r}')$ as

$$\rho_x(r,r') = -|\gamma(r,r')|^2/2\rho(r) \quad , \tag{31}$$

where in turn $\gamma(r,r') = \sum \phi_i^*(r)\, \phi_i(r')$ with $\gamma(r,r) = \rho(r)$. The Fermi hole satisfies the constraints of charge neutrality $\int \rho_x(r,r')dr' = -1$, value at electron position $\rho_x(r,r) = \rho(r)/2$ and negativity $\rho_x(r,r') \leq 0$. The local potential representing electron correlations is then $W(r) = W_H(r) + W_x(r)$, where the exchange potential $W_x(r)$ is the work done in the force field $\mathcal{E}_x(r)$ of the Fermi hole charge:

$$W_x(r) = -\int_\infty^r \mathcal{E}_x(r') \cdot dl' \quad \text{where} \quad \mathcal{E}_x(r) = \int \frac{\rho_x(r,r')(r-r')}{|r-r'|^3} dr' \quad , \tag{32}$$

and the differential equation governing the system is

$$\left[-\frac{1}{2} \nabla^2 + v_{ext}(r) + v_H(r) + W_x(r) \right] \phi_i(r) = \epsilon_i \, \phi_i(r) \quad . \tag{33}$$

The electron interaction energy is the sum of the Hartree E_H and exchange E_x energies, where

$$E_x = \frac{1}{2} \int \int \frac{\rho(r)\, \rho_x(r,r')}{|r-r'|} dr\, dr' \quad , \tag{34}$$

is the energy of interaction between the density and the Fermi hole charge. Equations (32)-(34) constitute the work formalism Hartree-Fock approximation, the source charge distribution giving rise to both the local exchange potential $W_x(r)$ and the exchange energy E_x being the Fermi hole.

As in the Hartree approximation, the differential equation Eq. (33) of the work formalism in the Pauli-correlated approximation is a Sturm-Liouville equation in which all the electrons move in the same local potential. As such its orbitals are not the same as those of Hartree-Fock theory. The reason for this is that although the starting point of each theory is a Slater determinant wavefunction, and that in each there is no self-interaction, the physical interpretation as well as the mathematical structure of the Hartree-Fock theory equations is different. To contrast the two we note that in Hartree-Fock theory according to Slater [13], each electron is surrounded by its ***orbital-dependent*** Fermi hole charge

$$\rho_{x,i}(r,r') = - \sum_{\substack{j \\ j \neq i}} \frac{\phi_j^*(r')\, \phi_i(r')\, \phi_j(r)}{\phi_i(r)} \quad , \tag{35}$$

which in turn gives rise to an ***orbital-dependent*** exchange potential

$$v_{x,i}(r) = \int \frac{\rho_{x,i}(r,r')}{|r - r'|}\, dr' . \tag{36}$$

Thus, in Hartree-Fock theory, as in Hartree theory, each electron can be thought of as moving in an ***orbital-dependent-potential*** of its own. The Hartree-Fock theory differential equation, derived by application of the variational principle, can then be written as

$$\left[-\frac{1}{2} \nabla^2 + v_{ext}(r) + v_H(r) + v_{x,i}(r) \right] \phi_i(r) = \epsilon_i\, \phi_i(r) \quad , \tag{37}$$

and leads to the best single-particle orbitals within the Pauli-correlated approximation. These orbitals, however, do not possess [19] the properties of the Sturm-Liouville theory. The expression for the exchange energy is, of course, the same as Eq. (34) of the work formalism.

Although Hartree-Fock theory has been provided a physical interpretation by rewriting its differential equation as given above, the concept of an orbital-dependent Fermi hole, though appealing, is physically unrealistic. This charge is singular at the nodes of the orbitals and leads to a singular orbital-dependent exchange potential. Furthermore, it can be both positive and negative, and thus no probabilistic quantum-mechanical interpretation can be given to it. The orbital-dependent hole, however, does satisfy the constraints of charge neutrality $\int \rho_{x,i}(r,r')\,dr' = -1$ and value at the electron position $\rho_{x,i}(r,r) = \rho(r)/2$. In contrast, the local exchange potential $W_x(r)$ of the work formalism is derived by Coulomb's law from a ***real*** physical effect, viz. the Fermi hole. Note further that the Fermi hole $\rho_x(r, r')$, which physically is the reduction in density about each electron in the distribution of electrons of the same spin, is mathematically [13] the weighted average of the orbital-dependent Fermi hole $\rho_{x,i}(r, r')$ weighted by the probability of occupation of that state. Thus, whereas in Hartree-Fock theory the orbital-dependent potentials $v_{x,i}(r)$ are determined from the orbital-dependent Fermi holes, the local exchange potential $W_x(r)$ is the work done in the force field of the ***weighted average*** of these orbital-dependent Fermi holes.

The question which then arises is whether one can reinterpret Hartree-Fock theory from the physical perspective of the work formalism. Based on the results for atoms to be discussed below, the answer is in the affirmative. Ground-state energies obtained via the work formalism lie within a few parts per million of those of Hartree-Fock theory. As such the orbitals of Hartree-Fock theory can also be thought of as being generated by a ***local*** exchange operator which has the physical interpretation of the work done to move an electron in the force field of the Fermi hole. The exchange energy in turn is the energy of interaction between the electronic density and the Fermi hole charge, which is the same interpretation as in Hartree-Fock theory.

2.4. THE WORK FORMALISM CONFIGURATION-INTERACTION APPROXIMATION

As noted previously, the differential equation Eq. (6) of the work formalism is a Sturm-Liouville equation, and as such the system wavefunction can be written as an infinite linear combination of Slater determinants of the spin-orbitals corresponding to the occupied and virtual states of the system. In practice, however, one can generate only a finite set of basis functions with the coefficients of the expansion being determined by energy minimization. We thereby have the configuration interaction (CI) approximation within the work formalism. In this manner the wavefunction, and consequently the pair-correlation density $g(\boldsymbol{r}, \boldsymbol{r}')$ and local potential $W(\boldsymbol{r})$, now incorporate both Pauli and Coulomb correlation effects.

The construction of the configuration-interaction wavefunction as described within the framework of the work formalism via Eq. (6) differs in fundamental ways from those of conventional [20] CI calculations. The most significant of these is that the orbitals of the wavefunction implicitly include in their structure the effects of ***both*** Pauli and Coulomb correlations since they are generated self-consistently from the pair-correlation density, which in turn is expressed in terms of the wavefunction. The basis functions could also be generated as in standard CI calculations, where the Hartree-Fock theory ground state is the reference state, from the differential equation Eq. (35) of the Pauli-correlated approximation. In either case the orbitals have the correct asymptotic behavior since they are generated by a local effective potential which decays as (-1/r) in atoms and molecules. In contrast the Hartree-Fock theory orbitals [21] all have the same asymptotic structure decaying exponentially with an exponent corresponding to the highest-occupied-orbital eigenvalue. Although yet not proved, it is expected that as a consequence of the ***intrinsic*** inclusion of Coulomb correlation effects and the correct asymptotic structure of the orbitals, the number of configurations necessary to achieve a certain accuracy for the energy will be smaller than in standard CI calculations.

2.5. ASYMPTOTIC STRUCTURE OF EXCHANGE-CORRELATION POTENTIAL

An important attribute of the work formalism is that for ***all*** nonuniform electron gas systems whether in atoms, molecules, metallic surfaces or metallic clusters, the asymptotic structure of the exchange-correlation potential $W_{xc}(r)$ in the classically forbidden region is that of the exchange potential $W_x(r)$, and therefore exactly determinable. To understand this we first define the Coulomb hole charge $\rho_c(r, r')$ as the difference between the Fermi-Coulomb $\rho_{xc}(r, r')$ and Fermi $\rho_x(r, r')$ hole charge distributions:

$$\rho_c(r, r') = \rho_{xc}(r, r') - \rho_x(r, r') \quad , \tag{38}$$

where the Fermi hole is determined within the Pauli-correlated approximation. The Coulomb hole charge gives rise to a correlation potential $W_c(r)$ which is the work done to move an electron in its force field $\mathcal{E}_c(r)$:

$$W_c(r) = -\int_{\infty}^{r} \mathcal{E}_c(r') \cdot dl' \quad \text{where} \quad \mathcal{E}_c(r) = \int \frac{\rho_c(r, r')(r - r')}{|r - r'|^3} dr' \quad , \tag{39}$$

and to a correlation energy E_c which is the interaction energy between it and electronic density:

$$E_c = \frac{1}{2} \int \int \frac{\rho(r)\rho_c(r, r')}{|r - r'|} dr\, dr' \quad . \tag{40}$$

Now since the total charge of both the Fermi-Coulomb and Fermi holes is (negative) unity, the total charge of the Coulomb hole is $\int \rho_c(r, r') dr' = 0$. Thus, for asymptotic positions of the electron beyond where this charge exists there is no force field $\mathcal{E}_c(r)$ due to it, and therefore the correlation potential $W_c(r)$ vanishes in this region. The exchange-correlation potential $W_{xc}(r)$ then reduces to $W_x(r)$ which arises due to the finite charge of the Fermi hole. Therefore, the asymptotic structure of the fully-correlated system potential $W_{xc}(r)$, i.e. when both Pauli and Coulomb correlations are present, can be determined ***exactly*** by solving the problem in the Pauli-correlated approximation.

There is yet another important consequence of the above conclusion. In local effective potential theories, the highest occupied eigenvalue is governed principally by the asymptotic structure of the exchange-correlation potential. Since the asymptotic structure of $W_{xc}(r)$ is given by $W_x(r)$, it is meaningful to compare the highest occupied eigenvalue of the Pauli-correlated approximation differential equation to the experimental ionization potential and electron affinity. For the same reason it is also meaningful to compare transition energies obtained within the Pauli-correlated approximation of the work formalism to experiment. In contrast we note that the highest occupied eigenvalue of Hartree-Fock theory has the meaning of a removal energy only within the context of Koopmans' theorem [22] which requires the orbitals of the neutral and ionized systems to be the same.

3. Comparisons with Other Local Potential Theories

In the work formalism, the potential representing electron conditions is local and derived by physical arguments based on Coulomb's law. The formalism thus leads naturally to insights into other local-potential theories. In the following subsections we compare the work formalism to Kohn-Sham and Slater theory, and thereby provide a physical understanding of the corresponding local many-body potentials of these theories. We also rederive the popular local density approximation of Kohn-Sham theory via the work formalism. The derivation, which is rigorous, then leads to a fundamental understanding of how electrons are correlated within this approximation.

3.1. COMPARISON WITH KOHN-SHAM THEORY

In Kohn-Sham (KS) theory [3], the ground-state energy $E[\rho]$, which is a universal functional of the density $\rho(r)$, is partitioned as follows:

$$E[\rho] = T_s[\rho] + \int v_{ext}(r)\rho(r)dr + E_H[\rho] + E_{xc}^{KS}[\rho] \quad , \tag{41}$$

where $T_s[\rho]$ is the kinetic energy of a system of ***noninteracting*** electrons having the same density distribution as the interacting system, the second term on the right is the expectation of the external potential $V_{ext}(r)$, $E_H[\rho]$

the Coulomb self-energy of Eq. (14), and $E_{xc}^{KS}[\rho]$ the Kohn-Sham theory exchange-correlation energy. Thus, $E_{xc}^{KS}[\rho]$ differs from the quantum-mechanical definition of the exchange-correlation energy E_{xc} of Eq. (14) in that as a result of the above partition it also accounts for the correlation contribution $T_c[\rho]$ to the kinetic energy.

On application of the variational principle for the energy to the functional of Eq. (41) for arbitrary variations of the density, one obtains the Kohn-Sham differential equation

$$\left[-\frac{1}{2}\nabla^2 + v_{ext}(\mathbf{r}) + v_H(\mathbf{r}) + v_{xc}(\mathbf{r})\right]\phi_i(\mathbf{r}) = \epsilon_i\phi_i(\mathbf{r}) \quad , \tag{42}$$

where the DFT Hartree $v_H(\mathbf{r})$ and exchange-correlation $v_{xc}(\mathbf{r})$ potentials are the functional derivatives of $E_H[\rho]$ and $E_{xc}^{KS}[\rho]$, respectively:

$$v_H(\mathbf{r}) = \frac{\delta E_H[\rho]}{\delta\rho(\mathbf{r})} = \int \frac{\rho(\mathbf{r}')}{|\mathbf{r}-\mathbf{r}'|}d\mathbf{r}' \quad , \tag{43}$$

and

$$v_{xc}(\mathbf{r}) = \frac{\delta E_{xc}^{KS}[\rho]}{\delta\rho(\mathbf{r})} \quad . \tag{44}$$

The ground-state density $\rho(\mathbf{r})$ of the system and the non-interacting kinetic energy are obtained from a ***single*** Slater determinant $\Phi\{\phi_i\}$ of the lowest occupied orbitals $\phi_i(\mathbf{r})$ of the Kohn-Sham differential equation:

$$\rho(\mathbf{r}) = \langle\Phi\{\phi_i\}|\sum_i \delta(\mathbf{r}_i-\mathbf{r})|\Phi\{\phi_i\}\rangle = \sum_i |\phi_i(\mathbf{r})|^2 \quad , \tag{45}$$

and

$$T_s[\rho] = \sum_i \int \phi_i^*(\mathbf{r})\left[-\frac{1}{2}\nabla_i^2\right]\phi_i(\mathbf{r})d\mathbf{r} \quad . \tag{46}$$

With the orbitals $\phi_i(\mathbf{r})$ and the density $\rho(\mathbf{r})$, the ground-state energy is then determined via the functional of Eq. (41). Furthermore, the highest occupied eigenvalue of the Kohn-Sham differential equation is [3] the negative of the removal energy. However, the exchange-correlation energy functional $E_{xc}^{KS}[\rho]$, and therefore its functional derivative $v_{xc}(\mathbf{r})$ and the highest occupied eigenvalue are at present unknown.

Analogous to the quantum-mechanical definitions, $E_{xc}^{KS}[\rho]$ can also be thought of as the energy of interaction between the density $\rho(r)$ and the Kohn-Sham Fermi-Coulomb hole charge $\rho_{xc}^{KS}(r, r')$. This hole charge differs from the quantum-mechanical hole in that its structure also incorporates the correlation contribution to the kinetic energy. The Kohn-Sham hole charge is defined in terms of the electron-electron-interaction coupling constant λ integral as

$$\int_0^1 d\lambda g_\lambda(r, r') = \rho(r') + \rho_{xc}^{KS}(r, r') \quad , \tag{47}$$

where $g_\lambda(r, r')$ is the pair-correlation density of a hypothetical system in an external potential $v_{ext,\lambda}(r)$ chosen such that the ground-state density $\rho_\lambda(r) = <\Psi_\lambda| \sum_i \delta(r_i - r) |\Psi_\lambda>$ is identical with the true density $\rho(r)$ for all values of the constant λ.

It is evident from the above description of Kohn-Sham theory that a point of commonality between it and the work formalism is the equivalence of the DFT Hartree potential $v_H(r)$ of Eq. (43) to the potential $W_H(r)$ of Eq. (12). Thus, the functional derivative of the Coulomb self-energy $E_H[\rho]$ has the physical interpretation of being the work done to move an electron in the force field $\mathcal{E}_H(r)$ of the electronic density $\rho(r)$.

Before discussing the exchange-correlation potentials $W_{xc}(r)$ and $v_{xc}(r)$, we compare the work formalism with Kohn-Sham theory in the Pauli-correlated approximation. In this approximation the functional $E_{xc}^{KS}[\rho]$ in Eq. (41) and potential $v_{xc}(r)$ of Eq. (42) are replaced by the exchange energy functional $E_x^{KS}[\rho]$ and potential $v_x(r) = \delta E_x^{KS}[\rho]/\delta\rho(r)$, respectively. The exchange energy $E_x^{KS}[\rho]$ can also be defined as the energy of interaction between the corresponding density and the Kohn-Sham Fermi hole, where in turn the latter is defined in terms of the indempotent Dirac density matrix formed from the exchange-only Kohn-Sham orbitals. Since the functional dependence of the density matrix $\gamma(r, r')$ on $\rho(r)$ is unknown, the potential $v_x(r)$ cannot be determined as a functional derivative. In contrast the exchange potential $W_x(r)$ is known precisely since it is determined directly from the Fermi hole itself.

The exchange potential $v_x(r)$ of Kohn-Sham theory also satisfies the following three conditions [23]. These are (i) the virial sum rule relating the exchange energy to its functional derivative:

$$E_x^{KS}[\rho] + \int dr \rho(r) r \cdot \nabla v_x(r) = 0 \quad , \tag{48}$$

(ii) the scaling condition

$$v_x(r;[\rho_\lambda]) = \lambda v_x(\lambda r;[\rho]) \quad , \tag{49}$$

where $\rho_\lambda(r) = \lambda^3 \rho(\lambda r)$, and (iii) the second derivative condition

$$\delta v_x(r)/\delta\rho(r') = \delta v_x(r')/\delta\rho(r) \quad , \tag{50}$$

which is one of symmetry in an interchange of r and r'. As a consequence of this same symmetry of the pair-correlation ***function*** $h_x(r,r') = g_x(r,r')/\rho(r')$, it can be ***analytically*** shown [6] that the potential $W_x(r)$ satisfies the virial sum rule. It is also ***analytically*** evident [6] that $W_x(r)$ satisfies the scaling condition. Again, since the dependence of the Dirac density matrix on the density is unknown, whether $W_x(r)$ satisfies the second derivative condition cannot be determined.

The Kohn-Sham exchange potential $v_x(r)$ is also defined in the literature [10] as that obtained by the optimized potential method [24] (OPM). In this method the electrons in an external potential $v_{ext}(r)$ are assumed to move in a local effective potential $V^{OPM}(r)$ which is then varied till the Hartree-Fock theory energy is minimized. The exchange potential $v_x(r)$, which is the difference between $V^{OPM}(r)$ and the DFT Hartree potential $v_H(r)$, satisfies the equations

$$\left[-\frac{1}{2}\nabla_i^2 + v_{ext}(r) + V^{OPM}(r)\right]\phi_i(r) = \epsilon_i \phi_i(r) \quad , \tag{51}$$

and

$$\sum_i \int dr' \left[v_x(r') - v_{x,i}(r')\right] G_i(r', r) \phi_i(r') \phi_i^*(r) = 0 \quad , \tag{52}$$

where $G_i(r, r')$ is the Green function

$$G_i(r, r') = \sum_j{}' \phi_j(r) \phi_j^*(r') / \epsilon_j - \epsilon_i \quad , \tag{53}$$

and where $v_{x,i}(r)$ is the orbital-dependent potential Eq. (36) of Hartree-Fock theory. For atoms in the central field approximation, the exchange potential determined by the optimized potential method has been shown [25] numerically to satisfy the virial sum rule. (The potentials of the original OPM calculation [26] do not satisfy [6] the sum rule. The OPM code was subsequently [25] refined). The total ground-state energies obtained, closely approximate but are an upper bound to the Hartree-Fock theory values. The local exchange potential which leads to the Hartree-Fock theory ground-state energy and density as defined [3] by the integral equation relating the potential to the irreducible self-energy has yet not been determined.

As will be shown in the section on results, the potential $W_x(\mathbf{r})$ and the exchange potential $v_x(\mathbf{r})$ of the optimized potential method are ***essentially equivalent*** for the nonuniform electron systems in atoms and at metallic surfaces. Together with the fact that $W_x(\mathbf{r})$ satisfies the virial sum rule and the scaling condition, the exchange potential of Kohn-Sham theory can then be accurately interpreted as the work done to move an electron in the force field of the Fermi hole charge.

The Kohn-Sham theory exchange-correlation energy functional $E_{xc}^{KS}[\rho]$ and its functional derivation $v_{xc}(\mathbf{r})$ satisfy in turn the virial sum rule

$$E_{xc}^{KS}[\rho] + \int d\mathbf{r}\rho(\mathbf{r})\mathbf{r} \cdot \nabla v_{xc}(\mathbf{r}) = -T_c \leq 0 \quad . \tag{54}$$

(Note that there is no exchange contribution T_x to the kinetic energy in Kohn-Sham theory. Thus, the sum rule satisfied by the Kohn-Sham theory correlation energy $E_c^{KS}[\rho]$ and its functional derivative $v_c(\mathbf{r}) = \delta E_c^{KS}[\rho]/\delta\rho(\mathbf{r})$ is the same as Eq. (54)). Once again as a result of the symmetry of the pair-correlation ***function*** $h(\mathbf{r}, \mathbf{r}') = g(\mathbf{r}, \mathbf{r}') / \rho(\mathbf{r}')$, the left hand side of this sum rule can be shown [27] to vanish when $W_{xc}(\mathbf{r})$ is substituted for $v_{xc}(\mathbf{r})$. Thus, the exchange-correlation potential of the work formalism is not the Kohn-Sham theory potential. The reason for this is that $W_{xc}(\mathbf{r})$ is determined from the quantum-mechanical Fermi-Coulomb hole charge $\rho_{xc}(\mathbf{r}, \mathbf{r}')$ which does not incorporate the correlation contribution to the kinetic energy $T_c[\rho]$ in its structure. [Even if one were to employ the Kohn-Sham Fermi-Coulomb hole charge $\rho_{xc}^{KS}(\mathbf{r}, \mathbf{r}')$, (which does include this contribution), to determine $W_{xc}(\mathbf{r})$, the left hand side of the sum rule would still vanish due to the symmetry of the pair-correlation function. This implies that the component of $\rho_{xc}^{KS}(\mathbf{r}, \mathbf{r}')$ due to the correlation-kinetic-energy does not contribute to the force field. Equivalently, Coulomb's law cannot account for the kinetic energy contributions to the potential]. Thus, the potential $W_{xc}(\mathbf{r})$ can be thought of as representing the exchange and purely Coulomb correlation components of the Kohn-Sham potential $v_{xc}(\mathbf{r})$.

3.2. RIGOROUS INTERPRETATION OF THE LOCAL DENSITY APPROXIMATION

In this subsection we rederive the equations of Kohn-Sham theory in the local density approximation (LDA) via the work formalism, and thereby provide a ***rigorous*** physical interpretation for this approximation scheme. Furthermore, the derivation through the work formalism provides additional insight into how electrons are correlated in this approximation. From the perspective of Dirac [28] who originally proposed the LDA, or of Kohn-Sham theory, in which the exchange-correlation energy functional is first approximated and the many-body potential then obtained as its functional derivative, the electrons within the LDA are assumed correlated as in a ***uniform*** electron gas. The derivation via the work formalism shows that in addition to these correlations there exist in this approximation correlations between electrons which ***explicitly*** account for the ***nonuniformity*** of the electronic density of the system. This more physically realistic description of electron correlations is the reason for the accuracy of the LDA.

To explain the above remarks let us compare the derivations of the LDA via Kohn-Sham theory and the work formalism. Let us also initially consider the Pauli-correlated approximation for which case the derivation within each theory is entirely analytical. In Kohn-Sham theory a pair-correlation density $g_x^{(0)}\{r, r'; \rho(r)\}$ is obtained by first determining the expectations of Eq. (2) with a wavefunction that is a Slater determinant of plane waves, and then assuming the resulting expression to be valid locally. With the superscript (0) indicating that the result is derived from uniform electron gas theory, we have

$$g_x^{(0)}(r,r') = \rho(r') + \rho_x^{(0)}(r,r') \quad , \tag{55}$$

where

$$\rho_x^{(0)}\{r,r';\rho(r)\} = -\frac{1}{2}\,\rho(r)\left[\frac{9j_1^2(x)}{x^2}\right] \quad , \tag{56}$$

$j_1(x)$ is the first-order spherical Bessel function, $x = k_F R$, k_F is the Fermi momentum, $k_F(r) = [3\pi^2\rho(r)]^{1/3}$ and $\boldsymbol{R} = \boldsymbol{r}' - \boldsymbol{r}$. The term $\rho_x^{(0)}\{r, r'; \rho(r)\}$ is the Fermi hole for the uniform electron gas which is ***spherically symmetric*** about the electron irrespective of its position. This hole charge satisfies all the constraints of the exact Fermi hole. The electron interaction energy in the LDA $E_{ee}^{(0)}$, which is the energy of interaction between the density $\rho(r)$ and the pair-correlation density $g_x^{(0)}\{r, r'; \rho(r)\}$, is then

$$E_{ee}^{(0)} = \frac{1}{2} \iint \frac{\rho(r)\, g_x^{(0)}\{r,r';\rho(r)\}}{|r-r'|}\, dr\, dr' \tag{57}$$

$$= E_H[\rho] + E_x^{LDA}[\rho] \quad , \tag{58}$$

where the exchange energy in the LDA

$$E_x^{LDA}[\rho] = \frac{1}{2} \iint \frac{\rho(r)\rho_x^{(0)}\{r,r';\rho(r)\}}{|r-r'|}\, dr\, dr' \quad , \tag{59}$$

is the energy of interaction between the density $\rho(r)$ and the Fermi hole charge $\rho_x^{(0)}\{r, r'; \rho(r)\}$. The expression for $E_x^{LDA}[\rho]$ can equivalently be written as

$$E_x^{LDA}[\rho] = \int \epsilon_x^{(0)}\{\rho(r)\}\, \rho(r) dr \quad , \tag{60}$$

where $\epsilon_x^{(0)}\{\rho(r)\} = -3k_F(r)/4\pi$ is the average exchange energy per electron for the uniform electron gas. The electron interaction potential of Kohn-Sham theory is then obtained as the functional derivative of the interaction energy:

$$v_{ee}^{LDA}(r) = \frac{\delta E_{ee}^{(0)}[\rho]}{\delta \rho(r)} = v_H(r) + v_x^{LDA}(r) \quad , \tag{61}$$

where the exchange potential in the LDA is

$$v_x^{LDA}(r) = \frac{\delta E_x^{LDA}[\rho]}{\delta \rho(r)} = -\frac{k_F(r)}{\pi} \quad . \tag{62}$$

Equations (58)-(62) constitute the LDA as derived by Kohn-Sham theory. Based on the fact that the starting point of the derivation is the pair-correlation density $g_x^{(0)}\{r, r'; \rho(r)\}$ of Eq. (55), it has been ***assumed*** that in the LDA the correlations between electrons are those of the uniform electron gas.

In the work formalism, the force field due to the pair-correlation density must first be determined, and then the potential representing electron correlations obtained as the work done to move an electron in this field. If the electrons are assumed correlated as defined by the pair-correlation density $g_x^{(0)}\{r, r'; \rho(r)\}$, then the corresponding force field and potential are the Hartree field $\mathcal{E}_H(r)$ and potential $v_H(r)$ respectively. This is because there can be no contribution to the field at the electron position from the spherically symmetric Fermi hole $\rho_x^{(0)}\{r, r'; \rho(r)\}$ term. The field arises only due to the density $\rho(r')$ which is a charge distribution that is ***not spherically symmetric*** about the electron at r. Thus, the pair-correlation density $g_x^{(0)}\{r, r'; \rho(r)\}$ leads via Coulomb's law to an electron interaction potential that is the Hartree potential $v_H(r)$ rather than $v_{ee}^{LDA}(r)$ of the LDA as given by Eq. (61). Therefore, $g_x^{(0)}\{r, r'; \rho(r)\}$ cannot be the pair-correlation density in the LDA. To obtain the correct LDA pair-correlation density $g_x^{LDA}\{r, r'; \rho(r)\}$, we expand the general expression $g_x(r, r')$ of Eq. (30) in gradients of the density about the uniform electron gas result, and then assume the resulting expressing to be valid locally. To obtain this expansion one requires the corresponding expansion of the Dirac density matrix $\gamma(r, r')$ whose diagonal matrix element is the density $\rho(r)$. In this manner the expansion for both terms of $g_x(r, r')$ are simultaneously obtained. The expansion [3] to lowest order in ∇ is

$$g_x^{LDA}\{r,r';\rho(r)\} = \rho(r') + \rho_x^{(0)}\{r,r';\rho(r)\} + \rho_x^{(1)}\{r,r';\rho(r)\} \quad , \tag{63}$$

where

$$\rho_x^{(1)}\{r,r';\rho(r)\} = \frac{9}{4}\,\rho(r)\left[\frac{j_0(x)j_1(x)}{k_F^3}\,\hat{R}\cdot\nabla k_F^2\right] \quad , \tag{64}$$

$j_0(x)$ is the zeroth-order spherical Bessel function, $\hat{R} = R/R$, and where the superscript (1) indicates the expression to be of $O(\nabla)$. [Note that the lowest-order correction term in the expansion for the density $\rho(r)$ is of $O(\nabla^2)$]. As with the density $\rho(r')$, the term $\rho_x^{(1)}\{r, r'; \rho(r)\}$ is ***not spherically symmetric*** about the electron and contributes [25] to the force field so that

$$\begin{aligned}\mathcal{E}^{LDA}(r) &= \int \frac{g_x^{LDA}\{r,r';\rho(r)\}(r-r')}{|r-r'|^3}\,dr' \\ &= \mathcal{E}_H(r) + \nabla\left[\frac{k_F(r)}{\pi}\right] \quad .\end{aligned} \tag{65}$$

The curl of this force field vanishes, i.e. $\nabla \times \mathcal{E}^{LDA}(r) = 0$. Thus, the work $W^{LDA}(r)$ required to move an electron in this field is path-independent, and given by

$$W^{LDA}(r) = v_H(r) - \frac{k_F(r)}{\pi} \quad , \tag{66}$$

which is the same as Eq. (61) for $v_{ee}^{LDA}(r)$ of Kohn-Sham theory. The electron interaction energy E_{ee}^{LDA} in turn is the energy of interaction between the density $\rho(r)$ and the pair-correlation density $g_x^{LDA}\{r, r'; \rho(r)\}$. However, the non-spherically symmetric component $\rho_x^{(1)}\{r, r'; \rho(r)\}$ does not contribute to this integral, so that

$$E_{ee}^{LDA} = \frac{1}{2} \int\int \frac{\rho(r) g_x^{LDA}\{r, r'; \rho(r)\}}{|r - r'|} dr\, dr' \tag{67}$$

$$= E_H[\rho] + E_x^{LDA}[\rho] \quad , \tag{68}$$

which is the same as Eq. (58) for the electron interaction energy of Kohn-Sham theory. Thus, the work formalism leads to the same expressions for the potential and energy as derived by the Kohn-Sham scheme. It thereby provides a rigorous physical interpretation for this approximation: the local potential representing electron correlations in the LDA is the work done in the force field of the pair-correlation density $g_x^{LDA}(r, r')$, and the electron interaction energy is the energy of interaction between the resulting electronic density and this pair-density. Furthermore, the work formalism derivation also shows that the pair-correlation density $g_x^{LDA}\{r, r'; \rho(r)\}$, which is the quantum-mechanical source charge in the approximation, contains a term proportional to the gradient of the density. Therefore, the nonuniformity of the electronic density is explicitly accounted for in the representation of electron correlations within the LDA. The existence of the additional correlations, and the fact that it is these correlations which generate the LDA exchange potential $v_x^{LDA}(r)$ and orbitals, cannot be gleaned from Kohn-Sham theory.

It is also evident from Eq. (63) for $g_x^{LDA}\{r, r'; \rho(r)\}$ that the Fermi hole in the LDA is given by the expression

$$\rho_x^{LDA}\{r, r'; \rho(r)\} = \rho_x^{(0)}\{r, r'; \rho(r)\} + \rho_x^{(1)}\{r, r'; \rho(r)\} \quad . \tag{69}$$

This charge distribution contains a term proportional to the gradient of the density and is not spherically symmetric about the electron. Although it satisfies the constraint of charge neutrality and value at the electron position, it does not satisfy the negativity constraint, and this is one source of error in the LDA. In spite of this, the structure of $\rho_x^{LDA}\{r, r'; \rho(r)\}$ is far more accurate than its spherically symmetric component $\rho_x^{(0)}\{r, r'; \rho(r)\}$. We refer the reader to reference 11 for examples of the structure of LDA hole in atoms and at metallic surfaces.

When both Pauli and Coulomb correlations are considered, the corresponding LDA Fermi-Coulomb hole charge $\rho_{xc}^{LDA}\{r, r'; \rho(r)\}$ is also not spherically symmetric about the electron, and is the sum of a spherically symmetric component $\rho_{xc}^{(0)}\{r, r'; \rho(r)\}$ obtained from uniform electron gas theory, and a non-spherically symmetric component $\rho_{xc}^{(1)}\{r, r'; \rho(r)\}$ proportional to the first-order in the gradient of the density determined by an expansion about the uniform gas value. Since in the calculations for the uniform electron gas, the kinetic energy is treated as that of the interacting system, the $\rho_{xc}^{LDA}\{r, r'; \rho(r)\}$ thus derived will not contain any correlation-kinetic-energy contributions. Its non-spherically symmetric component will give rise to a LDA potential and orbitals, and its spherically symmetric component to the energy. An expression for the non-spherically symmetric component has not yet been derived. However, it has been possible [29] to study the spherically symmetric component of the LDA Fermi-Coulomb and Coulomb holes to explain why the error in the LDA correlation energy for atoms is so large.

In concluding we learn from the work formalism that the representation of electron correlations within the LDA of Kohn-Sham theory is far more physically realistic than previously understood to be the case. The approximation does not correspond to one in which the nonuniform electron gas is considered uniform at each point, but rather one in which the non-uniformity of the electronic density is explicitly accounted for at each electron position. The LDA pair-correlation density $g^{LDA}\{r, r'; \rho(r)\}$ in fact contains a term proportional to the gradient of the density, and may be written as

$$g^{LDA}\{r, r'; \rho(r)\} = g^{(0)}\{r, r'; \rho(r)\} + O(\nabla_\rho) \quad , \tag{70}$$

where $g^{(0)}\{r, r'; \rho(r)\}$ is the uniform electron gas pair-correlation density. From this fact and its definition via Eq. (2) in terms of the corresponding LDA wavefunction $\Psi^{LDA}\{r_1, .. r_N; \rho(r)\}$ which is

$$g^{LDA}\{r,r';\rho(r)\} = \frac{\left\langle \Psi^{LDA} \left| \sum_{i,j}{}' \delta(r_i - r)\delta(r_j - r') \right| \Psi^{LDA} \right\rangle}{\left\langle \Psi^{LDA} \left| \sum_i \delta(r_i - r) \right| \Psi^{LDA} \right\rangle} , \tag{71}$$

we learn that Ψ^{LDA} explicitly incorporates elements of the physics appropriate to regions of space where the potential is rapidly varying as well as of the classically forbidden region. The fact that the LDA wavefunction possesses these properties is the fundamental reason for the accuracy achieved by the approximation.

3.3. COMPARISON WITH SLATER THEORY

After interpreting Hartree-Fock theory in terms of orbital-dependent Fermi holes $\rho_{x,i}(r,r')$ and exchange potentials $v_{x,i}(r)$ (see Eqs. (35-37)), Slater [13] attempted to simplify the theory by replacing the latter by a local exchange potential $V_x^s(r)$. He argued that since ***each*** orbital-dependent Fermi hole satisfied the same constraints of charge neutrality and value at the electron position, they could not be very different from each other. As such, no great error would be made if a weighted average of these orbital-dependent Fermi holes, weighted by the probability of occupation of each state, was considered instead. This weighted average, as noted previously, is simply the Fermi hole $\rho_x(r,r')$. Analogous to the expression Eq. (36) for the orbital-dependent potential $v_{x,i}(r)$, Slater then proposed a local exchange potential in terms of the Fermi hole defined similarly as

$$V_x^s(r) = \int \frac{\rho_x(r,r')}{|r-r'|} dr' . \tag{72}$$

However, the Slater potential $V_x^s(r)$ does not take into consideration the dynamic nature of the Fermi hole. The expression for the potential is valid only for static charge distributions. In other words, the effect of the electron on the charge distribution to which it gives rise is not accounted for in Slater theory as it is in the work formalism. Thus, the results of Slater theory are not as accurate [6]. This is also reflected by the fact that although the Slater potential satisfies the scaling condition, it does not satisfy the virial sum rule. The functional derivative of the Slater potential can also be written [30] as the sum of a local and non-local term. If this functional derivative is approximated by its local term, then the Slater potential satisfies the second derivative condition. With the inclusion of the nonlocal term, it is unclear whether it does.

Slater also further approximated his potential by its local density version $V_x^{S,LDA}(r)$. In other words he substituted for the Fermi hole $\rho_x(r, r')$ in Eq. (72) the expression for the uniform electron gas as given by $\rho_x^{(0)}\{r, r'\}$ of Eq. (56) and then assumed the resulting potential to be valid locally. However, since the dynamic nature of the approximate Fermi hole is not accounted for in Slater theory, the potential obtained is $V_x^{S,LDA}(r) = -3k_F(r)/2\pi$, which differs by a factor of 2/3 from the Kohn-Sham theory and work formalism value. In the Slater $X\alpha$ method [14], the local exchange-correlation potential is defined as $\alpha V_x^{S,LDA}(r)$, where α is a parameter that is varied to determine the energy minimum. Thus, in the $X\alpha$ method, the dynamic nature of the Fermi hole is once again not taken into consideration.

4. Results of Application to Atoms

In order to explain the work formalism further and to demonstrate its accuracy we present in this section a few results of its application to atoms.

4.1. STRUCTURE OF THE FERMI HOLE AND EXCHANGE POTENTIAL

Since the quantum-mechanical source charge for the exchange potential in the Pauli-correlated approximation is the Fermi hole $\rho_x(r, r')$, we begin by studying its structure for the spherically-symmetric Neon atom (Fig. 1) as a function of electron position. The orbitals employed in its construction are those generated self-consistently by the work formalism in this approximation. The three electron positions considered are at the nucleus r=0 a.u. (Fig. 1(a)), at the maximum of the radial probability density in the L shell at r=0.651 a.u. (Fig. 1(b)), and at r=4.00 a.u. in the classically forbidden region (Fig. 1(c)). Observe that for an electron at the nucleus, the Fermi hole is spherically symmetric about it. Thus the force field at this position due to the charge vanishes. The exchange potential $W_x(r)$ must therefore approach the origin quadratically, having zero slope at the nucleus. For other electron positions the Fermi hole is not spherically symmetric about the electron (see Figs. 1(b) and (c)), and further the structure changes significantly. For asymptotic positions of the electron (r $\geq$ 10 a.u.) the hole

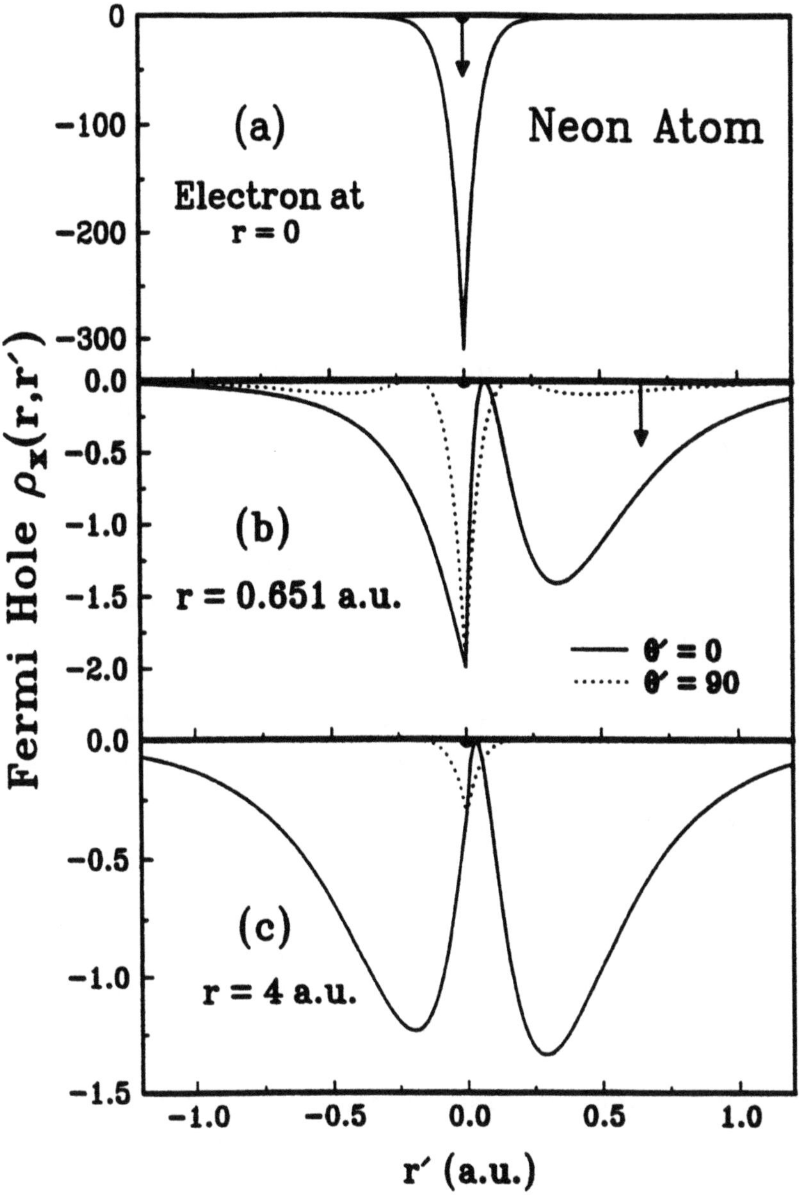

Figure 1. Cross-sections of the Fermi hole $\rho_x(r,\ r')$ for the Ne atom in the electron nucleus plane ($\theta' = 0^\circ$) (solid line) and in a perpendicular plane passing through the nucleus ($\theta' = 90^\circ$) (dashed line) for electron positions at (a) r=0 a.u., (b) r=0.651 a.u., and (c) r=4 a.u. on the z$'$ axis.

charge stabilizes and becomes essentially static [6]. Since the total charge of the Fermi hole is (negative) unity, the asymptotic structure of $W_x(r)$ must be -1/r.

The above discussion on the structure of the Fermi hole and exchange potential $W_x(r)$ though specific to Neon is entirely general and the same for all atoms. Thus, in Fig. 2 we plot $W_x(r)$ for the Argon atom. Observe that shell structure is clearly delineated by the potential. It is monotonic with positive slope everywhere, indicating thereby that positive work must be done to move an electron in the force field of the Fermi hole. We have also plotted in the figure the exchange potential [26] of the optimized potential method [24] (OPM). Observe that the two potentials are essentially equivalent, differing only in the intershell region where the OPM potential possesses bumps. From the perspective [6] of the work formalism the negative slope of these bumps implies that the electron is being pushed out by the Fermi hole charge in this region. However, although these bumps help lower [6] the total energy by confining the electrons, there is thus far no physical explanation for their existence. The OPM potential also has zero slope at the nucleus. Why this is the case is not explained by Kohn-Sham theory. On the other hand, the work formalism explanation based on the structure of the Fermi hole and resulting zero force field at the nucleus makes clear why this must be so. Further, by considering the contribution of only the highest occupied orbital, it can be shown [24] that the OPM potential goes as -1/r asymptotically. This, as discussed above, must be the case since for these electron positions the Fermi hole is essentially a static charge distribution. In Fig. 2 we also plot the Slater potential $V_x^s(r)$ which differs significantly from both the work formalism $W_x(r)$ and OPM exchange potentials. It does, however, delineate the shells, and also decays asymptotically as (-1/r) since for these electron positions as noted the Fermi hole is static, and the expression Eq. (72) for the Slater potential therefore correct. Thus, the Slater potential has the correct asymptotic structure in atoms and molecules. As a point of interest we note that the asymptotic structure of the Slater potential at metal surfaces is incorrect [31,32], decaying as (-1/2x), whereas both the work formalism [32,33] and OPM [34] exchange potentials have the correct image potential structure of (-1/4x).

As noted in section 2.1, the exchange potential $W_x(r)$ can also be determined from a ***static*** effective exchange charge distribution $\rho_x^{eff}(r)$ such that

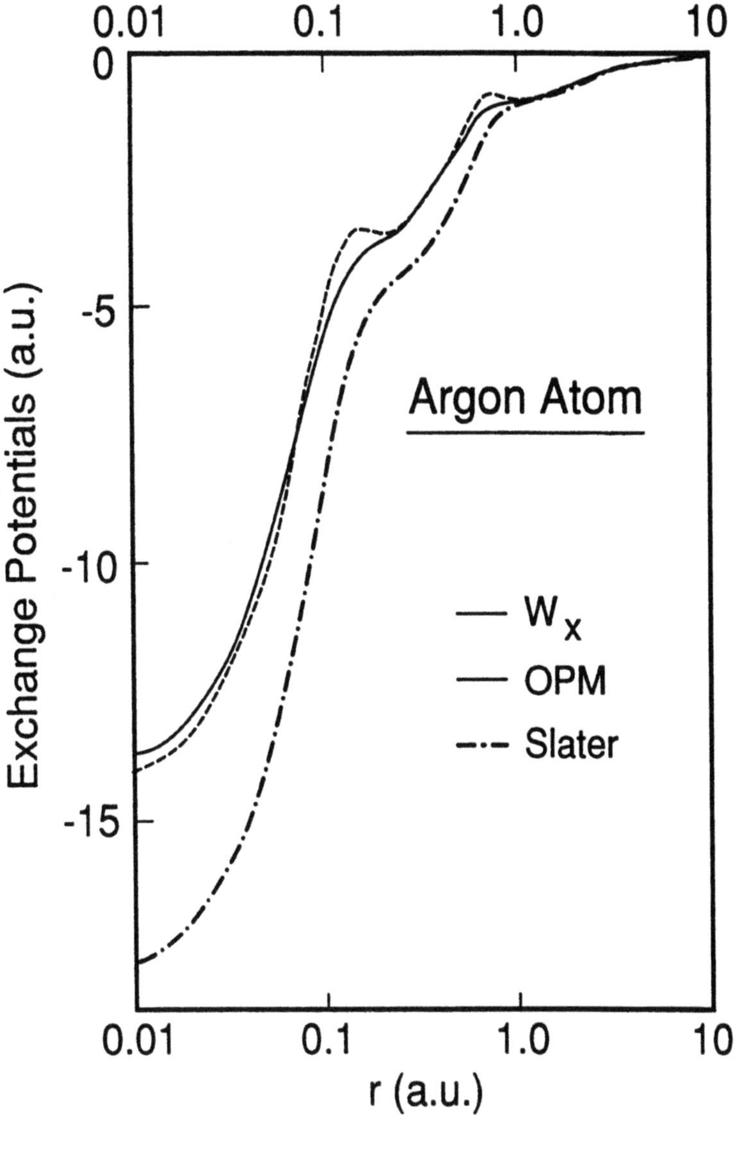

Figure 2. Exchange potentials for Ar as determined by the work formalism $W_x(r)$, the optimized potential method (OPM), and Slater theory $V_x^s(r)$.

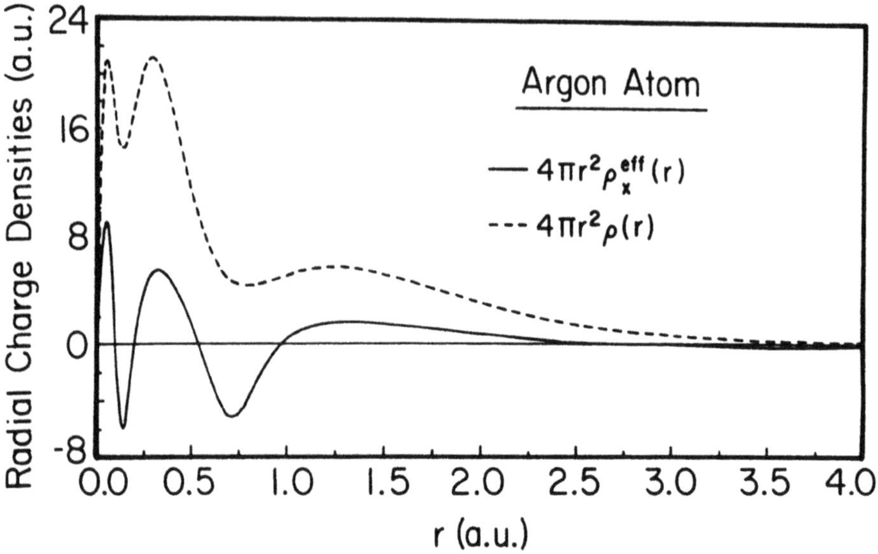

Figure 3. The effective radial exchange charge density $4\pi r^2 \rho_x^{eff}(r)$ and the radial probability density $4\pi r^2 \rho_x^{eff}(r)$ for Ar atom.

$$W_x(r) = \int \frac{\rho_x^{eff}(r)}{|r - r'|} dr' \tag{73}$$

where $\rho_x^{eff}(r)$ is the divergence of the field $\mathcal{E}_x(r)$ due to the Fermi hole $\rho_x(r, r')$. This effective exchange charge can also be written as

$$\rho_x^{eff}(r) = \frac{1}{2}\rho(r) + \frac{1}{4\pi}\int \nabla \rho_x(r, r') \cdot \frac{(r - r')}{|r - r'|^3} dr' \quad . \tag{74}$$

In Fig. 3 we plot the function $4\pi r^2 \rho_x^{eff}(r)$ together with the radial probability density $4\pi r^2 \rho(r)$ for the Argon atom. Observe that unlike the Fermi hole which is negative, the effective exchange charge is both positive and negative. It, however, closely follows the probability density in exhibiting the shell structure of the atom. This, of course, must be the case since it is this static charge which gives rise to the potential $W_x(r)$ of Fig. 2 with its shell structure.

4.2. STRUCTURE OF THE COULOMB HOLE AND CORRELATION POTENTIAL

We next discuss the structure [35] of the correlation potential $W_c(r)$ for the Helium atom. As was the case for the exchange potential $W_x(r)$, the structure of $W_c(r)$ can be understood by studying the corresponding source charge distribution, viz. the Coulomb hole charge $\rho_c(r, r')$. To determine the Coulomb hole requires knowledge of the exact or an accurate approximate wavefunction. For the present, instead of solving the differential equation Eq. (13) self-consistently for the orbitals $\phi_i(r)$ of the occupied and virtual states to construct the wavefunction, we assume the wavefunction known. The wavefunction employed in these calculations is the 39-parameter correlated wavefunction of Kinoshita. [36] This wavefunction leads to a ground-state energy which is the same as that of the Pekeris [37] 1078-parameter correlated wavefunction to seven significant figures, and to single particle expectation values which are the same to four-to-five significant figures.

In Fig. 4(a) we plot the structure of the Fermi-Coulomb hole $\rho_{xc}(r, r')$, and its Fermi $\rho_x(r, r')$ and Coulomb $\rho_c(r, r')$ hole components for an electron at the nucleus. (Within local potential theories of the Helium atom, the Fermi hole is equal to the negative of the self-interaction term in the density and is independent of electron position.) Note that for this electron position, these charge distributions are all spherically symmetric about the electron. Consequently, the exchange-correlation $W_{xc}(r)$, the exchange $W_x(r)$ and correlation $W_c(r)$ potentials all have zero slope at this position. In the figure we also plot the Coulomb hole for an electron within the atom at r=0.8 a.u. (Fig. 4(b)), and for an electron at r=5.0 a.u. in the classically forbidden region (Fig. 4(c)). Observe that in contrast to the Fermi hole (see also Fig. 1) which is negative for all electron positions, the Coulomb hole can be both positive and negative, and substantially positive about the nucleus for asymptotic positions of the electron. Of course, at the position of the electron, the Fermi-Coulomb hole is always more negative than the Fermi hole. This is because as a consequence of the additional correlations due to Coulomb repulsion the electron digs a deeper hole. Also observe the cusp in the Coulomb hole (Figs. 4(a) and (b)) at the electron position which is a reflection of the satisfaction of the electron-electron cusp condition by the wavefunction.

The potentials $W_{xc}(r)$, $W_x(r)$ and $W_c(r)$ resulting from the above described charge distributions are plotted in Fig. 5. All the potentials approach the nucleus quadratically having zero slope there. The correlation potential $W_c(r)$ is negative and monotonic with positive slope, thereby once again indicating that positive work must be done to move an electron in the force field of the Coulomb hole. It is significant only within the atom, with a value

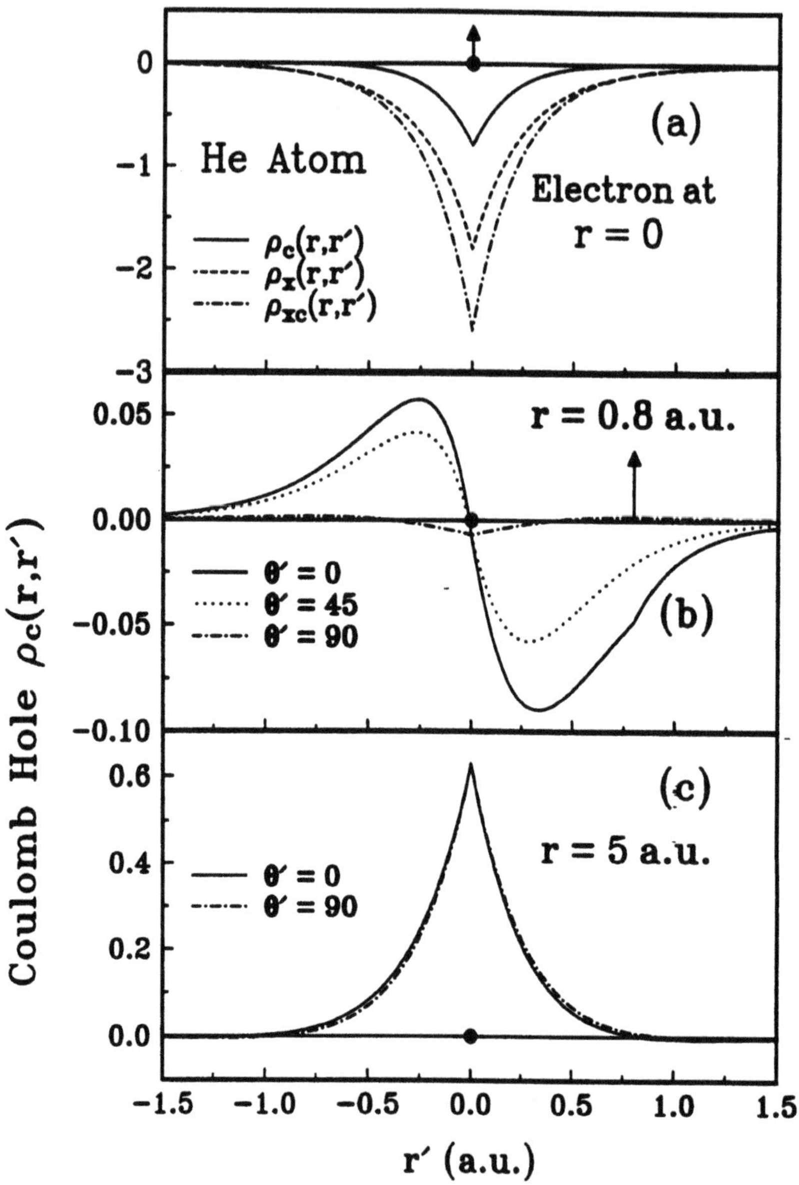

Figure 4. (a) The Fermi-Coulomb $\rho_{xc}(r, r')$, Fermi $\rho_x(r, r')$ and Coulomb $\rho_c(r, r')$ hole charge distributions for the He atom for an electron at the nucleus. (b) and (c) The Coulomb hole $\rho_c(r, r')$ for electron positions at r=0.8 a.u. and 5 a.u. respectively. (From Ref. 35).

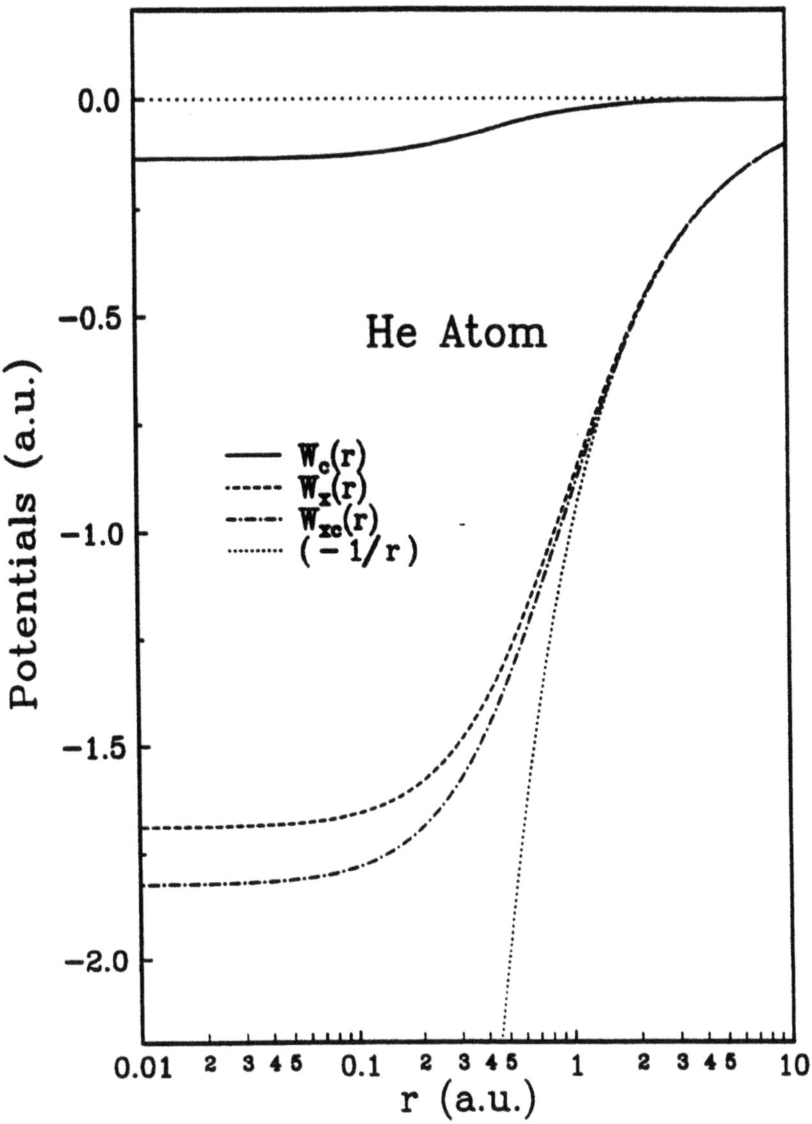

Figure 5. The exchange-correlation $W_{xc}(r)$, exchange $W_x(r)$ and correlation $W_c(r)$ potentials for the He atom. The function -1/r is also plotted. (From Ref. 35).

that is an order of magnitude smaller than the exchange potential $W_x(r)$. Further, as a result of the facts that the total Coulomb hole charge is zero, and significantly positive and localized about the nucleus for asymptotic positions

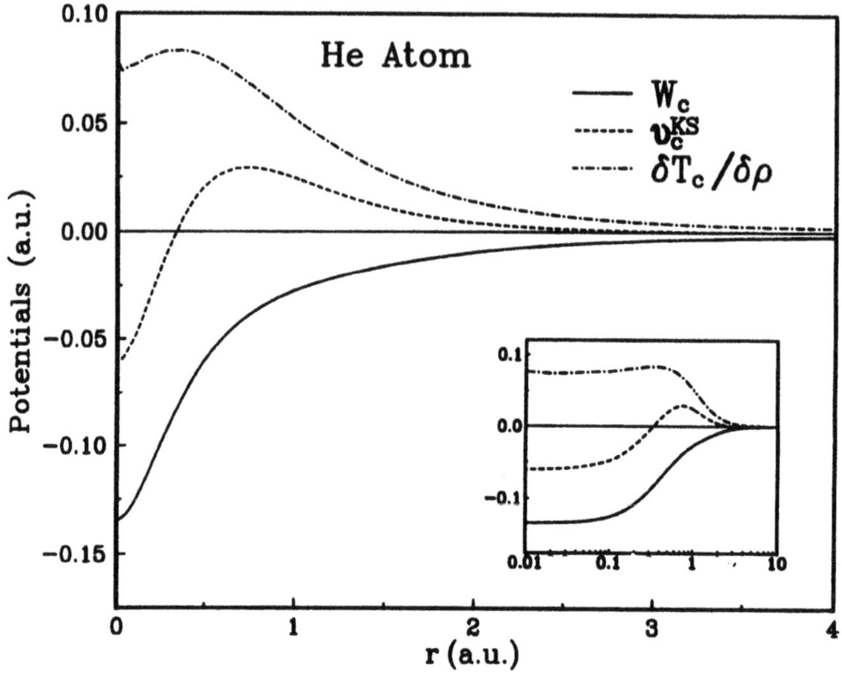

Figure 6. The correlation potentials $W_c(r)$ and $v_c(r)$ of the work formalism and Kohn-Sham theory respectively for the He atom. The functional derivative $\delta T_c/\delta\rho$, where T_c is the correlation contribution to the kinetic energy, as obtained from the ansatz $\delta T_c/\delta\rho = v_c(r) - W_c(r)$ is also plotted. (From Ref. 35).

of the electron, the correlation potential $W_c(r)$ decays rapidly becoming negligible by about r≃4 a.u. Thus as may be observed from the figure, the asymptotic structure of the exchange-correlation potential $W_{xc}(r)$ is that of the exchange potential $W_x(r)$ which is -1/r. This confirms the tenet that the highest occupied eigenvalue of the work formalism differential equation when both Pauli and Coulomb correlations are considered is determined principally by the exchange potential, and that it is therefore meaningful to compare the highest occupied eigenvalue in the Pauli-correlated approximation to experimental ionization potentials, electron affinities and transition energies. The results for this eigenvalue in atoms within the latter approximation are given in the following subsection.

For the Helium atom, the exchange potential $W_x(r)$ and $v_x(r)$ of the work formalism and Kohn-Sham theory, respectively, are the same. However, the correlation potentials $W_c(r)$ and $v_c(r)$ are significantly different. The correlation potential $v_c(r)$ can be determined [38] by inverting the Kohn-Sham equation. The result [35] for $v_c(r)$ employing the density obtained from the Kinoshita wavefunction is plotted together with $W_c(r)$ in Fig. 6. Observe that since the potential $W_c(r)$ due to the Coulomb hole $\rho_c(r, r')$ is strictly representative of Coulomb correlations whereas $v_c(r)$ contains in addition the correlation contribution $T_c[\rho]$ to the kinetic energy, the potential $v_c(r)$ is less attractive than $W_c(r)$. The inset in the figure shows that $v_c(r)$ too has zero slope at the origin. However $v_c(r)$ is not a monotonic function and goes positive at about r~0.3 a.u., decaying asymptotically as a positive function. Recent work [39] employing a 491-term correlated wavefunction, however, shows that $v_c(r)$ once again goes negative at about r~4 a.u. and vanishes as a negative function. These negative values of $v_c(r)$ are three-orders of magnitude smaller than its value at the origin. Finally, once again because $W_c(r)$ is strictly representative of Coulomb correlations, it is reasonable to assume that the difference $[v_c(r) - W_c(r)]$ is a good approximation to $\delta T_c[\rho]/\delta\rho(r)$, the contribution of $T_c[\rho]$ to $v_c(r)$. The result of this ***ansatz*** for $\delta T_c/\delta_\rho$ is also plotted in Fig. 6. Observe that it is of the same order of magnitude as $v_c(r)$, and that it is a positive but not monotonic function.

4.3. GROUND-STATE PROPERTIES IN THE PAULI-CORRELATED APPROXIMATION

In this subsection we demonstrate the accuracy of the physics invoked in the work formalism by presenting fully-self-consistent results of a few ground-state properties of atoms in the Pauli-correlated approximation. In Table 1 we quote [40] the ground-state energy of noble gas and closed s-subshell atoms together with those of Hartree-Fock theory [41]. (For the results of open-shell atoms as determined within the central-field approximation, we refer the reader to reference 40). Observe that the results are a rigorous upper bound to and lie within 50 ppm of Hartree-Fock theory. From ^{36}Kr to ^{71}Lu, the difference lies between (10-5) ppm, and from ^{72}Hf to ^{86}Rn the difference is less than 5 ppm. (The corresponding differences in the results [1,42] of Slater theory are an order of magnitude greater, and of the OPM [25] about half those of the work formalism). Thus, in essence the work formalism reproduces the ground-state energies of Hartree-Fock theory.

The accuracy of the electronic density in the interior of atoms as determined by the work formalism is demonstrated by the satisfaction of the

TABLE 1. Self-consistent ground-state energies of noble gas and closed s-subshell atoms within the Pauli-correlated approximation as obtained by the work formalism and Hartree-Fock theory. The negative values of the energies in Rydbergs are quoted.

Atom	Work formalism[a]	Hartree-Fock theory[b]
^{2}He	2.86168	2.86168
^{4}Be	14.5714	14.5730
^{10}Ne	128.542	128.547
^{12}Mg	199.606	199.615
^{18}Ar	526.804	526.818
^{20}Ca	676.743	676.758
^{30}Zn	1777.820	1777.848
^{36}Kr	2752.030	2752.055
^{38}Sr	3131.519	3131.546
^{48}Cd	5465.093	5465.133
^{54}Xe	7232.101	7232.138
^{56}Ba	7883.506	7883.544
^{70}Yb	13391.389	13391.456
^{80}Hg	18408.932	18408.991
^{86}Rn	21866.721	21866.772

a. See Ref. 40 b. See Ref. 41

Kato-Steiner [15] electron-nucleus cusp condition

$$- \lim_{r \to 0} \left[\frac{d\rho(r)/dr}{2Z\rho(r)}\right] = C_{en} = 1 \quad , \tag{75}$$

where C_{en} is the cusp ratio. In Table 2 we quote [43] this cusp ratio for the noble gas atoms, and observe that the cusp condition is satisfied to within 2 ppm up to Xe. For the calculations of the cusp ratio a modified Herman-Skillman program was employed [44,45]. Comparisons with the results of Hartree-Fock theory are not meaningful in this case because in the work of Fischer [41] the cusp ratio is fixed for the outward integration.

TABLE 2. The electron-nucleus cusp ratio C_{en} for noble gas atoms as determined by the work formalism in the Pauli-correlated approximation. (From Ref. 43).

Atom	C_{en}
He	1.000000
Be	0.999999
Ne	0.999999
Ar	1.000000
Kr	0.999999
Xe	0.999998
Rn	0.999945

TABLE 3. The core charge q and cutoff radius r_c of atoms in atomic units as determined by the work formalism in the Pauli-correlated approximation and Hartree-Fock theory.

Atom	Work formalism[a]		Hartree-Fock theory[b]	
	r_c	q	r_c	q
$^{19}K(4s)^1)$	2.97	18.00	2.91	17.94
$^{20}Ca(4s^2)$	2.36	17.94	2.33	17.89
$^{21}Sc(3d^1 4s^2)$	2.25	18.77	2.26	18.78
$^{36}Kr(4s^2 4p^6)$	0.96	27.48	0.97	27.49
$^{37}Rb(5s^1)$	3.43	36.03	3.34	35.95
$^{38}Sr(5s^2)$	2.78	36.00	2.76	35.93
$^{39}Y(4d^1 5s^2)$	3.02	37.10	2.70	36.73
$^{54}Xe(5s^2 5p^6)$	1.34	45.62	1.34	45.61

a. See Ref. 46 b. See Ref. 47

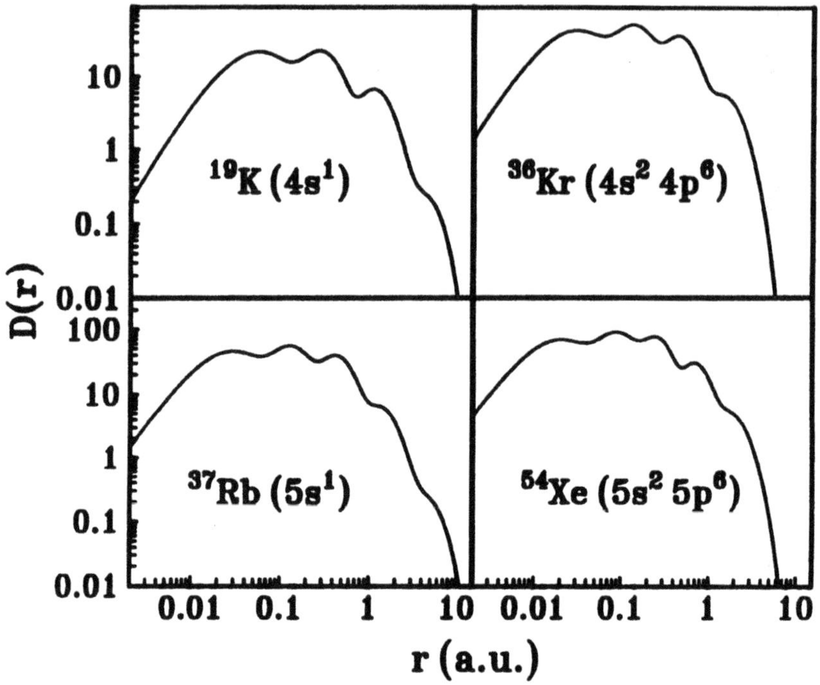

Figure 7. A log-log plot of the radial probability density D(r) for K, Kr, Rb and Xe as determined by the work formalism in the Pauli-correlated approximation. (From Ref. 46).

The densities of atoms determined by the work formalism also clearly exhibit [46] atomic shell structure. In Fig. 7 we plot the radial probability density $D(r) = 4\pi r^2 \bar{\rho}(r)$, where $\bar{\rho}(r)$ is the spherical average of $\rho(r)$, for $^{19}K(4s^2)$, $^{36}Kr(4s^2 4p^6)$, $^{37}Rb(5s^2)$ and $^{54}Xe(5s^2 5p^6)$. The existence of ***all*** the shells and in particular the valence shell is clearly evident. Employing the criterion of the midpoint of the last two successive points of inflection in the distribution function to define the core-valence cutoff radius r_c, we quote in Table 3 the resulting core charge q. The Hartree-Fock theory results [47] employing the same cutoff criterion are also given. The two sets of results are essentially equivalent.

4.4. IONIZATION POTENTIALS, ELECTRON AFFINITIES AND TRANSITION ENERGIES

As explained and shown previously, the asymptotic structure of the exchange-correlation potential $W_{xc}(r)$ is that of the exchange potential $W_x(r)$, so that it is meaningful to compare the highest occupied eigenvalue in the Pauli-correlated approximation of the work formalism to experiment. On the other hand, the corresponding eigenvalue of Hartree-Fock theory has the meaning of a removal energy only in the context of Koopmans' theorem [22] in which the neutral and ionized atom orbitals are assumed to be the same.

In Table 4 we compare the highest occupied eigenvalues of a few neutral atoms as determined [40] by the work formalism with those of Hartree-Fock theory [41] and experimental [48] ionization potentials. Observe that for atoms where the last closed subshell is an s-subshell and the third transition group, the experimental results lie ***below*** those of Hartree-Fock theory. The corresponding work formalism eigenvalues lie between these two results, thus being consistently superior to those of Hartree-Fock theory in comparison to experiment. Further, the work formalism results eliminate a substantial fraction of the difference between the Hartree-Fock theory results and experiment. For noble gas atoms, the experimental ionization potentials lie ***above*** those of Hartree-Fock theory. Once again, with the exception of ^{10}Ne for which the difference between the Hartree-Fock theory and work formalism eigenvalues is a hundredth of a Rydberg, the results of the latter are consistently superior to the former in comparison to experiment.

In Table 5 we quote the highest occupied eigenvalues of negative ions as determined [44,49] by the work formalism in the Pauli correlated approximation together with the Hartree-Fock theory results and experimental [50] electron affinities. Note that for the alkali ions, the experimental results lie ***below*** those of Hartree-Fock theory, and the work formalism eigenvalues lie in between. For the Halogens and negative ions with incomplete p-shells, the experimental results lie ***above*** those of Hartree-Fock theory. Once again, with the exception of S^-, the work formalism eigenvalues lie between the two. Thus the results of the work formalism are again consistently superior to those of Hartree-Fock theory in comparison to experiment.

TABLE 4. Highest-occupied-orbital eigenvalues of atoms in the central-field model as obtained by the work formalism in the Pauli-correlated approximation. The corresponding Hartree-Fock-theory results as well as the experimental ionization potentials are also given. The negative values of the eigenenergies in rydbergs are quoted. (From Ref. 40).

Atom	Structure	Hartree-Fock[a]	Work Formalism	Experiment[b]
Atoms with last closed subshell an s subshell				
^{2}He	$1s^2$	1.836	1.836	1.807
^{4}Be	$[He]2s^2$	0.619	0.626	0.685
^{12}Mg	$[Ne]3s^2$	0.506	0.521	0.562
^{20}Ca	$[Ar]4s^2$	0.391	0.402	0.449
^{30}Zn	$[Ar]3d^{10}4s^2$	0.585	0.646	0.690
^{38}Sr	$[Ar]3d^{10}4p^6 5s^2$	0.357	0.369	0.419
^{48}Cd	$[Kr]4d^{10 5}5s^2$	0.530	0.583	0.661
^{56}Ba	$[Xe]6s^2$	0.315	0.325	0.383
^{70}Yb	$[Xe]4f^{14}6s^2$	0.365	0.383	0.460
^{80}Hg	$[Xe]4f^{14}5d^{10}6s^2$	0.522	0.580	0.767
Noble-gas atoms				
^{10}Ne	$[He]2s^2 2p^6$	1.701	1.713	1.585
^{18}Ar	$[Ne]3s^2 3p^6$	1.182	1.178	1.158
^{36}Kr	$[Ar]3d^{10}4s^2 4p^6$	1.048	1.035	1.029
^{54}Xe	$[Kr]4d^{10}5s^2 5p^6$	0.915	0.899	0.892
^{86}Rn	$[Xe]4f^{14}5d^{10}6s^2 6p^6$	0.856	0.838	0.790
Third transition group				
^{72}Hf	$[Xe]4f^{14}5d^n 6s^2$ $n = 2$	0.418 (6s) 0.649 (5d)	0.470 (6s) 0.426 (5d)	0.515
^{73}Ta	$n = 3$	0.435	0.494	0.580
^{74}W	$n = 4$	0.450	0.513	0.587
^{75}Re	$n = 5$	0.461	0.531	0.579
^{76}Os	$n = 6$	0.478	0.540	0.639
^{77}Ir	$n = 7$	0.491	0.551	0.669
^{78}Pt	$n = 8$	0.503	0.561	0.662

a. See Ref. 41 b. See Ref. 48

TABLE 5. Highest occupied eigenvalues of negative ions in the central field model as obtained by the work formalism in the Pauli-correlated approximation. The corresponding Hartree-Fock theory results as well as the experimental electron affinities are also given. The negative values of the eigenenergies in Rydbergs are quoted. (From Refs. 44 and 49).

Atomic Number	Ion	State	Hartree-Fock Theory	Work Formalism	Experiment[(a)]
			Hydrogen		
1	H^-	$1s^2\ ^1S$	0.09244	0.09244	0.0554
			Alkalis		
3	Li^-	$2s^2\ ^1S$	0.02908	0.03017	0.04542
11	Na^-	$3s^2\ ^1S$	0.02670	0.02918	0.04027
19	K^-	$4s^2\ ^1S$	0.02064	0.02280	0.03686
37	Rb^-	$5s^2\ ^1S$	0.01906	0.02136	0.03571
			Halogens		
9	F^-	$2p^6\ ^1S$	0.3620	0.3576	0.2498
17	Cl^-	$3p^6\ ^1S$	0.3006	0.2884	0.2658
35	Br^-	$4p^6\ ^1S$	0.2787	0.2612	0.2473
53	I^-	$5p^6\ ^1S$	0.2583	0.2401	0.2248
			Ions with incomplete p Shells		
5	B^-	$2p^2\ ^3P$	0.05314	0.04933	0.0204
6	C^-	$2p^3\ ^4S$	0.1562	0.1511	0.0928
8	O^-	$2p^5\ ^2P$	0.1434	0.1414	0.1074
13	Al^-	$3p^2\ ^3P$	0.04124	0.03259	0.0324
14	Si^-	$3p^3\ ^4S$	0.1241	0.1099	0.1018
16	S^-	$3p^5\ ^2P$	0.1551	0.1457	0.1527

(a) See Ref. 50

In Table 6 we quote the total energy for the ground and various excited states of the Sodium atom as determined [51] by the work formalism in the Pauli-correlated approximation, and by Hartree-Fock theory. Note that as is the case for the ground state, the work formalism results for the excited states constitute an upper bound to the Hartree-Fock theory results, the two being essentially equivalent and within 55 ppm of each other. The corresponding highest occupied eigenvalues of these states are also given in the table. From these eigenvalues of the work formalism we obtain the

TABLE 6. Total ground and excited state energies, and highest occupied eigenvalues of the Sodium atom in the central field model as determined by the work formalism in the Pauli-correlated approximation and Hartree-Fock theory. The negative values in Rydbergs are quoted. (From Ref. 51).

Highest Occupied State	Work Formalism		Hartree-Fock Theory	
	Total Energy	Highest Occupied Eigenvalue	Total Energy	Highest Occupied Eigenvalue
$3s^1$	323.700	0.390	323.718	0.364
$4s^1$	323.476	0.160	323.494	0.140
$5s^1$	323.424	0.084	323.428	0.074
$3p^1$	323.554	0.232	323.572	0.220
$4p^1$	323.450	0.108	323.454	0.106

TABLE 7. Transition energies for sodium in Rydbergs as determined by the work formalism via highest occupied eigenvalue differences, and by Hartree-Fock theory via total energy differences. (From Ref. 51).

Transitions	Hartree-Fock Theory	Work Formalism	Experiment[a]
$3^2S \rightarrow 4^2S$	0.224	0.230	0.234
$3^2S \rightarrow 5^2S$	0.290	0.306	0.302
$3^2S \rightarrow 3^2P$	0.146	0.158	0.154
$3^2S \rightarrow 4^2P$	0.264	0.282	0.276

(a) See Ref. 52

transition energies quoted in Table 7. The Hartree-Fock theory values quoted, however, are obtained from total energy differences since this is the most accurate way of determining transition energies by this theory. Once again in comparison to experiment [52] the work formalism results are superior to those of Hartree-Fock theory.

It is evident from the results of Tables 4-7 that the structure of the exchange potential $W_x(\mathbf{r})$ is an accurate representation of the effective potential in atoms for asymptotic positions of the electron. The reason why the highest occupied eigenvalues of Hartree-Fock theory are not as accurate is because, as noted previously, in this theory all the orbitals contribute [21] to the asymptotic structure of the density rather than just the highest occupied or most energetic orbital as must physically be the case. In the work formalism, not only is the asymptotic structure of the effective potential obtained correctly, but also only the highest occupied orbital contributes to the asymptotic density. Thus, the resulting highest occupied eigenvalue is an accurate approximation to experiment.

The work formalism has also been applied to atoms for the determination [53] of static dipole and quadropole polarizabilities, to metal clusters [54] for a study of the size dependence of the ionization potential and electron affinity, and to metal surfaces [32,33,55] for a study of the image potential. For the results and details of these calculations we refer the reader to the original literature.

5. Conclusions and Future Work

In conclusion we reiterate the principal points made in this article on the work formalism. As described this is a physically based theory of electronic structure. The arguments of the theory for the construction of a local potential representing electron correlations via a quantum-mechanical source charge distribution and Coulomb's law are general and valid for both ground and excited states of nonuniform density systems. The application to few-electron atomic and many-electron metallic surface systems has lead to results that are accurate both in comparison to other theories as well as to experiment. This accuracy then speaks to the correctness of the physics invoked. Other theories of electronic structure can as a consequence then be reinterpreted in terms of this physics. For example, Hartree-Fock theory can be described in terms of a quantum-mechanical source charge distribution which is the Fermi hole. The ***integral*** exchange operator of the theory can in effect be replaced by a ***multiplicative*** operator which has the physical interpretation of the work done to move an electron in the force field of the Fermi hole. The exchange energy in turn is the energy of interaction between the electronic density and the Fermi hole. This same physical interpretation is also an accurate description of the exchange potential of Kohn-Sham theory. Thus the functional derivative of the exchange energy functional of the density can now be understood in purely physical terms. The work formalism also provides physical insights into other local-potential approximation schemes. Thus, we now understand that in Slater theory and its approximations, it is the dynamic nature of the Fermi hole charge that is not accounted for in the construction of the potential. By deriving the source charge distribution of the local density approximation for exchange and correlation of Kohn-Sham theory, we learn that the correlations between electrons in this approximation are not solely those of the uniform electron gas as previously assumed, but that the nonuniformity of the electronic density is in fact also explicitly incorporated in its representation of these correlations. These insights then provide a more meaningful evaluation of the results of the approximations. Finally, we note the important attribute of the work formalism that the asymptotic structure of the exchange-correlation potential of ***all*** nonuniform density systems arises solely due to the Fermi hole charge, and therefore is exactly determinable by solution within the Pauli-correlated approximation.

The current and future research envisaged with regard to the work formalism falls into the categories of fundamental studies and applications. Within the former, an issue being investigated is the question of path-dependence of the work for non-symmetrical density systems for which the external potential is physically realistic. We have, however, recently [7] investigated the accuracy of the approximation of determining a potential for non-symmetrical density systems from the irrotational component of the force field. For a highly non-symmetrical-density model atom in which a hypothetical external potential leads to a force field whose curl does not vanish, the solenoidal component is observed to be negligible and two-orders of magnitude smaller than the irrotational component. Thus, the latter accounts for essentially all the correlations between the electrons. As a consequence, the resulting potential is accurate, as is the approximation. We further propose to investigate the work formalism for spin-uncompensated systems as well as for nonuniform density systems in the presence of external magnetic fields. We are also presently studying the structure of the Fermi-

Coulomb and Coulomb hole charge distributions and the resulting work formalism exchange-correlation and correlation potentials for the ground and excited states of light atoms and small molecules employing accurate Monte Carlo generated wavefunctions. Simultaneously, we are working towards determining atomic wavefunctions and the corresponding local many-body potentials of the same accuracy by self-consistent solution of the work formalism Sturm-Liouville equation.

The author thanks Eugene Kryachko and Jean-Louis Calais for their invitation to contribute to this volume.

6. References

1. E. Schrödinger, Ann. Phys. **79**, 361 (1926).
2. P. Hohenberg and W. Kohn, Phys. Rev. **136**, B864 (1964); W. Kohn and L.J. Sham, Phys. Rev. **140**, A1133 (1965).
3. R.G. Parr and W. Yang, *Density Functional Theory of Atoms and Molecules* (Oxford University Press, Oxford, 1989); R.M. Dreizler and E.K.U. Gross, *Density Functional Theory* (Springer-Verlag, Berlin, 1990); E.S. Kryachko and E.V. Ludeña, *Energy Density Functional Theory of Many-Electron Systems* (Kluwer, Dordrecht, 1990); N.H. March, *Electron Density Theory of Atoms and Molecules* (Academic, London 1992).
4. *Density-Functional Theory*, vol. 336 of NATO Advanced Study Institute, series B: Physics, edited E.K.U. Gross and R.M. Dreizler (Plenum, New York 1994).
5. Proceedings of the Symposium on *Thirty Years of Density-Functional Theory: Concepts and Applicatons* held at Cracow, Poland (June 13-16, 1994) to be published in the Int. J. Quantum Chem. (1995).
6. M.K. Harbola and V. Sahni, Phys. Rev. Lett. **62**, 489 (1989); V. Sahni and M.K. Harbola, Int. J. Quantum Chem. Symp. **24**, 569 (1990); M.K. Harbola and V. Sahni, J. Chem. Educ. **70**, 920 (1993).
7. M.K. Harbola, M. Slamet and V. Sahni, Phys. Lett. A**157**, 60 (1991); M. Slamet, V. Sahni and M.K. Harbola, Phys. Rev. A**49**, 809 (1994).
8. V. Sahni, Int. J. Quantum Chem. (in press).
9. V. Fock, Z. Physik **61**, 126 (1930); ibid. **62**, 795 (1930); J.C. Slater, Phys. Rev. **35**, 210 (1930).
10. V. Sahni, J. Gruenebaum and J.P. Perdew, Phys. Rev. B**26**, 4371 (1982); V. Sahni and M. Levy, Phys. Rev. B**33**, 3869 (1986).

11. V. Sahni, in Ref, 4; V. Sahni and M. Slamet, Phys. Rev. B48, 1910 (1993); M. Slamet and V. Sahni, Phys. Rev. B45, 4013 (1992).

12. M. Slamet and V. Sahni, Int. J. Quantum Chem. Symp. **26**, 333 (1992); ibid **25**, 235 (1991).

13. J.C. Slater, Phys. Rev. **81**, 385 (1951).

14. J.C. Slater, T.M. Wilson, and J.H. Wood, Phys. Rev. **179**, 28 (1969).

15. T. Kato, Commun. Pure Appl. Math. **10**, 151 (1957); E.J. Steiner, J. Chem. Phys. **39**, 2365 (1963).

16. P.O. Löwdin, Adv. Chem. Phys., **2**, 207 (1959).

17. H. Ou-Yang and M. Levy, Phys. Rev. A41, 4038 (1990); M. Rasolt and D.J.W. Geldart, Phys. Rev. Lett. **65**, 276 (1990); M.K. Harbola and V. Sahni, Phys. Rev. Lett. **65**, 277 (1990).

18. D.R. Hartree, Proc. Cambridge Philos. Soc. **24**, 39 (1928); ibid **24**, 111 (1928); ibid **24**, 426 (1928).

19. M.M. Morrell, Ph.D. thesis, The John Hopkins University 1974 (unpublished).

20. A. Szabo and N.S. Ostlund, *Modern Quantum Chemistry*, (McGraw Hill, New York, 1989).

21. N.C. Handy, M.T. Marron, and H.J. Silverstone, Phys. Rev. **180**, 45 (1969).

22. T. Koopmans, Physica **9**, 104 (1933).

23. H. Ou-Yang and M. Levy, Phys. Rev. Lett. **65**, 1036 (1990); Phys. Rev. A44, 54 (1991).

24. R.T. Sharp and G.K. Horton, Phys. Rev. **90**, 3876 (1953); J.D. Talman and W.F. Shadwick, Phys. Rev. A14, 36 (1976).

25. Y. Wang, J.P. Perdew, J.A. Chevary, L.D. Macdonald and S.H. Vosko, Phys. Rev. A41, 78 (1990).

26. K. Aashamar, T.M. Luke and J.D. Talman, Phys. Rev. A19, 6 (1979); At. Data Nucl. Data Table **22**, 443 (1978).

27. A. Nagy, Phys. Rev. Lett. **65**, 2608 (1990); M.K. Harbola and V. Sahni, Phys. Rev. Lett. **65**, 2609 (1990).

28. P.A.M. Dirac, Proc. Cambridge Philos. Soc. **26**, 376 (1930).

29. M. Slamet and V. Sahni, Bull. Am. Phys. Soc. **39**, 394 (1994) (manuscript in preparation).

30. A. Solomatin, V. Sahni, and N.H. March, Phys. Rev. B49, 16856 (1994).

31. M.K. Harbola and V. Sahni, Phys. Rev. B36, 5024 (1987).

32. V. Sahni, Surf. Sci. **213**, 226 (1989).

33. M.K. Harbola and V. Sahni, Phys. Rev. B39, 10437 (1989).

34. M.K. Harbola and V. Sahni, Int. J. Quantum Chem. Symp. **27**, 101 (1993).

35. M. Slamet and V. Sahni, Bull Am. Phys. Soc. **38**, 139 (1993); (submitted to Phys. Rev. A).

36. T. Kinoshita, Phys. Rev. **105**, 1490 (1957).

37. C.L. Pekeris, Phys. Rev. **115**, 1216 (1959).

38. C.O. Almbladh and A.C. Pedroza, Phys. Rev. **A29**, 2322 (1984).

39. C.J. Umrigar and X. Gonze, Phys. Rev. A (in press).

40. V. Sahni, Y. Li and M.K. Harbola, Phys. Rev. **A45**, 1434 (1992); Y. Li, M.K. Harbola, J.B. Krieger and V. Sahni, Phys. Rev. **A40**, 6084 (1989).

41. C.F. Fischer, *The Hartree-Fock Method for Atoms* (Wiley, New York, 1977).

42. L. Kleinman, Phys. Rev. **B49**, 14197 (1994).

43. K.D. Sen, (submitted to Chem. Phys. Lett.).

44. K.D. Sen and M.K. Harbola, Chem. Phys. Lett. **178**, 347 (1991).

45. K.D. Sen., L.J. Bartolotti, M.K. Harbola, and V. Sahni, Comp. Phys. Com. (manuscript in preparation).

46. M. Slamet, K.D. Sen and V. Sahni (manuscript in preparation).

47. K.D. Sen, M. Slamet and V. Sahni, Chem. Phys. Lett. **205**, 313 (1993).

48. C.E. Moore, *Ionization Potentials and Ionization Limits Derived from the Analysis of Optical Spectra*, Natl. Bur. Stand. Ref. Data Ser., Natl. Bur. Stand. (U.S.) (U.S. GPO, Washington D.C., 1970), Vol. 34.

49. Y. Li, J.B. Krieger and G.J. Iafrate, Chem. Phys. Lett. **191**, 38 (1992).

50. H. Hotop and W.C. Lineberger, J. Phys. Chem. Ref. Data **14**, 731 (1985).

51. K.D. Sen, Chem. Phys. Lett. **188**, 510 (1992).

52. S. Bashkin and J.O. Stoner, *Atomic energy levels and Grotarian diagrams*, Vol. 1 (North-Holland, Amsterdam, 1975).

53. K.D. Sen, Phys. Rev. A **44**, 756 (1991); J. Samuel and K.D. Sen, Int. J. Quantum Chem. **44**, 1041 (1992).

54. M.K. Harbola, J. Chem. Phys. **97**, 2578 (1992).

55. L. Orosz, Phys. Rev. **B47**, 12806 (1993).

INDEX

Absorption line shape 186
Arrhenius law 17
Auger double ionization 172
avoided crossing 100

Bath of harmonic oscillators 13
Born-Huang expansion 99
Born-Oppenheimer approximation 99, 142
Bohr-Sommerfeld theory 155, 173

Conditional probability distribution 162
complete active space self-consistent field 39
Coulomb correlation 216

Density matrix 9
diabatic representation 106
diagonalization of valence shell Hamiltonian 38
dimensional scaling 133
dipole interaction 101
Dirac density matrix 222
double ionization 171
dynamic coupling 101

Eckart potential 3
electron-nuclear cusp 248
effective valence shell Hamiltonian 30

Fermi-Coulomb hole 218
Fermi hole 223
Fermi momentum 232
Fock-Dirac density matrix 84

Genuine primitive function 78

Hamiltonian
reaction path 7
Hartree approximation 220
Hartree-Fock approximation 82
Hartree-Fock model 157
Helmholtz theorem 219
Hirschfelder-Silbey theory 81
Hohenberg-Kohn-Sham density functional 213
Huzinaga equation 87
hydrogen transfer 1
hyperpolarizability 187

Intermolecular perturbation theory 74

Kepler orbit 156
Kohn-Sham theory 213

Langmuir orbit 174
Langmuir vibration 139
level shift operator 79
Lewis structure 139
localized orbital 83
localized wave function 74
local density approximation 232
local exchange operator 225
local Morse oscillator 175
Löwdin orthogonalization 59
Löwdin's natural orbital 77

Malonaldehyde 5
minimum reaction path 7
model Hamiltonian 184
multiphoton absorption 108

Nonlocal operator 213

Oscillator strength 168

Pauli correlation 216
point charge model 63
pseudo-spectral approach 105

Quadrupole moment 168
quantum beat 14
quantum time evolution 104

Rectangular barrier 3
relaxation rate 14
rotor-vibrational model 166
Runge-Lenz vector 156

Separability concept 73
separable N-electron wave function 70
Slater Xα method 237
strong orthogonality 78
supermultiplet 160

Time-dependent Schrödinger equation 105
transmission coefficient 3
tunneling 1, 9
tunneling splitting 5
two-level system 9
two-level system relaxation 11ff

Valence orbitals of trans-butadiene 40
valence space perturbation treatment 32

Unphysical state 69,71

Zero differential overlap approximation 51

GPSR Compliance
The European Union's (EU) General Product Safety Regulation (GPSR) is a set of rules that requires consumer products to be safe and our obligations to ensure this.

If you have any concerns about our products, you can contact us on

ProductSafety@springernature.com

In case Publisher is established outside the EU, the EU authorized representative is:

Springer Nature Customer Service Center GmbH
Europaplatz 3
69115 Heidelberg, Germany

www.ingramcontent.com/pod-product-compliance
Ingram Content Group UK Ltd.
Pitfield, Milton Keynes, MK11 3LW, UK
UKHW021238300926
15471UKWH00013B/533
* 9 7 8 9 4 0 1 1 0 2 6 4 3 *